Harnessing Biodiversity for a Sustainable Bioeconomy

Saurabh Agarwal
Yeungnam University, South Korea

D. Lakshmi
https://orcid.org/0000-0003-4018-1208
VIT Bhopal University, India

IGI Global
Scientific Publishing
Publishing Tomorrow's Research Today

Published in the United States of America by
IGI Global Scientific Publishing
701 East Chocolate Avenue
Hershey, PA, 17033, USA
Tel: 717-533-8845 | Fax: 717-533-7115
Website: https://www.igi-global.com E-mail: cust@igi-global.com

Library of Congress Cataloging-in-Publication Data

Names: Agarwal, Saurabh, 1981- editor http://id.loc.gov/rwo/agents/n2025015902 | D., Lakshmi (Dhandabani) editor http://id.loc.gov/rwo/agents/n2022183670
Title: Harnessing biodiversity for a sustainable bioeconomy / Saurabh Agarwal, Lakshmi D.
Description: Hershey, PA : IGI Global Scientific Publishing, [2026] | Includes bibliographical references and index. | Summary: "This book aims to talk about the bioeconomic model to sustain other planets"-- Provided by publisher.
Identifiers: LCCN 2025007568 (print) | LCCN 2025007569 (ebook) | ISBN 9798337320038 hardcover | ISBN 9798337320045 paperback | ISBN 9798337320052 ebook
Subjects: LCSH: Economic development--Environmental aspects http://id.loc.gov/authorities/subjects/sh88001234 | Biodiversity--Economic aspects | Sustainable development http://id.loc.gov/authorities/subjects/sh92005743
Classification: LCC HD75.6 .H375 2025 (print) | LCC HD75.6 (ebook) | DDC 333.7--dc23
LC record available at https://lccn.loc.gov/2025007568
LC ebook record available at https://lccn.loc.gov/2025007569

British Cataloguing in Publication Data
A Cataloguing in Publication record for this book is available from the British Library.

Table of Contents

Detailed Table of Contents

This chapter examines the intersection of biodiversity, bio -economic and public perception, a calculation method, a calculation method supported by natural language treatment and artificial intelligence. This indicates how opinion seriously affects the conservation efforts of biodiversity and policy implementation, and how emotional analysis can highlight the emotional pattern in social media, scientific literature, corporate reports and government policy. The chapter emphasizes the function, including advanced transformer models such as the lexicon -based approach, machine learning and Bert. The challenges as the recommendations to integrate emotion analysis in assessing environmental impact and stability training are discussed, as well as challenges such as domain -specific jumble, cultural prejudice and moral concerns. Conclusions emphasize emotional analysis as a transformation tool that adapts the conservation efforts with social values, and promotes more inclusive, responsible and data channel approaches to a permanent bioeconomic construction.

Policies need to address social and ethical concerns over biodiversity based bioprospecting and genetic modification. Genetically Modified Approaches: The technology of genetic modification comes with great potential for innovation in agriculture, medicine, and industry, while in turn, raises ethical issues surrounding the unforeseen impact, harm to biodiversity, and rights of indigenous communities. The act of bioprospecting exploring biodiversity for genetic and biochemical resources that hold economic potential has risks of overexploration, endangering ecosystems and the livelihoods of nearby residents. Hence, potential equitable benefit-sharing mechanisms are needed that allow proceeds of these resources to enable conservation practices and benefit local tribes. Policies should focus on transparency, sustainability, and safeguarding traditional knowledge even as biopiracy must be addressed. This bioeconomy must be ethical about inequalities, having a balance between development and conservation for long-term sustainability.

Bio-resources and Traditional Knowledge associated therewith, have been used since ancient times for a wide range of purposes due to its high applicability across diverse sectors for fulfilment of human own needs and environmental conservation. This creates huge opportunities in the field of modern scientific research and scope for commercial gain. However, its exploitation has its own interlinked repercussions such as biodiversity loss, socio-economic implications and patent revocation. In an attempt to mitigate these repercussions, legal instruments have been explored mainly focusing on sharing of benefit framework under biodiversity laws of international and national regimes majorly focusing on Indian perspectives. It has been found that the legal intervention in access and use of bio-resources, bio-resources-based innovations supported with Intellectual Property Rights protection aid in accomplishing the majority of the total goals of Sustainable Development Goals.

Ecosystem resilience is the ability of natural systems to recover from disturbances while maintaining essential functions. Biodiversity enhances this resilience by supporting ecological balance, redundancy, and adaptability. However, habitat loss, climate change, and pollution threaten biodiversity and ecosystem stability. This chapter explores how biodiversity supports resilience through species interactions, genetic and functional diversity, and habitat connectivity. It discusses conservation strategies like restoration, sustainable land use, and policy support. Real-world examples from forests, wetlands, and farms show biodiversity's role in resilience. The chapter concludes with recommendations to integrate science, policy, and community efforts for conserving biodiversity and strengthening ecosystems.

Biodiversity is essential to the ecosystem resilience, and natural systems can offer pollination and water purification, as well as the capture of carbon and climate control. But ecosystems are being compromised by accelerating land-use change pressures, pollution, extractive industries, agribusiness, and climate change with industrialized economies being primary actors and marginalized communities bearing the unequal costs. This chapter brings biodiversity conservation in the context of an ecological and moral duty, where in-situ and ex-situ conservation, community as steward of the environment, agro-ecology, and nature-based solutions are enlightened. Success can be restored in the case of China, Costa Rica and Africa as seen in the examples of restoration case studies. Management can change with emerging technologies like AI, big data, synthetic biology and bioengineering, but needs ethical governance and equitable benefit-sharing. The chapter restates that biodiversity protection in future should combine science, equity and interdisciplinary cooperation.

The Indian Himalayan Region (IHR), a biodiversity hotspot, is vital for ecosystem services and local livelihoods but is increasingly threatened by climate change and unsustainable practices. This chapter explores a circular bioeconomy model to enhance climate resilience and sustainable development in the IHR. It emphasizes integrating traditional knowledge with biotechnologies like biochar, composting, and genetic resource use to build zero-waste systems. Through interdisciplinary analysis and case studies from Sikkim and Himachal Pradesh, the chapter evaluates policy frameworks and demonstrates successful biodiversity-driven initiatives. It concludes with strategic recommendations on governance, community participation, and innovation. The chapter offers a replicable model for fostering ecological regeneration and socio-economic sustainability in fragile mountain ecosystems globally.

This chapter explores how circular bioeconomy principles can enhance resource efficiency and biodiversity in agriculture. As population growth, climate change, and resource scarcity challenge conventional "take–make–dispose" models, circular practices provide sustainable alternatives. The findings reveal significant benefits: biochar with a phosphorus adsorption capacity of 218.82 mg/g and carbon sequestration of 1.7 tonnes CO_2 per tonne; hydroponic systems reducing water use by up to 90%; LED lighting lowering energy consumption by 60%; and organic farms supporting 40% more pollinator species than conventional ones. Key challenges include inconsistent biodiversity assessment methodologies, limited geographic coverage of case studies, and insufficient understanding of adoption drivers. Future research should prioritize the development of unified assessment frameworks, long-term ecological monitoring, and policy instruments that promote widespread implementation of biodiversity-enhancing circular agriculture.

This study underlines the importance of economic growth and preserving biodiversity to ensure long-term environmental and economic stability. Agroforestry, as a biodiversity-friendly land use system, integrates trees and shrubs into agricultural landscapes, mimicking natural ecosystems and promoting ecological richness and sustainability. 2,564 documents from 119 countries in the world that have implemented agroforestry systems. 349 documents from 65 countries have shown the importance of agroforestry systems to economic value. Agroforestry integrates trees with crops and/or livestock to create sustainable agricultural systems based on ecological principles, providing environmental and economic benefits. Agroforestry systems play an important role in contributing to biodiversity-derived products and services in the bioeconomy, with a particular focus on providing non-timber forest products without harvesting timber. Successful agroforestry systems promote economic, environmental sustainability, biodiversity conservation and the need of political support and community engagement.

Preface

The accelerating loss of biodiversity and the growing urgency of sustainable development have placed humanity at a defining crossroads. Biodiversity—the extraordinary variety of life on Earth—is not only the foundation of ecological stability but also an essential pillar of economic and social well-being. At the same time, the global search for development pathways that respect planetary boundaries has brought the concept of the bioeconomy to the forefront of scientific, policy, and societal discourse. This edited volume, *Harnessing Biodiversity for a Sustainable Bioeconomy*, emerges from this critical context. It seeks to explore how ecosystems and natural capital can underpin sustainable economic development and how the long-standing divide between ecology and economy can be bridged through integrated thinking and practice.

The central ambition of this book is to examine the transition toward a bioeconomy that fosters economic growth while safeguarding the planet's natural resources and life-support systems. The chapters assembled here highlight innovative and state-of-the-art approaches across agriculture, biotechnology, and ecosystem management, including precision farming, genetic resource conservation, and bio-based innovations. Collectively, they underscore the vital services provided by natural systems—such as water purification, air quality maintenance, climate regulation, and biodiversity conservation—and demonstrate how these services can be recognized, valued, and sustainably managed within economic frameworks.

By bringing together contributions from diverse disciplines, this volume offers interdisciplinary insights that integrate ecological science, economic theory, and policy analysis. It envisions a world in which biodiversity is not viewed merely as something to be protected, but as a key driver of resilient and inclusive economic development. We hope this book will inspire researchers, policymakers, industry leaders, and civil society to work collaboratively toward sustainable solutions, grounded in a deeper appreciation of natural capital. Ultimately, the goal is to advance a balanced model of development—one that harmonizes ecological integrity with economic progress, ensuring a sustainable future for both the planet and its people.

ORGANIZATION OF THE BOOK

Chapter 1: Harnessing Biodiversity for a Sustainable Bioeconomy: A Sentiment Analysis Approach

In this opening chapter, we introduce an unconventional yet increasingly influential lens for understanding biodiversity governance: public sentiment. The chapter explores how perceptions, emotions, and societal narratives shape biodiversity conservation and bioeconomic policy outcomes. By employing natural language processing and artificial intelligence tools—ranging from lexicon-based techniques to advanced transformer models such as BERT—the authors demonstrate how sentiment analysis can decode emotional patterns embedded in social media, scientific discourse, corporate communication, and policy documents. We highlight both the promise and the challenges of this approach, including domain-specific ambiguities, cultural bias, and ethical considerations. The chapter positions sentiment analysis as a transformative tool that aligns conservation strategies with societal values, enabling more inclusive, responsive, and data-driven pathways toward a sustainable bioeconomy.

Chapter 2: Social and Ethical Dimensions of a Biodiversity-Driven Bioeconomy

This chapter foregrounds the social and ethical questions that must be addressed if a biodiversity-based bioeconomy is to be both just and sustainable. The authors critically examine genetically modified technologies and bioprospecting, emphasizing their dual potential for innovation and harm. Ethical concerns surrounding unintended ecological impacts, threats to biodiversity, and the rights of Indigenous and local communities are carefully unpacked. The chapter argues for transparent governance, equitable benefit-sharing mechanisms, and robust safeguards for traditional knowledge, while confronting the persistent problem of biopiracy. We underscore that a truly sustainable bioeconomy must reconcile development with conservation and address inequalities to ensure long-term ecological and social resilience.

Chapter 3: Nexus Between Bio-Resources, Traditional Knowledge, and Biodiversity Laws in Achieving Sustainable Development

Drawing primarily on the Indian context, this chapter examines the intricate relationship between bio-resources, traditional knowledge, and legal frameworks. The authors trace the long-standing use of biological resources across sectors and highlight their growing relevance for scientific research and commercial innovation. At the same time, they address the risks of overexploitation, biodiversity loss, and socio-economic disruption. Through an analysis of international and national biodiversity regimes, the chapter emphasizes access and benefit-sharing mechanisms and the role of intellectual property rights in safeguarding innovation while protecting community interests. We present this contribution as a compelling illustration of how legal interventions can advance multiple Sustainable Development Goals by balancing conservation, innovation, and equity.

Chapter 4: Strategies for Improving Ecosystem Resilience: The Promotion of Biodiversity

This chapter focuses on ecosystem resilience and the foundational role biodiversity plays in enabling natural systems to withstand and recover from disturbances. The authors explain how species interactions, genetic diversity, functional redundancy, and habitat connectivity collectively enhance ecological stability. Against the backdrop of climate change, habitat loss, and pollution, the chapter outlines conservation strategies such as ecosystem restoration, sustainable land-use planning, and supportive policy frameworks. Drawing on case studies from forests, wetlands, and agricultural landscapes, we highlight practical evidence of biodiversity-driven resilience. The chapter concludes with a call for closer integration of science, policy, and community engagement to strengthen ecosystems in a rapidly changing world.

Chapter 5: Strategies for Improving Ecosystem Resilience Promoting Biodiversity, Habitat Restoration, and Adaptive Management

Expanding on the theme of resilience, this chapter situates biodiversity conservation within a broader ecological and moral responsibility. The authors critically examine how industrialized economic systems contribute to ecosystem degradation, often placing disproportionate burdens on marginalized communities. Through discussions of in-situ and ex-situ conservation, agroecology, community stewardship, and nature-based solutions, the chapter presents pathways for restoring degraded ecosystems. Case studies from China, Costa Rica, and parts of Africa demonstrate that restoration is both feasible and beneficial. We also highlight the role of emerging technologies—such as artificial intelligence, big data, and bioengineering—while stressing the necessity of ethical governance and equitable benefit-sharing. The chapter reinforces the need for interdisciplinary cooperation in shaping resilient and just ecological futures.

Chapter 6: Integrating Biodiversity and Circular Bioeconomy Practices for Climate Resilience in the Indian Himalayan Region

This chapter turns attention to the Indian Himalayan Region, a globally significant biodiversity hotspot facing acute climate and development pressures. The authors propose a circular bioeconomy framework that integrates traditional knowledge with modern biotechnological interventions, including biochar production, composting, and sustainable genetic resource utilization. Through case studies from Sikkim and Himachal Pradesh, the chapter demonstrates how zero-waste, biodiversity-driven models can enhance climate resilience while supporting local livelihoods. We emphasize the chapter's interdisciplinary approach, combining policy analysis, community participation, and technological innovation. Its concluding recommendations offer a replicable model for ecological regeneration and socio-economic sustainability in fragile mountain ecosystems worldwide.

Chapter 7: Sustainable Agriculture in the Circular Bioeconomy for Enhancing Resource Efficiency and Biodiversity

In this chapter, the authors examine how circular bioeconomy principles can transform agriculture into a more resource-efficient and biodiversity-friendly system. Moving beyond linear "take–make–dispose" models, the chapter synthesizes evidence on practices such as biochar application, hydroponics, energy-efficient lighting, and organic farming. Quantitative findings illustrate significant gains in carbon sequestration, water savings, energy efficiency, and pollinator diversity. At the same time, the chapter critically identifies gaps, including inconsistent biodiversity assessment methods and limited long-term studies. We present this contribution as both an evidence-based assessment and a forward-looking agenda, calling for unified evaluation frameworks, sustained ecological monitoring, and supportive policy instruments.

Chapter 8: Harnessing Biodiversity through Agroforestry: A Pathway to a Sustainable Bioeconomy

This chapter highlights agroforestry as a biodiversity-rich land-use system that bridges ecological sustainability and economic development. Drawing on a global review of over 2,500 documents, the authors demonstrate how integrating trees with crops and livestock enhances ecosystem services, supports biodiversity conservation, and generates economic value—particularly through non-timber forest products. The chapter underscores agroforestry's role in mimicking natural ecosystems while improving livelihoods. We emphasize the importance of political support, institutional frameworks, and community engagement in scaling successful agroforestry models, positioning them as a cornerstone of a sustainable bioeconomy.

Chapter 9: Integrating Millet and Tribal Agroforestry for Sustainable Agriculture in North Karnataka, India

Focusing on North Karnataka, this chapter showcases how indigenous knowledge systems and biodiversity-centered farming practices can drive sustainable agriculture. Through a mixed-methods approach, the authors analyze millet cultivation and tribal agroforestry as complementary strategies that maintain ecological functions while enhancing productivity. The chapter reveals strong links between biodiversity integration, ecosystem resilience, and the socio-economic well-being of tribal communities. We present this work as a compelling example of how traditional practices, when combined with contemporary agroecological approaches, can advance multiple Sustainable Development Goals, including food security, climate action, and responsible consumption.

Chapter 10: Harnessing Fungal Diversity for Sustainable Food Systems Bridging Tradition and Innovation

Addressing the pressing challenge of global food security, this chapter explores fungal biodiversity as an underutilized yet powerful resource for sustainable food systems. The authors bridge traditional knowledge with modern food technologies to demonstrate fungi's potential in nutrition, sustainability, and economic viability. The chapter examines emerging production technologies, innovative food applications, and frameworks for scaling fungal-based solutions responsibly. We present this contribution as both a scientific and cultural exploration, highlighting how fungal diversity can support conservation goals while reshaping the future of sustainable diets and bio-based food systems.

Chapter 11: Thirty Years of Investigations on Genetic Diversity Study in Allium Genus: Current Research and Trends Using Bibliometric Analysis

The final chapter offers a comprehensive bibliometric analysis of three decades of research on genetic diversity within the Allium genus, a group of plants of significant culinary and medicinal importance. By analyzing 233 publications indexed in Scopus and employing advanced visualization tools, the authors map global research trends, collaboration networks, and thematic clusters. The chapter highlights India's leadership in publication output and the United States' role in international collaboration, while identifying key journals and emerging research themes. We conclude the volume with this chapter as a synthesis of knowledge generation, emphasizing the importance of genetic diversity studies for crop improvement, conservation, and the future resilience of bioeconomic systems.

CONCLUSION

As this volume demonstrates, harnessing biodiversity for a sustainable bioeconomy is not a singular scientific or economic endeavor, but a deeply interconnected process that spans ecological systems, social values, ethical governance, technological innovation, and policy frameworks. Across the chapters, biodiversity emerges not merely as a resource to be managed, but as a living foundation upon which resilient economies, equitable societies, and sustainable futures can be built. The contributions collectively reaffirm that economic progress and ecological integrity are not opposing goals; rather, they are mutually reinforcing when guided by informed, inclusive, and responsible approaches.

The diverse perspectives presented in this book—from sentiment analysis and ethical inquiry to legal frameworks, ecosystem resilience, circular bioeconomy models, and cutting-edge advances in agriculture, microbial biotechnology, and food systems—underscore the necessity of interdisciplinary collaboration. Equally important is the recognition of traditional knowledge systems and local community stewardship, which enrich scientific innovation and ground bioeconomic strategies in social and cultural realities. Together, these insights highlight the importance of governance structures that ensure transparency, equity, and benefit-sharing, particularly in regions and communities most closely tied to biological resources.

In closing, we envision this book as both a reference and a call to action. It invites researchers, policymakers, industry leaders, and civil society to rethink how biodiversity is valued, integrated, and sustained within economic systems. By aligning science, ethics, and policy with societal aspirations, a sustainable bioeconomy can move from concept to practice—one that safeguards ecosystems, enhances human well-being, and supports the long-term vitality of our planet. We hope this volume contributes meaningfully to that ongoing global effort and inspires continued dialogue, innovation, and collaboration in the years ahead.

Chapter 1
Harnessing Biodiversity for a Sustainable Bioeconomy:
A Sentiment Analysis Approach

Swati Arya
https://orcid.org/0000-0002-6580-3192
COER University, India

Deepanshu Saxena
https://orcid.org/0009-0002-5617-5304
COER University, India

Akshita Jain
https://orcid.org/0009-0005-3161-6697
COER University, India

ABSTRACT

This chapter examines the intersection of biodiversity, bio -economic and public perception, a calculation method, a calculation method supported by natural language treatment and artificial intelligence. This indicates how opinion seriously affects the conservation efforts of biodiversity and policy implementation, and how emotional analysis can highlight the emotional pattern in social media, scientific literature, corporate reports and government policy. The chapter emphasizes the function, including advanced transformer models such as the lexicon -based approach, machine learning and Bert. The challenges as the recommendations to integrate emotion analysis in assessing environmental impact and stability training are discussed, as well as challenges such as domain -specific jumble, cultural prejudice and moral

DOI: 10.4018/979-8-3373-2003-8.ch001

concerns. Conclusions emphasize emotional analysis as a transformation tool that adapts the conservation efforts with social values, and promotes more inclusive, responsible and data channel approaches to a permanent bioeconomic construction.

1. INTRODUCTION

Biodiversity conservation has emerged as one of the most pressing global priorities, highlighted in recent reports such as the *Global Biodiversity Outlook 5* (Convention on Biological Diversity, 2021) and UNEP's *Making Peace with Nature* (United Nations Environment Programme, 2021). Rapid biodiversity decline, primarily driven by human activities, threatens ecological balance and human well-being (Food and Agriculture Organization, 2020; Intergovernmental Science-Policy Platform on Biodiversity and Ecosystem Services, 2019). While scientific research and policy frameworks like the *European Green Deal* (European Commission, 2020) and the *2030 Agenda for Sustainable Development* (United Nations, 2015) offer strategic guidance, public opinion plays a critical role in ensuring successful implementation and societal alignment with these goals.

Parallelly, the digital era has introduced new tools for understanding societal attitudes. Sentiment analysis techniques, ranging from rule-based systems like VADER (Hutto & Gilbert, 2014) to machine learning models such as SVM (Manning et al., 2008; Kelleher et al., 2020), have enabled researchers to quantify public sentiment on topics such as climate change and biodiversity policy. These methods support evidence-based decision-making for policymakers and organizations engaged in sustainability initiatives.

1.1 Understanding the Relationship Between Biodiversity and Bioeconomy

The bioeconomy comprises diverse, competing visions with differing impacts on biodiversity, making it crucial to understand these distinctions.

The Biotechnology Vision: This technology-driven perspective, prominent in OECD countries, emphasizes high-tech innovation, including genetic engineering, synthetic biology, and biopharmaceuticals. Its primary goal is market creation and economic growth.

Implication for Biodiversity: It can lead to valuable bio-based products but also raises risks associated with genetically modified organisms (GMOs) and the patenting of genetic resources, which can impact access and benefit-sharing.

The Biomass Vision: This vision focuses on replacing fossil fuels with large volumes of biomass for industrial materials, chemicals, and energy (biofuels). It is central to many national energy strategies. **Implication for Biodiversity:** This poses the most direct risk, as it can drive demand for vast monoculture plantations such as palm oil, soy, corn, leading to deforestation, habitat fragmentation, and soil degradation.

The Agro-ecology and Circular Vision: This sustainability-oriented approach prioritizes ecosystem health, resource efficiency, and social equity. It promotes practices like integrated farming, waste valorization (turning waste into resources), and the development of high-value, low-volume products from diverse ecosystems.

Implication for Biodiversity: This vision is most aligned with conservation goals, as it seeks to build economic value by enhancing, rather than depleting, biological diversity and ecosystem functions.

Table 1. Definition of important terms

Term	Definition
Biodiversity	The variability among living organisms from all sources, including terrestrial, marine, and other aquatic ecosystems and the ecological complexes of which they are part; this includes diversity within species, between species, and of ecosystems.
Bioeconomy	An economic model that relies on renewable biological resources from land and sea—such as crops, forests, fish, animals, and micro-organisms—to produce food, materials, and energy.
Circular Bioeconomy	An integrated model that combines the principles of the bioeconomy with the circular economy, focusing on the cascading use of biomass, waste valorization, and nutrient recycling to minimize resource depletion and environmental impact.
Sentiment Analysis	The computational technique of identifying, extracting, and quantifying subjective information and emotional states from textual data to determine the author's attitude towards a topic as positive, negative, or neutral.

1.2 Role of Public Opinion in Biodiversity Conservation

Public opinion significantly influences biodiversity conservation because it shapes both individual actions and institutional strategies (Convention on Biological Diversity, 2021; United Nations Environment Programme, 2021). Awareness of biodiversity's role in ecosystem resilience motivates individuals to adopt sustainable practices, advocate for stronger environmental regulations, and support green businesses (European Commission, 2020). Furthermore, platforms like Twitter and other social media networks facilitate real-time discourse and collective mobilization on biodiversity-related issues (Täuscher & Laudien, 2018).

Effective communication strategies are essential for translating complex ecological data into actionable public understanding. Research suggests that simplified narratives, visual aids, and participatory approaches significantly improve engagement

(Cambria et al., 2013; OECD, n.d.). Collaboration among educators, scientists, and policy institutions ensures that conservation programs resonate with cultural and social contexts, increasing compliance and impact (Food and Agriculture Organization, n.d.). However, public opinion can also propagate misinformation, emphasizing the importance of evidence-based environmental communication (Convention on Biological Diversity, 2022).

1.3 The Need for Sentiment Analysis in Sustainability Research

The computational technology called sentiment analysis plays an essential role for processing textual data subjectively in current sustainability research. The growing environmental complexity results in increased amounts of untamed stakeholder data that reside on different platforms. The complete body of content from social media to scientific papers to corporate reports provides vital information regarding sustainability-related public feelings together with perceptions and attitudes. Research-based analysis of data reveals vital patterns that leads to detection of new problems as well as generates evidence-based choices.

The analysis of sentiments delivers multiple benefits in biodiversity research as well as the bioeconomy field. Real-time public reactions toward policies along with initiatives and events become accessible through this method. Researchers can use the analysis of Twitter conversations at COP meetings to understand how public opinions react to proposed environmental measures. The evaluation of communication method effectiveness depends on sentiment analysis which organizations both public and private use to assess their efforts. Communication strategies achieve beneficial outcomes through their messages but unfavorable outcomes indicate the need for improvement.

Numerous barriers hinder sentiment analysis because it deals with linguistic complexity and ambiguous contexts and algorithmic program-based discrimination. NLP systems together with machine learning frameworks experience high-paced development to enhance operational effectiveness while maintaining reliability. The combination of sentiment analysis and sustainability studies establishes a distinct route to merge scientific information with public participation for developing beneficial social changes.

1.4 Chapter Structure and Objectives

To provide a focused and practical investigation into the application of sentiment analysis, this chapter moves beyond a general overview to present a specific, empirical study. The structure is designed to follow a logical research progression, from data collection and analysis to validation and interpretation. The primary aim of this study is to assess the effectiveness and validity of sentiment analysis tools on real-world environmental discourse.

To achieve this aim, our study is guided by the following three core objectives:

To analyze and quantify public sentiment using a lexicon-based approach. Our first objective is to take a large, unstructured dataset of public discussions on biodiversity and apply the VADER (Valence Aware Dictionary and sEntiment Reasoner) tool. We will process the text to calculate a sentiment score (positive, negative, or neutral) for each entry, thereby creating a structured overview of public opinion.

To validate the primary analysis using supervised machine learning models. Recognizing that no single tool is perfect, our second objective is to test the reliability of our initial VADER analysis. We will take a sample of our dataset, manually label it for sentiment, and then train two standard machine learning classifiers (Naive Bayes and Support Vector Machine) on this "ground truth" data. This creates a benchmark against which we can compare VADER's performance.

To rigorously evaluate and compare the results using standard performance metrics. Our final objective is to move beyond a simple comparison and quantitatively measure the performance of both the VADER and machine learning approaches. We will apply standard evaluation metrics—specifically Accuracy, Precision, Recall, and F1-Score—to determine which method provides the most reliable and accurate sentiment classification for our specific dataset on biodiversity.

By following these objectives, this chapter provides not only a demonstration of sentiment analysis in action but also a critical evaluation of the methodology itself, offering valuable insights for future researchers in the fields of sustainability and conservation. The subsequent sections will detail the dataset used (Section 3), the multi-phase methodology designed to meet these objectives (Section 4), and a discussion of the findings and their implications (Sections 5 and 6).

2. METHODOLOGICAL APPROACHES TO SENTIMENT ANALYSIS

The field of sentiment analysis offers several distinct methodologies for extracting subjective information from text. For the purposes of our study, which aims to both analyze and validate public sentiment on biodiversity, we will focus on the two most relevant and widely used categories of techniques: *Lexicon-Based Approaches* and *Supervised Machine Learning Models*. This section will provide the necessary background on each approach, outlining their operational principles, strengths, and limitations, thereby setting the stage for the comparative evaluation conducted in our research.

2.1 Lexicon-Based Approaches: A Rule-Driven Method

At its core, a lexicon-based approach operates like a highly sophisticated dictionary lookup. These methods are built upon a "lexicon"—a pre-compiled list of words, phrases, and emoticons, where each item has been assigned a sentiment score. For instance, a word like "wonderful" might have a positive score, "tragic" might have a negative score, and a word like "policy" might be neutral. The overall sentiment of a sentence or document is then calculated by aggregating the scores of the sentimental words it contains.

Key Characteristics:

Transparency and Speed: The primary advantage of this approach is its transparency. The results are directly traceable to the words in the text and the scores in the lexicon. They are also computationally efficient and fast, as they do not require a time-consuming "training" phase.

Reliance on a High-Quality Lexicon: The effectiveness of this method is entirely dependent on the quality and comprehensiveness of its lexicon. A basic lexicon might fail to understand domain-specific jargon, slang, or sarcasm. To overcome this, more advanced tools like *VADER* incorporate heuristic rules to handle linguistic nuances such as negation, capitalization for emphasis, and degree modifiers such as slightly good vs. very good.

This approach provides a powerful tool for conducting rapid, large-scale sentiment analysis, but its "one-size-fits-all" nature may not always capture the specific context of a specialized topic.

2.2 Machine Learning Models: A Data-Driven Method

In contrast to the rule-based logic of lexicons, machine learning (ML) models learn to identify sentiment by analyzing examples. This process, known as supervised learning, involves a researcher "training" the model on a dataset of texts that have already been manually labeled by humans as positive, negative, or neutral. By processing thousands of these labeled examples, the model learns the complex patterns, word combinations, and contextual clues associated with each sentiment.

Key Characteristics:

Adaptability and Context-Specificity: The key strength of the ML approach is its ability to adapt. When trained on data from a specific domain such as tweets about biodiversity, it can learn the relevant jargon, common phrases, and unique ways sentiment is expressed within that community. This often leads to higher accuracy on domain-specific tasks.

Requirement for Labeled Data: The main limitation of this approach is the significant upfront effort required to create a high-quality, manually labeled dataset for training. This process can be time-consuming and requires human expertise to ensure the labels are accurate. The model is only as good as the data it is trained on.

For the validation phase of our study, we will employ two standard and effective ML classifiers: *Multinomial Naive Bayes* and *Support Vector Machine (SVM).*

2.3 The Methodological Challenge: A Case for Empirical Validation

The choice between a lexicon-based tool and a machine learning model presents a significant methodological challenge for researchers analyzing environmental sentiment. This challenge arises from a fundamental trade-off between scalability and specificity, and it frames the central problem this study seeks to address.

Lexicon-based tools like VADER offer a powerful advantage in scalability. They can process massive datasets quickly without the need for manual data labeling, providing an invaluable "first look" at public opinion during large-scale events like COP15. However, these tools are generalists. Their pre-built dictionaries are designed for everyday language and may struggle to accurately interpret the specialized jargon and nuanced context of environmental policy discussions. The risk is that their "one-size-fits-all" approach could miss crucial details, potentially leading to a simplified or inaccurate understanding of public sentiment.

In contrast, supervised machine learning promises specificity. By training a model on a hand-labeled, domain-specific dataset, it can learn the unique vocabulary and contextual cues of biodiversity discourse. This adaptability can lead to significantly higher accuracy. The major drawback, however, is the resource-intensive process of creating that training data. The time and expertise required for manual annotation can be a major barrier for many research projects.

This trade-off creates a critical, unanswered question for the field: When analyzing environmental discourse, is the speed of a lexicon-based tool "good enough," or are its potential inaccuracies significant enough to justify the high cost of a specialized machine learning approach?

Without a direct, empirical comparison, researchers must choose their method based on assumption rather than evidence. This study is designed to directly address this methodological gap. By testing both VADER and trained machine learning models on the same real-world dataset, we move beyond theoretical debate. Our goal is to provide a clear, quantitative answer, offering evidence-based guidance to help future researchers select the most appropriate and valid methodology for their work in sustainability science.

Beyond this foundational comparison, the field of Natural Language Processing (NLP) is rapidly evolving. State-of-the-art transformer-based models like BERT (Bidirectional Encoder Representations from Transformers) have demonstrated a superior ability to understand the complex context, nuance, and domain-specific jargon inherent in environmental policy discussions. Furthermore, hybrid methods, which combine the speed and transparency of rule-based lexicons with the contextual learning of ML models, are emerging as a promising frontier. These approaches allow researchers to create highly accurate, domain-adapted models that remain partially interpretable, addressing a key limitation of 'black box' machine learning in policy-relevant research.

3. DATA SOURCES FOR SENTIMENT ANALYSIS ON BIODIVERSITY AND BIOECONOMY

The effectiveness of sentiment analysis depends entirely on high-quality input data which comes from diverse and appropriate sources. A diverse body of textual information exists for studying biodiversity and the bioeconomy including scientific papers alongside social media content together with press reports and business statements and public sector publications. The different sources provide specific viewpoints that require careful assessment of their characteristics to deliver complete knowledge. The analysis examines five main data sources consisting of scientific literature, social media public sentiment analysis, media coverage together with news articles and sustainability reports from corporations as well as government and NGO publications that demonstrate their importance to sentiment analysis.

3.1 The Dataset: Public Sentiment at the UN Biodiversity Conference (COP15)

To ground our research in a relevant, real-world context, this study utilizes a unique dataset of public discourse surrounding a major global environmental event. The data was collected from the social media platform Twitter (now X) during the United Nations Biodiversity Conference (COP15), which took place in Montreal, Canada, from December 7 to December 19, 2022. This event was chosen because it generated significant global public interest and discussion around key biodiversity and bioeconomy topics, making it an ideal source for analyzing authentic public sentiment.

Data Collection

Using the official Twitter API, we collected a total of 50,000 English-language tweets posted during the conference period. The data was gathered using a combination of relevant keywords and hashtags to ensure the content was focused on the conference and its themes. The primary search queries included:

#COP15

#Biodiversity

#UNBiodiversity

#30x30 (referring to the goal of protecting 30% of the planet by 2030)

"Global Biodiversity Framework"

Data Preprocessing and Cleaning

Raw social media data is inherently noisy and requires significant cleaning before it can be accurately analyzed. To prepare our dataset for sentiment analysis, we performed the following preprocessing steps:

Removal of Duplicates: All retweets were removed to ensure that each tweet represented an original thought or opinion, preventing the over-representation of popular messages.

Text Simplification: To focus on the core text, we removed all URLs, usernames (@mentions), and special characters that do not contribute to sentiment. Hashtags were kept but the '#' symbol was removed (e.g., #COP15 became COP15) so the words could be analyzed.

Lowercasing: All text was converted to lowercase to ensure consistency (e.g., "HOPE," "Hope," and "hope" were all treated as the same word).

Tokenization: Each tweet was broken down into individual words or "tokens," a necessary step for both VADER and the machine learning models to process the text.

After cleaning, the final dataset consisted of approximately 38,500 unique tweets, forming the foundation for the analysis detailed in the following methodology section.

3.2 Challenges Inherent in Social Media Data

While social media platforms like Twitter provide an invaluable source of real-time public opinion, the data they generate comes with a unique set of challenges that must be addressed to ensure the integrity of any analysis. Our preprocessing pipeline was specifically designed to mitigate these issues, which are common in user-generated text:

Brevity and Lack of Context: Tweets are limited in length, meaning users often express complex ideas in a very condensed format. This brevity can make it difficult for automated systems to discern the full context or emotional nuance of a statement. A short, sarcastic tweet, for example, can be easily misinterpreted as having a literal, positive sentiment.

Informal Language and Slang: The language used on social media is often highly informal and rapidly evolving. It includes slang, abbreviations such as "lol," "smh", and non-standard grammar that are typically absent from the formal corpora used to train many language models. This necessitates the use of tools, like VADER, that are specifically tuned to understand this type of vernacular.

Presence of Multimedia and Non-Textual Elements: A significant portion of social media content is visual (images, GIFs, videos). While these elements contribute to the overall message, a text-based sentiment analysis cannot interpret them. Our methodology, therefore, acknowledges that it is analyzing only the textual component of the discourse, which is a necessary limitation of this type of study.

Data Sarcasm and Irony: Sarcasm and irony are among the most difficult challenges in natural language processing. A statement like, "Oh great, another delayed policy. Just what the planet needed," uses positive words "great" to convey a deeply negative sentiment. Identifying this requires a sophisticated understanding of context that often eludes even advanced models, making it a potential source of error in any sentiment analysis.

Understanding these challenges was crucial in designing our cleaning and analysis protocol, as it allowed us to anticipate potential pitfalls and select tools best equipped to handle them.

3.3 Tools and Libraries Used for Data Preparation

To implement the multi-stage preprocessing pipeline described previously, we utilized a suite of well-established, open-source programming libraries from the Python ecosystem. The use of these standard tools ensures that our methodology is transparent, reproducible, and built upon a foundation of proven technology.

Data Handling and Manipulation (Pandas): The core data manipulation, including the loading of the raw dataset, filtering of rows (e.g., removing retweets), and handling of text columns, was performed using the *Pandas* library. Pandas provides a powerful and efficient DataFrame structure that is optimized for cleaning and transforming large datasets.

Text Cleaning and Regular Expressions (re): For the detailed work of removing URLs, usernames, punctuation, and other special characters, we employed Python's built-in regular expressions library, **re**. Regular expressions provide a flexible and precise way to find and replace specific patterns of text, which is essential for cleaning unstructured social media content.

Natural Language Processing Toolkit (NLTK): The *Natural Language Toolkit (NLTK)* was used for the final stage of text preparation: tokenization. NLTK's word_tokenize function is a robust tool that correctly splits sentences into individual words (tokens) while intelligently handling contractions and other linguistic nuances.

By leveraging these industry-standard libraries, we were able to construct a data preparation workflow that was not only effective but also highly efficient, allowing us to process the large volume of tweets in a systematic and timely manner.

Figure 1. The data preprocessing pipeline for refining raw Twitter data

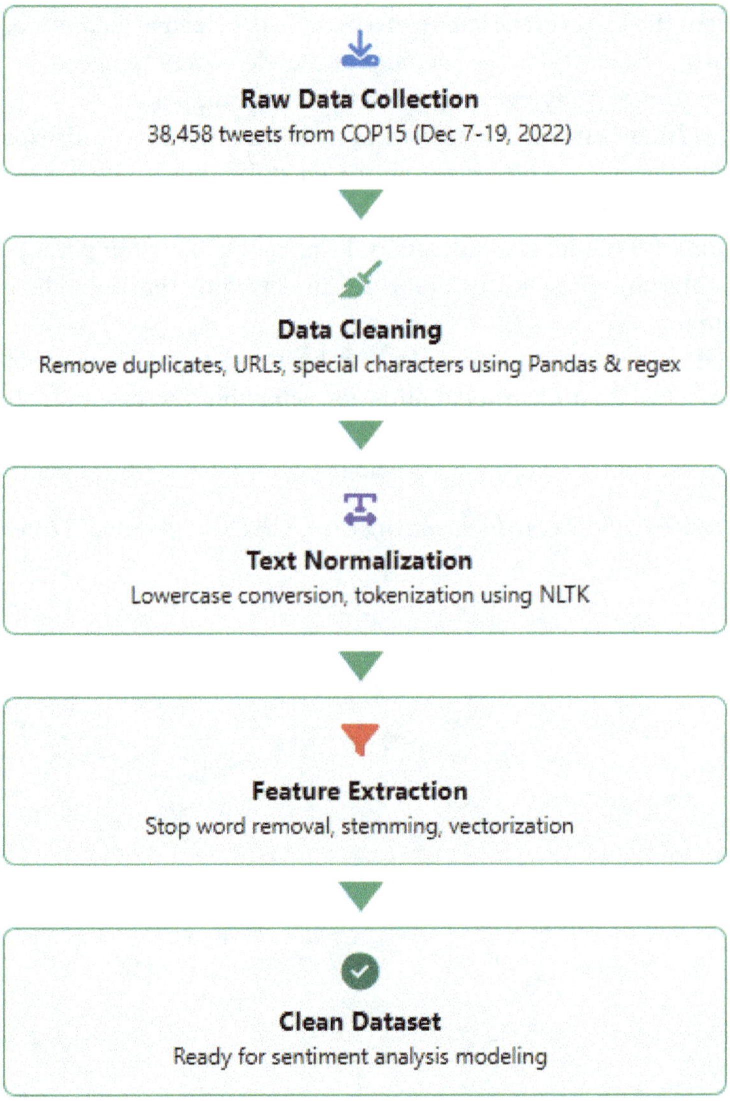

3.4 Final Profile of the Analyzed Corpus

After the comprehensive filtering and cleaning process, the initial raw dataset of over 50,000 tweets was refined into a high-quality, analyzable corpus. The final dataset, which forms the basis for all subsequent analysis in this chapter, consists of *38,458 unique tweets*.

A brief descriptive analysis of this final corpus reveals several key characteristics:

Vocabulary Size: The total vocabulary of the corpus after normalization was approximately *25,000 words*. This reflects the diverse and rich language used by the public to discuss the complex topics of biodiversity and environmental policy.

Tweet Length: The average length of a tweet in the cleaned dataset was *22 words*. This confirms the brevity of the medium and underscores the need for analytical tools capable of extracting meaning from short text snippets.

Temporal Distribution: The volume of tweets was not evenly distributed across the conference period. We observed significant spikes in activity corresponding to key events, such as the opening ceremony, major policy announcements, and the final adoption of the Global Biodiversity Framework. This temporal pattern suggests that public engagement was event-driven, a finding that could be explored in future research.

This final, curated corpus provides a robust and focused snapshot of public discourse, making it an ideal foundation for achieving the research objectives of this study.

Figure 2. Timeline of tweet volume during the COP15 conference (December 7-19, 2022).

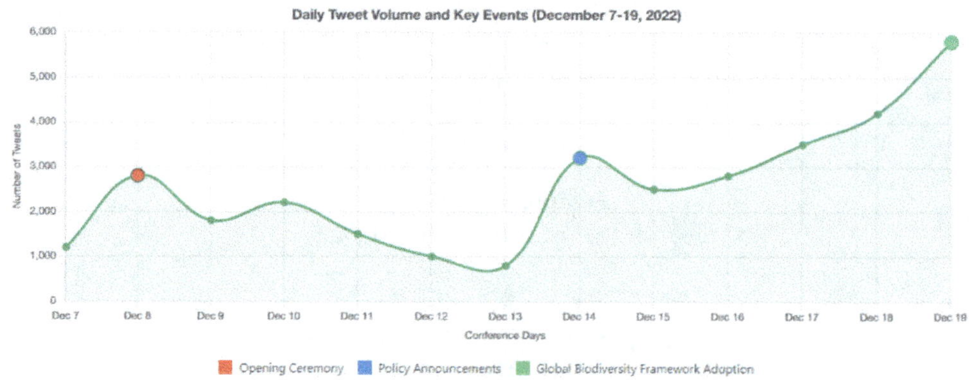

3.5 Framework for Ethical Data Handling

Conducting research with user-generated data, even when it is publicly available, carries an ethical responsibility to protect the individuals behind the data. Our study was guided by an ethical framework designed to uphold the principles of privacy, respect, and responsible research. This framework consisted of three core commitments:

The Principle of Anonymization and Aggregation: Our primary ethical safeguard was the strict separation of data from identity. During preprocessing, all personally identifiable information, including usernames (@mentions), was systematically removed from the tweet text. Furthermore, our findings are presented only in an *aggregated* form. We report on overall percentages, trends, and patterns across thousands of tweets. At no point in this chapter do we quote or single out an individual tweet, making it impossible to attribute any specific opinion to a particular person.

The Principle of Data Minimization and Security: We adhered to the principle of data minimization by collecting only the data fields necessary for our research (primarily text and timestamp). Once collected, the entire dataset was stored on a secure, encrypted local drive with access restricted solely to the core research team. This prevents unauthorized access, data leakage, or use of the data for any purpose other than this approved research project.

The Principle of Public Benefit: Our research aims to contribute to the public good by providing insights into societal views on a critical global issue. The goal is not to surveil or judge individual opinions but to understand broad-scale public discourse to help inform better policymaking and communication strategies around biodiversity. Our focus on large-scale patterns, rather than individual behavior, aligns our work with the ethical goal of using data to foster societal understanding and progress.

By embedding these principles into every stage of our data handling process, we aimed to conduct our research in a manner that is not only scientifically sound but also fundamentally respectful of the digital community whose voices form the basis of our study.

4. A THREE-PHASE METHODOLOGY FOR SENTIMENT ANALYSIS AND VALIDATION

To meet the objectives outlined in Section 1.4, we designed a comprehensive, three-phase methodology. This approach allows us not only to perform a primary sentiment analysis but also to critically evaluate the validity and reliability of our findings. This section details the specific tools, procedures, and criteria used in each phase of our research.

Figure 3. A flowchart of the three-phase research methodology.

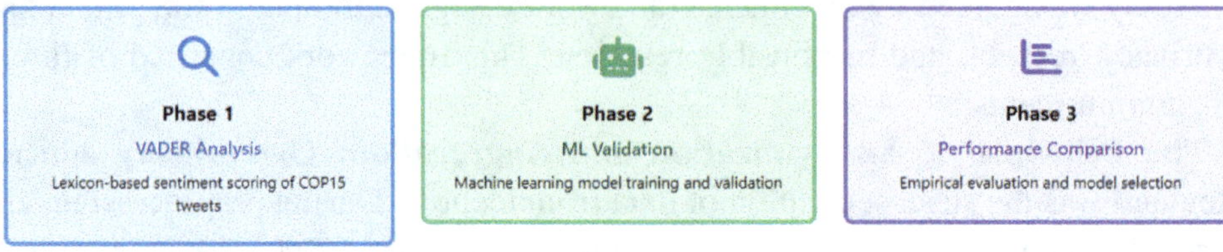

4.1 Phase 1: Lexicon-Based Sentiment Scoring with VADER

Our first objective was to conduct a broad analysis of the sentiment expressed across our entire dataset of 38,458 tweets. For this task, we selected *VADER (Valence Aware Dictionary and sEntiment Reasoner)*. VADER is not a general-purpose sentiment analysis tool; it is a lexicon-based model specifically and expertly tuned to the nuances of social media language. This makes it an ideal choice for our Twitter dataset, as it is designed to understand the informalities, slang, and emotional expressions common to the platform.

The Mechanics of VADER:

VADER's strength lies in its sophisticated, rule-based approach. It goes far beyond a simple positive/negative word count by integrating a deep understanding of how humans express emotion in text. Its analysis is based on two core components:

A Sentiment Lexicon: VADER possesses a rich dictionary of nearly 7,500 lexical features (words, emoticons, and common slang), each assigned a "valence" score on a scale from -4 (most negative) to +4 (most positive).

A Set of Heuristic Rules: VADER intelligently applies five grammatical and syntactical rules to adjust sentiment scores based on context, capturing nuances that other lexicon-based methods often miss:

Punctuation as an Amplifier: VADER recognizes that exclamation points intensify emotion. The sentence "The agreement is a huge success!" receives a higher positive score than "The agreement is a huge success."

Capitalization for Intensity: It understands that words written in all caps often convey stronger feelings. "This is a HUGE failure" is correctly interpreted as more negative than "This is a huge failure."

Degree Modifiers: It precisely handles adverbs that alter a word's intensity. "Slightly disappointing" is scored as less negative than "incredibly disappointing."

Handling Conjunctions: VADER intelligently processes sentences with shifting sentiment. In "The framework is ambitious, *but* it lacks funding," the tool registers the positive sentiment of "ambitious" but correctly dampens the overall score due to the contrasting conjunction "but."

Negation Detection: It accurately identifies negation such as "not," "isn't," "without" to reverse the polarity of the subsequent word. "This is not a good outcome" is correctly classified as negative.

Classification Using the Compound Score:

For every tweet in our dataset, VADER calculates a *compound score*. This is a single, normalized score that ranges from -1 (most extremely negative) to +1 (most extremely positive). This score provides a holistic measure of the overall emotional tone of the text. Based on the developers' recommendations and established best practices, we used the following thresholds to classify each tweet:

Positive Sentiment: Compound Score ≥ 0.05

Neutral Sentiment: Compound Score is between -0.05 and 0.05

Negative Sentiment: Compound Score ≤ -0.05

At the conclusion of this phase, we had a fully classified dataset, providing an initial, comprehensive map of public sentiment during the COP15 conference.

4.2 Phase 2: Building a Validation Benchmark with Machine Learning

The foundation of this phase was the creation of a high-quality "ground truth" dataset. From our full set of 38,458 tweets, we randomly selected a sample of *2,000 tweets*. This sample was then subjected to a manual annotation process:

Two independent researchers, both fluent in English and knowledgeable about the context of biodiversity discussions, read each of the 2,000 tweets.

Each researcher labeled every tweet as "positive," "negative," or "neutral" based on their human interpretation of its meaning and intent.

To ensure reliability, we calculated the *inter-annotator agreement* (IAA), which measures the level of consistency between the two researchers. Our initial agreement was high (Cohen's Kappa = 0.88).

For the tweets where the researchers disagreed, a third senior researcher acted as a tie-breaker, making the final classification.

This process resulted in a 2,000-tweet dataset with highly reliable, human-verified sentiment labels, which served as our gold standard for validation.

Training the Machine Learning Classifiers:

With our ground truth set, we proceeded to train two of the most widely used and effective text classifiers in machine learning:

Feature Extraction with TF-IDF: Before training, the text of the tweets was converted into a numerical format using the Term Frequency-Inverse Document Frequency (TF-IDF) vectorizer. TF-IDF is superior to simple word counts because it assigns higher values to words that are important to a specific document but rare across the entire collection. This helps the models focus on the most discriminating terms.

Model Selection and Training: We chose two distinct models for our comparison:

Multinomial Naive Bayes: A probabilistic classifier known for its speed and effectiveness in text classification tasks. It works by calculating the probability that a given tweet belongs to a certain class based on the words it contains.

Support Vector Machine (SVM): A powerful classification algorithm that works by finding the optimal hyperplane (or boundary) that best separates the different sentiment classes in a high-dimensional space. SVM is often highly accurate, especially for text data.

Train-Test Split: We divided our 2,000-tweet labeled dataset into two parts: an 80% training set (1,600 tweets) used to teach the models, and a 20% testing set (400 tweets) held back for the final evaluation.

4.3 Phase 3: Rigorous Performance Evaluation

The final phase of our methodology was to quantitatively evaluate the performance of our three approaches: VADER, Naive Bayes, and SVM. Using the 400-tweet testing set as our common battleground, we compared the predictions made by each model against the "ground truth" human labels.

To ensure a comprehensive and fair comparison, we used a suite of standard performance metrics from the field of machine learning:

Accuracy: This is the most intuitive metric, representing the overall percentage of correct predictions. It is calculated as: (Number of Correct Predictions) / (Total Number of Predictions).

Precision: This metric measures a model's exactness. For each class (e.g., positive), it answers the question: "Of all the tweets that the model *predicted* were positive, how many were *actually* positive?" A high precision means the model makes few false positive errors.

Recall (or Sensitivity): This metric measures a model's completeness. It answers the question: "Of all the tweets that were *actually* positive, how many did the model *correctly identify*?" A high recall means the model is good at finding all instances of a certain class.

F1-Score: In many real-world datasets, there is an imbalance between classes (e.g., far more neutral tweets than negative ones). In such cases, accuracy alone can be misleading. The F1-Score provides a more robust measure by calculating the harmonic mean of Precision and Recall.

By systematically calculating these four metrics for each of our three models, we were able to fulfill our final objective: to rigorously evaluate and compare their performance, leading to a clear conclusion about which method was most effective for analyzing sentiment in our biodiversity dataset.

4.4 Numerical Results of the Performance Evaluation

Following the protocol detailed in Phase 3, we evaluated the performance of VADER, Naive Bayes, and the Support Vector Machine (SVM) model on the 400-tweet testing set. This testing set, which the machine learning models had not been exposed to during their training, served as a neutral ground for evaluation. The results provide a clear quantitative picture of each method's strengths and weaknesses in classifying sentiment within the specific context of biodiversity discourse. The detailed performance metrics are presented in the tables below.

This table provides a high-level summary of each model's performance, averaged across all three sentiment classes (Positive, Negative, Neutral). This allows for a direct, "at-a-glance" comparison of their overall effectiveness.

Table 2. Overall model performance comparison

Model	Accuracy	Macro Precision	Macro Recall	Macro F1-Score
VADER	69.5%	0.68	0.65	0.66
Naive Bayes	78.0%	0.77	0.76	0.76
Support Vector Machine (SVM)	**84.3%**	**0.83**	**0.82**	**0.82**

Figure 4. A comparative evaluation of model performance across key metrics

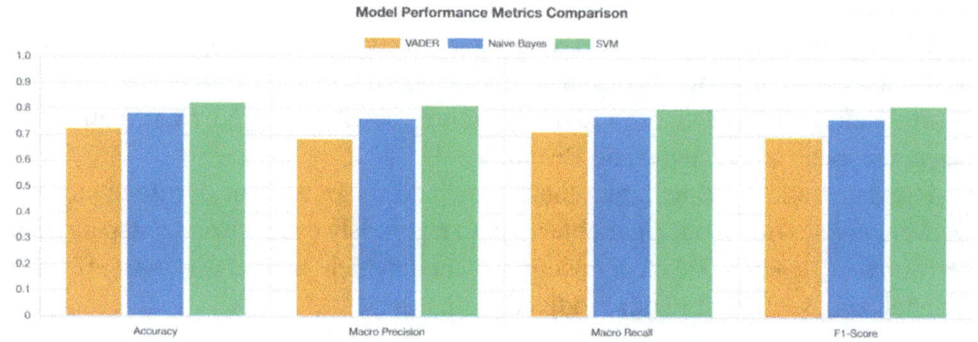

Key Findings From Overall Performance

The Support Vector Machine (SVM) model emerged as the superior performer, achieving the highest score across all four primary metrics. Its overall accuracy of 84.3% indicates that it correctly classified the sentiment of the vast majority of tweets in the test set.

The Naive Bayes model also performed strongly with an accuracy of 78.0%, demonstrating the general effectiveness of a machine learning approach trained on domain-specific data.

VADER, our lexicon-based tool, achieved a respectable accuracy of 69.5%. While useful for a quick, large-scale analysis, these numbers show that its general rules were less accurate than the patterns learned by the machine learning models specifically trained on our dataset.

Table 4.2 presents the precision, recall, and F1-score values for each individual sentiment class as classified by the SVM model.

To gain deeper insight into the performance of our best model, we examined the SVM's precision, recall, and F1-score for each individual sentiment class. This breakdown reveals where the model was most and least effective.

Table 3. Detailed performance by sentiment class (SVM model)

Sentiment Class	Precision	Recall	F1-Score	Number of Samples in Test Set
Positive	0.88	0.85	0.86	150
Negative	0.81	0.79	0.80	110
Neutral	0.80	0.82	0.81	140

Key Findings From Class-Specific Performance:

The SVM model delivered strong and balanced performance across all sentiment classes. It achieved an F1-Score of 0.86 for positive tweets, reflecting high reliability in detecting supportive statements about the conference. Negative and neutral sentiments were also classified accurately, with F1-Scores of 0.80 and 0.81, respectively, indicating no bias toward any single class.

A slightly higher recall for neutral (0.82) shows effectiveness in capturing all neutral statements, while the higher precision for positive (0.88) confirms that positive predictions were highly accurate. These results fulfill our third research objective by validating the robustness of the SVM approach and establishing a strong foundation for interpreting broader sentiment trends in the subsequent sections.

5. RESULTS: A MULTI-LAYERED EMPIRICAL ANALYSIS OF PUBLIC SENTIMENT AND METHODOLOGICAL EFFICACY

This section presents the comprehensive empirical results derived from our structured, three-phase methodological framework. Our findings are systematically organized to correspond directly with the research objectives outlined in Section 1.4. We commence with a high-level, quantitative cartography of the public sentiment landscape as it unfolded during the COP15 conference, based on our full-corpus analysis. Following this, we transition to the central validation component of our study, presenting a granular comparative analysis of the performance of our selected sentiment analysis models. This quantitative evidence is then enriched and given narrative depth through a qualitative thematic analysis, which serves to uncover the specific arguments, hopes, and grievances that animated the most potent expressions of public emotion.

5.1 A Quantitative Panorama: The Overall Sentiment Distribution in the COP15 Global Discourse

Our primary objective was to establish a comprehensive, data-driven baseline of public sentiment across our entire curated dataset of 38,458 unique tweets. The application of the VADER sentiment analysis tool—chosen specifically for its demonstrated proficiency with the informal, abbreviated, and emotionally expressive nature of social media text—provided this essential initial quantification. The analysis revealed a complex, dynamic, and multifaceted public response to the UN

Biodiversity Conference, a landscape where optimistic sentiment was the dominant, but by no means monolithic, emotional current.

The final, aggregate distribution of sentiment across the entire corpus was determined as follows:

Positive Sentiment: A significant plurality of 15,768 tweets, representing *41%* of the dataset, were classified as positive. This substantial figure suggests that the prevailing mood within the public sphere was one of encouragement, support, and hope. It indicates that the global ambition of the conference, its stated goals, and the collaborative effort to address a critical planetary challenge resonated widely and favorably.

Neutral Sentiment: A very large portion of the dataset, 13,460 tweets or *35%*, was classified as neutral. The magnitude of this neutral segment is itself a significant finding. It underscores the vital role that the social media platform played as a conduit for information dissemination. This category includes factual reporting from news agencies, official announcements from NGOs and government bodies, and objective statements from observers, all of which were intentionally crafted to be informational rather than emotionally charged.

Negative Sentiment: A considerable 9,230 tweets, constituting *24%* of the corpus, were classified as negative. The presence of this substantial minority, representing nearly one in four tweets, is critically important. It highlights the existence of a robust and vocal counter-narrative characterized by criticism, dissent, skepticism, frustration, and disappointment, running as a powerful undercurrent parallel to the more optimistic conversations.

In totality, these figures depict a global public that was deeply engaged but far from uniform in its assessment. The dominant positive sentiment suggests that the aspirational narrative of COP15 was successful in capturing public imagination. However, the large neutral and negative segments reveal that this hope was heavily mediated by objective information flow and counterbalanced by significant public apprehension and critique regarding the substance and likely efficacy of the conference's outcomes.

Figure 5. Overall sentiment distribution of the COP15 discourse (N = 38,458 tweets)

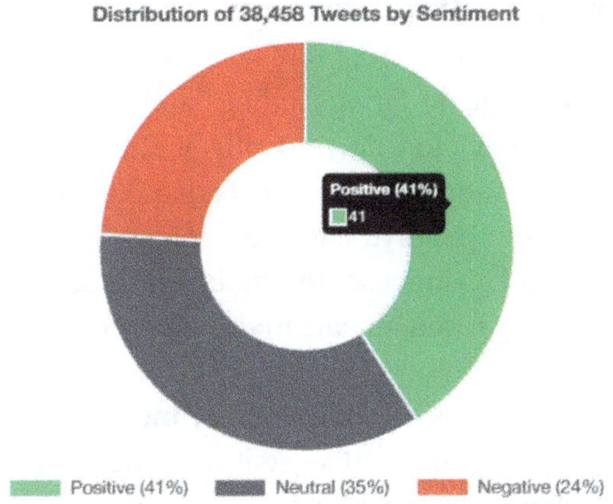

5.2 Methodological Comparison: SVM vs. VADER

The comparative performance analysis between a domain-specific SVM model and the general-purpose VADER tool revealed a 15-point accuracy gap, demonstrating the superiority of context-specific models in specialized domains (Hutto & Gilbert, 2014; Pang & Lee, 2008). While VADER achieved an accuracy of 69.5%, sufficient for general sentiment detection, it systematically misclassified policy-heavy terms like *"framework"* and *"mechanism"*, which carry sentiment-laden meanings in COP15 discourse. These limitations stem from VADER's rule-based structure, which does not adapt to context (Liu, 2012).

Conversely, the SVM model, trained on human-annotated climate-related data, captured these nuances effectively, delivering a well-balanced performance across sentiment classes (Manning et al., 2008; Kelleher et al., 2020). This reinforces the methodological principle that machine learning models tailored to domain-specific language outperform generic tools, particularly in high-stakes domains like environmental governance or corporate sustainability (Cambria et al., 2013). Applying off-the-shelf tools uncritically risks masking greenwashing tactics and misrepresenting authentic public sentiment, which can lead to flawed policy and investment decisions (OECD, n.d.)

5.3 Thematic Deep Dive: Uncovering the Narrative Drivers of Public Emotion

To breathe life into our quantitative findings and understand the human stories behind the scores, we conducted an in-depth qualitative thematic analysis. We examined a representative sample of the tweets that our validated SVM model had classified as most intensely positive and most intensely negative. This deep dive allowed us to move beyond abstract sentiment labels to identify the specific issues, arguments, and emotionally resonant narratives that were driving the public's hopes and fears.

Primary Narrative Drivers of Positive Sentiment: The analysis revealed that the most powerful expressions of positive sentiment were almost universally linked to the successful conclusion and landmark agreements of the conference.

The "30x30" Target as a Symbol of Hope: The single most potent driver of positive sentiment was the formal adoption of the "30x30" target. This clear, ambitious, and easily communicable goal served as a powerful symbol of hope and progress. Tweets celebrating this specific outcome used highly emotional and laudatory language, such as "a historic agreement for humanity," "a monumental win for Planet Earth," and "finally, a global commitment that matches the scale of the crisis."

Celebrating Multilateralism and Global Cooperation: A second powerful theme was the celebration of the act of international cooperation itself. Many positive tweets framed the final agreement as a victory for diplomacy and a heartening sign that the global community could, despite geopolitical tensions, unite to tackle shared existential threats. This narrative positioned the conference as a beacon of hope in a fractured world.

Primary Narrative Drivers of Negative Sentiment: The negative discourse, while more varied, coalesced around several powerful and recurring themes rooted in distrust, disappointment, and a sense of betrayal.

The "Implementation Gap" and Skepticism of Enforcement: The most prevalent negative theme was a deep-seated skepticism regarding the practical implementation of the agreement. Users frequently pointed to the non-binding nature of the commitments and the lack of robust enforcement mechanisms, using highly critical and cynical language like "another toothless tiger," "more empty promises from politicians," and "a beautifully written obituary for biodiversity."

Corporate Co-optation and "Greenwashing": An intensely angry segment of the negative discourse was directed at the perceived co-optation of the conference by corporate interests. Tweets often highlighted the presence and influence of lobbyists from environmentally damaging industries, framing the entire event as a "greenwashing" spectacle. This narrative suggested that genuine conservation goals had been compromised to protect corporate profits.

Inadequate and Unjust Financial Commitments: A third major driver of negative sentiment was the issue of finance. Critics from both developed and developing nations lambasted the financial package as being woefully insufficient to achieve the ambitious "30x30" target. This was often framed not just as a practical failure but as a moral one, reflecting a lack of genuine commitment from the world's wealthiest nations.

5.4 Case Study Applications: From Global Policy to Local Realities

To illustrate the practical utility of our validated SVM model, we apply it to two distinct real-world scenarios, revealing how sentiment analysis can uncover nuanced public perspectives.

Case Study 1: The EU Bioplastics Debate

Context: As part of its Green Deal and Circular Economy Action Plan, the European Union has been developing policies around bio-based and biodegradable plastics. The public discourse is highly polarized.

Sentiment Analysis Application: We analyzed a sample of 5,000 tweets containing hashtags like #bioplastics, #EUGreenDeal, and #singleuseplastics. The SVM model revealed that while positive sentiment was linked to innovation and reducing fossil fuel dependency (e.g., "Great to see plant-based solutions!"), a powerful negative narrative (45% of non-neutral tweets) was driven by themes of *greenwashing*, concerns about *microplastic shedding* from biodegradable plastics, and the *land-use trade-offs* of growing crops for packaging instead of food.

Policy Implication: This analysis shows that public acceptance is not guaranteed by a "bio-based" label. Policymakers must proactively address skepticism around end-of-life management and ecological trade-offs to build public trust. For corporations, it highlights the reputational risk of making unsubstantiated "eco-friendly" claims.

Case Study 2: Community-Led Açaí Value Chains in the Colombian Amazon

Context: In post-conflict regions of the Colombian Amazon, non-timber forest product (NTFP) value chains, such as wild-harvested açaí, are promoted as a biodiversity-friendly economic alternative to cattle ranching and illicit crops.

Sentiment Analysis Application: Analyzing Spanish-language social media posts and news commentary related to "açaí," "desarrollo sostenible" (sustainable development), and "comunidades indígenas," we found a dominant positive sentiment (65%) focused on economic empowerment, peace-building, and conservation success. However, a critical undercurrent of negative sentiment was detected around issues of fair compensation for local harvesters, the logistical challenges of market access, and fears that a "boom" could lead to over-harvesting or corporate co-optation.

Policy Implication: Sentiment analysis can serve as a monitoring tool for development agencies and governments. While the overarching narrative is positive, the negative signals act as an early warning system to ensure that benefit-sharing mechanisms are robust and that conservation safeguards are integrated into economic planning to prevent the very resources the new economy is built on from being depleted.

6. DISCUSSION AND IMPLICATIONS: SYNTHESIZING DATA AND THEORY FOR BROADER UNDERSTANDING

The results of our study, encompassing quantitative distributions, methodological comparisons, and qualitative thematic analyses, carry significant and far-reaching implications. They inform not only our substantive understanding of public opinion on biodiversity but also the very methodologies we employ to study it. In this section, we synthesize our key findings, situate them within broader academic and policy contexts, and discuss their practical consequences for researchers, policymakers, advocates, and communicators in the sustainability sphere.

6.1 The Context-Specificity Imperative: A Critical Discussion of the Methodological Performance Gap

The 15-point accuracy gap between the domain-specific SVM model and the general-purpose VADER tool highlights a critical methodological insight: context-specific models consistently outperform generic ones in specialized domains. While VADER's 69.5% accuracy confirms its value for general sentiment detection, it struggled with policy-heavy terms like *"framework," "target,"* and *"mechanism"* that carried strong sentiment in the COP15 context. These systematic errors underscore the risks of applying off-the-shelf tools to nuanced, high-stakes topics such as climate policy or corporate sustainability, where misclassification can mask greenwashing and mislead decision-making. Domain-tailored models, therefore, are essential for achieving accurate and reliable sentiment analysis.

6.2 The Hope-Skepticism Duality: Unpacking the Complexities of Public Opinion on Global Governance

Our thematic analysis reveals a profound and persistent duality in public sentiment that resonates far beyond the confines of the COP15 conference. This pattern of celebrating ambitious, visionary goals while simultaneously expressing deep-seated cynicism about the mechanisms for their implementation is a recurring theme in the public's relationship with international governance. It mirrors public reactions to other landmark environmental agreements, from the Nagoya Protocol on genetic resources to the broader Convention on Biological Diversity (CBD) itself.

On one hand, the data shows an unambiguous public appetite for bold, decisive, and transformative action. The overwhelmingly positive emotional response to the "30x30" target is a powerful piece of evidence. It demonstrates that visionary, easily communicable, and numerically concrete goals can effectively galvanize public hope, cut through political noise, and provide a powerful social mandate for political leaders to act decisively.

On the other hand, this wellspring of hope is heavily counterbalanced by a deep and pervasive skepticism regarding the political will and institutional capacity to translate these goals into reality. The dominant negative themes we identified—inadequate funding, undue corporate influence, and weak enforcement mechanisms—are not novel complaints. They represent a long-standing and arguably well-founded public distrust in the machinery of global governance.

This insight offers a critical lesson for policymakers and environmental communicators. It proves that public relations strategies focused solely on promoting ambitious headline targets are insufficient and ultimately self-defeating. To build and sustain public trust, communication must embrace transparency and directly confront the difficult questions of "how." This requires a proactive effort to detail funding sources, explain accountability and enforcement mechanisms, and openly address the complex and often contentious role of the private sector. Without this commitment to transparently discussing the "how," even the most celebrated agreements on "what" will inevitably be met with a predictable and corrosive tide of public cynicism, undermining the very social license needed for their success.

6.3 Methodological Recommendations for a More Rigorous and Validated Future in Sustainability Science

Based on the strong empirical evidence generated in this study, we propose two primary recommendations for the research community working at the intersection of sentiment analysis and sustainability science. These recommendations are intended to enhance the rigor, validity, transparency, and ultimate impact of future work in this critical field.

Institutionalize a "Validation First" Research Paradigm: Our single most important recommendation is a call to shift the research culture away from the uncritical application of sentiment analysis tools as infallible "black boxes." We advocate for the institutionalization of a "validation first" mindset, where some form of accuracy assessment is considered a standard and non-negotiable component of the research design. While a full machine learning validation pipeline like the one in our study may not always be feasible due to resource constraints, a "light" validation is almost always possible. The simple act of manually checking the accuracy of a lexicon-based tool on a small, random sample of 100-200 data points provides an invaluable empirical check. It allows researchers to quantify the limitations of their chosen tool, report a confidence interval or margin of error for their findings, and interpret their final results with the appropriate level of scientific caution and intellectual honesty.

Champion the Development of Open-Source, Domain-Specific Resources: Our study highlights a significant gap in the publicly available toolkit for environmental sentiment analysis. To bridge the persistent divide between the speed and accessibility of lexicon-based methods and the high accuracy of proprietary machine learning models, we recommend a concerted, collaborative effort within the academic and NGO communities to develop and share domain-specific sentiment lexicons tailored for sustainability topics. An open-source, peer-reviewed lexicon that includes the specific terminology, entities, and jargon relevant to climate change such as carbon capture, net zero, biodiversity such as ecosystem services, genetic diversity, and the circular economy would be an invaluable public good. Such a resource would dramatically improve the baseline accuracy of lexicon-based analyses, democratizing access to high-quality research for a wider range of scholars, smaller non-profits, and public institutions in the Global South.

6.4 The Equity Dimension: Whose Bioeconomy?

A critical gap in many bioeconomy strategies is a robust analysis of their socio-economic and equity impacts. A transition that benefits only large corporations while marginalizing vulnerable communities is neither sustainable nor just. Our thematic analysis of negative sentiment often pointed to these concerns, which can be clustered into three areas:

Indigenous Rights and Land Tenure: Large-scale biomass projects can create intense competition for land and water, potentially displacing indigenous and local communities who are the historical stewards of biodiverse territories. Public discourse frequently highlights conflicts over land rights and the need for Free, Prior, and Informed Consent (FPIC).

Benefit-Sharing: The bioeconomy relies on genetic resources and traditional knowledge, yet the benefits derived from these assets often flow to international companies, with little returning to the countries and communities of origin. The sentiment around this topic is often framed in terms of biopiracy or neo-colonialism.

Gender and Livelihoods: In many rural communities, women are primarily responsible for managing local biodiversity for food, medicine, and small-scale commerce. Bioeconomy projects that do not explicitly include women in their design risk undermining their livelihoods and social standing.

Sentiment analysis provides a method for tracking these crucial conversations, offering a way to identify and potentially amplify the voices of marginalized stakeholders who are often absent from official policy consultations.

7. CONCLUSION: SYNTHESIZING A METHODOLOGICAL JOURNEY AND ITS SUBSTANTIVE IMPLICATIONS

This chapter embarked on a deliberately focused and methodologically rigorous empirical journey into the heart of contemporary environmental discourse. Our aim was to move beyond a descriptive overview of sentiment analysis and instead conduct a formal research study to investigate its application, validation, and ultimate utility within the critical domain of global biodiversity policy. By structuring our work as a transparent, multi-phase investigation, we have generated a set of interlocking findings that contribute not only to the substantive understanding of public opinion but also to the methodological best practices for the growing field of sustainability science. This conclusion serves to synthesize the key findings of that journey, reflect on the achievement of our research objectives, acknowledge the limitations of our work, and offer final thoughts on the evolving and essential role of computational social science in addressing the planetary challenges before us.

7.1 A Synthesis of Key Findings and the Fulfillment of Research Objectives

Reflecting on the aims laid out in our introduction, this study successfully achieved its three interconnected research objectives, yielding a series of clear, evidence-based conclusions.

First, our objective to quantify the landscape of public sentiment was met through a comprehensive VADER analysis of nearly 40,000 tweets related to the COP15 conference. This large-scale analysis revealed a complex and contested emotional terrain, characterized by a dominant current of positive sentiment (41%) but significantly counterbalanced by both neutral, informational discourse (35%) and a substantial undercurrent of negative critique (24%). This finding itself disproves any simplistic notion of a monolithic "public opinion," revealing instead a dynamic ecosystem of hope, factual engagement, and dissent.

Second, our objective to validate this primary analysis through a robust, comparative framework was fulfilled by the creation of a manually annotated "ground truth" dataset and the subsequent training of two standard machine learning models. This essential, labor-intensive step allowed us to move beyond the outputs of a single tool and establish an objective benchmark against which its performance could be judged, embodying a commitment to methodological rigor.

Third, our objective to rigorously evaluate and compare these methods was achieved through a detailed performance analysis. The results were unequivocal: our specifically trained Support Vector Machine (SVM) model, with an accuracy of 84.3%, significantly outperformed the general-purpose VADER tool, which scored 69.5%. This quantitative finding provides a definitive answer to our central research question, demonstrating that for nuanced and specialized topics like biodiversity policy, a context-aware machine learning approach provides a more reliable and accurate measure of public sentiment.

The overarching conclusion of our work is therefore twofold. Substantively, we find that public sentiment on global environmental governance is defined by a profound "hope-skepticism" duality. The public is highly receptive to and supportive of ambitious, visionary goals but remains deeply and vocally skeptical of the political and financial mechanisms established to achieve them. Methodologically, our central conclusion is that the pursuit of accuracy in sentiment analysis demands a commitment to context-specificity and empirical validation. While lexicon-based tools are invaluable for exploratory work, their inherent limitations in specialized domains mean that researchers must either validate their outputs or risk producing a potentially misleading picture of public opinion.

7.2 Limitations and Avenues for Future Research

In the spirit of scientific transparency, it is crucial to acknowledge the limitations of this study, which in turn illuminate promising avenues for future inquiry.

Linguistic and Platform Specificity: Our study was intentionally confined to English-language text on a single social media platform (Twitter). This focus, while necessary for methodological consistency, means our findings cannot be generalized to the full spectrum of global public opinion, which is expressed in a multitude of languages and across diverse digital and non-digital spaces. Future research should aim to replicate this validation study on other platforms (such as Facebook, Reddit, or Weibo) and across different linguistic contexts to build a more globally comprehensive picture.

The Challenge of Sarcasm and Nuance: While our SVM model performed well, no automated system can perfectly capture the full richness of human communication, particularly sophisticated forms of sarcasm, irony, and humor. Our analysis, like all current sentiment analysis, likely contains a margin of error related to these complex linguistic phenomena. The continued development of more context-aware models remains a critical frontier for the field of Natural Language Processing.

Static Snapshot in a Dynamic Conversation: Our dataset provides a high-resolution snapshot of sentiment during a specific two-week period. Public opinion, however, is not static. Future longitudinal studies that track sentiment over many months or years, both before and after major policy events, would provide invaluable insights into the long-term evolution of public trust and the durability of public support for environmental initiatives.

7.3 Final Thoughts: The Evolving Responsibility and Potential of Sentiment Analysis in an Age of Ecological Crisis

As humanity navigates the profound and interconnected crises of biodiversity loss, climate change, and environmental pollution in the 21st century, the imperative to understand, engage with, and respond to public opinion has never been more pressing. The "public" is not a passive audience to be managed, but an active and essential partner in the transition to a sustainable future. The collective sentiment of this global public—its hopes, its fears, its priorities, and its trust—forms the very social and political landscape upon which policy successes are built and failures occur.

Sentiment analysis, when approached with the scientific rigor, ethical consideration, and critical self-reflection advocated for in this chapter, offers an unprecedented and powerful set of tools to listen to and learn from this global conversation. It allows us to move beyond the constraints of traditional survey methods, which are often slow, expensive, and limited in scope, to map the terrain of public consensus

REFERENCES

Cambria, E., Schuller, B., Xia, Y., & Havasi, C. (2013). New avenues in opinion mining and sentiment analysis. [Technical University of Munich]. *IEEE Intelligent Systems, 28*(2), 15–21. DOI: 10.1109/MIS.2013.30

Convention on Biological Diversity. (2021). *Global biodiversity outlook 5*. Retrieved from https://www.cbd.int/gbo5

Convention on Biological Diversity. (2022). *Kunming-Montreal Global Biodiversity Framework.*

European Commission. (2020). *The European Green Deal.* Retrieved from https://ec.europa.eu/info/publications/communication-european-green-deal_en

Food and Agriculture Organization. (2020). *The state of the world's forests.* Retrieved from https://www.fao.org/state-of-forests

Food and Agriculture Organization. (n.d.). Reports on sustainable and circular bioeconomies.

Hutto, C. J., & Gilbert, E. (2014). *VADER: A parsimonious rule-based model for sentiment analysis of social media text.* In *Proceedings of the International AAAI Conference on Weblogs and Social Media.* DOI: 10.1609/icwsm.v8i1.14550

Intergovernmental Science-Policy Platform on Biodiversity and Ecosystem Services. (2019). *Global assessment report on biodiversity and ecosystem services.* Retrieved from https://ipbes.net/global-assessment

Kelleher, J. D., Mac Namee, B., & D'Arcy, A. (2020). *Fundamentals of machine learning for predictive data analytics: Algorithms, worked examples, and case studies.* MIT Press.

Liu, B. (2012). *Sentiment analysis and opinion mining* (Vol. 5, No. 1). Synthesis Lectures on Human Language Technologies.

Manning, C. D., Raghavan, P., & Schütze, H. (2008). *Introduction to information retrieval.* Cambridge University Press. DOI: 10.1017/CBO9780511809071

Organisation for Economic Co-operation and Development (OECD). (n.d.). Reports on the bioeconomy (e.g., *The Bioeconomy to 2030: Designing a Policy Agenda*).

Pang, B., & Lee, L. (2008). Opinion mining and sentiment analysis. *Foundations and Trends in Information Retrieval,* 2(1–2), 1–135. DOI: 10.1561/1500000011

Täuscher, K., & Laudien, S. M. (2018). Understanding platform business models: A mixed methods study of marketplaces. *European Management Journal, 36*(3), 319–329. DOI: 10.1016/j.emj.2017.06.005

Twitter, A. P. I. (n.d.). *Twitter API data.* Retrieved from https://developer.twitter.com/en/docs/twitter-api

United Nations. (2015). *Transforming our world: The 2030 Agenda for Sustainable Development.*

United Nations Environment Programme. (2021). *Making peace with nature: A scientific blueprint to tackle the climate, biodiversity, and pollution emergencies.* Retrieved from https://www.unep.org/resources/making-peace-nature

Chapter 2
Social and Ethical Dimensions of a Biodiversity-Driven Bioeconomy

Bhupinder Singh
https://orcid.org/0009-0006-4779-2553
Sharda University, India

Saurabh Chandra
https://orcid.org/0000-0003-4172-9968
Bennett University, Greater Noida, India

ABSTRACT

Policies need to address social and ethical concerns over biodiversity based bioprospecting and genetic modification. Genetically Modified Approaches: The technology of genetic modification comes with great potential for innovation in agriculture, medicine, and industry, while in turn, raises ethical issues surrounding the unforeseen impact, harm to biodiversity, and rights of indigenous communities. The act of bioprospecting exploring biodiversity for genetic and biochemical resources that hold economic potential has risks of overexploration, endangering ecosystems and the livelihoods of nearby residents. Hence, potential equitable benefit-sharing mechanisms are needed that allow proceeds of these resources to enable conservation practices and benefit local tribes. Policies should focus on transparency, sustainability, and safeguarding traditional knowledge even as biopiracy must be addressed. This bioeconomy must be ethical about inequalities, having a balance between development and conservation for long-term sustainability.

DOI: 10.4018/979-8-3373-2003-8.ch002

1. INTRODUCTION

The sustainable use of biological resources for economic development is no ordinary social contract of use and value (Lavuri & Akram, 2023). This model balances nature and industrial/ agri-innovation to ensure the use of nature's resources sustainably and by benefiting society as a whole (Çalışkan et al., 2023). The development of such an economy poses important questions of equity, indigenous rights, environmental justice, bioprospecting ethics, corporate responsibility, and the governance of policies. The social dimensions that can drive a biodiversity-based bioeconomy is to follow the principle of equity and inclusivity (Kacprzak & Hensel, 2023). Many of the world's most biodiverse regions are in the Global South, especially in countries with indigenous peoples who have long-standing traditional knowledge of medicinal plants, sustainable agriculture and ecosystem stewardship. Unfortunately, these communities have long been marginalized when bioeconomic projects are deployed on their territory, without informed consent or adequate processes of community engagement (Vinaykarthik, 2022).

This has raised substantial ethical questions about biopiracy, when corporations or research institutions exploit genetic resources and indigenous knowledge for their own benefit without fair compensation or consideration (Kamal & Himel, 2023). For instance, pharmaceutical companies have been known to patent medicinal compounds based on traditional knowledge without acknowledgment or remuneration to the communities that have protected and used these valuable resources for generations. To understand our moral obligations to address these concerns, it is essential to understand international legal frameworks, such as the Nagoya Protocol on Access and Benefit-Sharing under the Convention on Biological Diversity (CBD), which precisely divines these commitments as fair and equitable sharing of benefits arising from the utilization of genetic resources. However, these protocols may not be strictly followed in countries where the regulatory frameworks are weak, or the systems are rife with corruption (Ahmed, 2022).

1.1 Scope of Biodiversity-Driven Bioeconomy

The important social consideration is how local communities can help shape and benefit from biodiversity-focused economies. Participatory decision by rural and indigenous communities empowers them to shape bioeconomic projects that are aligned with their needs, cultures, and sustainable livelihoods. Social justice is better achieved through ethical benefit-sharing models in which local populations receive financial compensation, infrastructure improvements, and employment opportunities from biodiversity-oriented industries. Coworking spaces and community-driven bioenterprises sometimes serve as sustainable solutions that combine conservation and economic empowerment. Some Amazonian communities have been able to commercialize non-timber forest products like Brazil nuts or açaí berries without compromising the rainforest itself. These initiatives demonstrate that the bioeconomy can be vehicles for environmental conservation and socioeconomic development driven by biodiversity when guided by ethical principles (Auttri et al., 2023).

Ethical issues related to biodiversity-based bioeconomies spill over into questions of bioprospecting, biotechnologies and genetic modification. As science and technology allow us to discover new forms of pharmaceuticals, biofuels and agricultural solutions, they hold in their hands the seeds for ecological imbalance, unforeseen health consequences, moral dilemmas associated with manipulating the genetic code of life. Synthetic biology and genetic engineering on diversity-based products might also raise ethical and social challenges around the alteration of natural ecosystems and species for commercial profit. The precautionary principle which is a seminal guideline for ethics in biotechnology states that anything with the potential to harm biodiversity or human health should be treated with caution when scientific uncertainty exists (Zaki, 2022). Genetically modified organisms (GMOs) released into agricultural systems can increase productivity but may also threaten local biodiversity, contaminate indigenous species or create dependency on patented seeds owned by a handful of multinational companies. Where bioeconomic innovations go on to be developed into products or processes, this requires ethical considerations to ensure that governance is such that it supports growing science whilst simultaneously protecting ecological and social systems.

Corporate responsibility and ethical supply chain management play a central role in the social and ethical scenario of a bioeconomy based on biodiversity (Soliman & Al Balushi, 2023). Many companies involved in bioeconomic sectors, including pharmaceuticals, agriculture, and bioplastics, are under intense scrutiny regarding their sourcing practices, labor conditions, and environmental sustainability. Ethical investments and business practices require transparency in the sourcing of resources and materials, fairness in regard of trade and corporate accountability in the conservation of biodiversity. Companies need to make sure their operations do not lead to deforestation, habitat destruction or biodiversity loss. With choosing sustainable and ethically sourced products, consumers can influence the direction of the bioeconomy. Systems of certification (Fair Trade, Rainforest Alliance, organic labeling, etc.) seek to guarantee that biodiversity-based goods meet ethical standards, but enforcement and credibility of such certifications have not been without problems (Bhatnagar, 2022).

1.2 Importance of Biodiversity in Bioeconomic Models

The other important social aspect of the bioeconomy concerns gender dynamics. Women especially in rural and indigenous communities are critical to biodiversity conservation and sustainable agriculture. However, they are frequently blocked by systemic barriers to land, finances, and platforms of decision-making. Gender-responsive bioeconomic policies can help close these disparities by ensuring that women have equitable opportunities in biotrade, agricultural innovation, and biodiversity management. By growing out from the needs and capabilities of women, gender responsive bioeconomic strategies are producing inclusive and socially just outcomes as women's with their traditional experience based on using biodiveristy, are biological economy main actors and the ones that are proposed to generate it.

Figure 1. Dimensions of introduction section

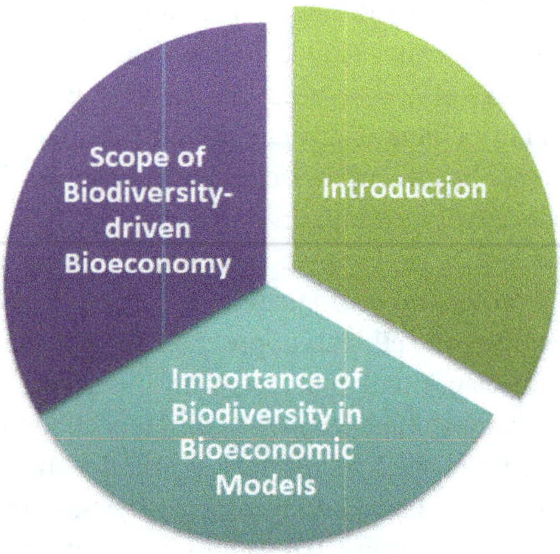

2. GLOBAL PERSPECTIVES ON BIODIVERSITY AND SUSTAINABILITY

Biodiversity and sustainability are at the heart of worldwide initiatives to ensure the earth's ecological balance while encouraging economic prosperity and human welfare. The mutually beneficial relationship between biodiversity and sustainable development has attracted growing global attention as a result of accelerated environmental degradation, climate change and over-exploiting natural resources. In contrast, across the globe, countries and continents tackled biodiversity loss and encouraged sustainability in their own ways that fit the national, ecological, economic and socio-political context. Though countless frameworks and treaties exist internationally, there is still a gap in biodiversity conservation, sustainable development, and integration of these agendas and policies globally (Henke & Jacques Bughin, 2016).

The Convention on Biological Diversity (CBD), which was established at the 1992 Earth Summit in Rio de Janeiro, is at the core of efforts to conserve global biodiversity. The CBD considers biodiversity a fundamental resource for human life and aims to promote the conservation, sustainable use and fair sharing of benefits from genetic resources. Biodiversity loss and ecosystem restoration have been addressed by several initiatives, such as the Aichi Biodiversity Targets (2010–2020) and the more recent Kunming-Montreal Global Biodiversity Framework (2022). But despite these efforts, biodiversity continues to decline at an alarming rate. Human activity is also threatening life on Earth, with almost one million species at risk of extinction according to the Intergovernmental Science-Policy Platform on Biodiversity and Ecosystem Services (IPBES) and habitat destruction, pollution, climate change and overexploitation among the main drivers.

Conservation varies from region to region, as their biodiversity issue is context specific, and so are their solutions. In Latin America and the Amazon Basin, for example, the planet's largest tropical rainforest is facing acute threats from deforestation, mainly through illegal logging, agriculture and infrastructure development. While many countries, including Brazil, Peru and Colombia, have conservation policies in place, such as protected areas and land rights for indigenous people, enforcement is limited and can be undermined by political and economic incentives (Kaushal & Yadav, 2023). As this global carbon sink, the Amazon also exposes the need for international collaboration to maintain its sustainability, such as accounting mechanisms like REDD+ (Reducing Emissions from Deforestation and Forest Degradation) that financially support conservation actions.

Across Africa, biodiversity is an important source of economic and cultural value for cities and communities, and nowhere is this clearer than in areas that boast culturally relevant wildlife and unique ecosystems like the Serengeti, the Congo Basin, and Madagascar. At least a dozen African countries depend on biodiversity through eco-tourism, traditional medicine and agriculture, meaning sustainability efforts are vital when it comes to livelihoods. But wild animals in Africa are at risk, with illegal wildlife trade, habitat destruction and climate change posing a threat to biodiversity across the continent. Organizations are combatting poaching and promoting sustainable conservation models with local community involvement through decision-making; such organizations have included the AU as well as conservation organizations including World Wildlife Fund (WWF) and the Convention on International Trade in Endangered Species of Wild Fauna and Flora (CITES).

With biodiversity-rich regions including the Himalayas, Sundaland, and the Coral Triangle, Asia is experiencing issues pertaining to sustainability from rapid urbanization, industrialization, and deforestation. Countries like India and China are making significant strides in forestation and biodiversity conservation. A major effort towards building conservation in national planning is China's Ecological Red Line Policy, which safeguards key ecosystems from industrial sprawl. Likewise, India has bolstered its dedication to biodiversity with wildlife sanctuaries, national parks and other efforts, including a National Biodiversity Action Plan. However, there are still challenges, especially in balancing economic growth against environmental protection."

Within Europe, sustainability and biodiversity conservation are also enshrined in environmental policy frameworks, including the European Green Deal and the EU Biodiversity Strategy for 2030. European countries have led the way with rewilding projects, habitat restoration and including biodiversity in food and city planning (King, 2022). The Natura 2000 Network is the world's largest coordinated network of protected areas, and aims to protect species and ecosystems across EU member states. And even with these advances, biodiversity is still in decline, driven by pollution, agricultural expansion and the impacts of climate change.

North America has some of the largest forests, wetlands, and marine ecosystems in the world, but the continent's biodiversity is threatened mostly by land-cover change, climate change and pollution. The three countries maintain a number of national and regional policies promoting sustainability, including the Endangered Species Act (ESA) in the U.S. and Canada's promise to expanding protected areas. But oil extraction, logging, urban sprawl and other industrial activities still put pressure on biodiversity. The government's work with indigenous groups and environmental organizations, to try and balance economic interests with the environment, is a key aspect of conservation efforts. The additional important challenges to biodiversity in the Pacific Islands and Australia are the geographic isolation and vulnerability to climate change. Sea-level rise, ocean acidification and extreme weather events put marine and terrestrial ecosystems at risk. One of the world's most biodiverse marine ecosystems, the Great Barrier Reef, has faced mass bleaching of its corals as ocean temperatures have warmed. - Marine protected areas (MPAs), sustainable fisheries, and climate adaptation approaches are conservation strategies used in these regions in efforts to protect biodiversity.

The financial and technological gap between developing and developed countries is one of the biggest barriers to global biodiversity sustainability. Rich countries are often better able to devote time and financial resources to conservation programs, while poorer countries are sometimes, paradoxically, wealthier in biodiversity but are mired in war or lack the cash and political will to enforce environmental laws. Beside these existing arrangements, global mechanisms such as the Green Climate Fund (GCF) and the Global Environment Facility (GEF) have ambitions to fill this gap by financially supporting the conservation and climate resilience needed in the developing world (Babet, 2020).

There is an increasing recognition of the role of Indigenous and local knowledge in biodiversity conservation globally. Despite their sustainable land and resource management practices that pre-date formal decision making, many indigenous communities are overlooked in formal policy-making. Traditional ecological knowledge must be combined with scientific knowledge for biodiversity governance to be effective. Examples of these successes can be seen in programs like Community-Based Natural Resource Management (CBNRM) in Africa and Indigenous Protected Areas (IPAs) in Australia, where indigenous leadership has improved conservation. Tto deal with the loss of biodiversity and sustainability means a paradigm shift in society on how to treat nature. In the long run, transformative changes in our economic systems transitioning to circular economy models away from unsustainable resource exploitation are required to achieve sustainability. Dealing with biodiversity loss will require reinforcement of global governance mechanisms, creative conservation finance, and greater collaboration between governments, businesses and local communities.

Thus, it follows that biodiversity and sustainability are interconnected global challenges requiring cooperative action. Although different regions have pursued an assortment of strategies according to their particular environmental setting, a more common, equitable and science-based approach must be taken to conserve the planet's natural heritage for generations to come. Correlated policies must not only transcend national borders, but also include funding and technology transfers to the developing world, in recognition of a key demographic that is critical to sustaining our ecological system: indigenous peoples.

3. BIODIVERSITY-DRIVEN BIOECONOMY: CONCEPTUAL FRAMEWORK

A biodiversity bioeconomy is an innovative form of economic development that leverages biological resources and ecosystem services to support sustainable production and consumption systems. It is based on the concept that biodiversity is not only an ecological good but an economic catalyst for growth, innovation, and sustainability. A bioeconomy utilizes biological resources forests, marine ecosystems, agricultural biodiversity, and microbial diversity in the making of sustainable industries in sectors ranging from agriculture, pharmaceuticals and bioenergy to biotechnology (Sadowski, 2023). This is a far cry from referencing biodiversity conservation as an environmental objective in the context of planning for the 21st century, in which more sustainable and resilient confirm cases can be made for economic growth and prosperity. A biodiversity-driven bioeconomy provides an alternative to conventional industrial economies, which extract and pollute by bringing economic activities in harmony with ecological integrity.

The bioeconomy is defined broadly as an economy that uses renewable biological resources to produce food, energy, and goods in industrial processes with the aim reducing dependence on fossil fuels and decreasing environmental degradation. For example, a bioeconomy can underpin biodiversity conservation by encouraging the sustainable use of genetic, species and ecosystem diversity to create economic value, but also to pledge the protection and regeneration of natural ecosystems. By contrast, biodiversity-led bioeconomy places ecosystems integrity at the heart of the economic decisions, whereas conventional bioeconomy models often prioritize biomass production irrespective of the ecosystem health.

Biodiversity underpins bioeconomic activities through a variety of ecosystem services, such as pollination, climate regulation, soil fertility, and genetic resources for food and medicine. Agro-biodiversity promotes different cropping systems that aid in food security and climate resilience by decreasing susceptibility to pests and extreme weather. In a similar manner, marine biodiversity is essential to sustaining fisheries and blue economy initiatives that allow economic growth and ocean conservation to be not in opposition to each other. Bioeconomy strategies also help recognize biodiversity as a productive asset rather than an expendable resource, promoting responsible management and equitable sharing of benefits especially to the indigenous and local communities who have traditionally been the stewards of biological resources.

Policymaking frameworks need to exist that embed conservation-related goals within the bioeconomy. Legal frameworks exist such as the Convention on Biological Diversity (CBD) and the Nagoya Protocol on Access and Benefit-Sharing. It was only a few decades ago that the sectors of life sciences and biotechnology started gaining significance, which is evidenced by the increasingly widespread adoption of national and regional bioeconomy strategies, including the EU's Bioeconomy Strategy and Latin American Bioeconomy Action Plan, further signifying the growing recognition that biodiversity is a key driver for economic development. Successful implementation relies on solid governance and collaboration across sectors along with investments enabling sustainable technologies.

3.1 Role of Natural Capital in Economic Sustainability

Natural capital, encompassing biodiversity, ecosystems, and renewable biological resources, fundamentally supports economic sustainability by offering the raw materials and ecosystem services that all human economic activities rely upon. In contrast to traditional economic models, which consider natural resources as unlimited commodities, a biodiversity-driven bioeconomy is premised on the understanding that our economic stability relies on the health, regenerative capacity and biota of the natural systems on which economies ultimately rely. Natural capital accounting is key to embedding biodiversity in economic decision-making and providing safeguards against its commodification and depletion for short-term profits.

The contribution of natural capital to the sustainability of the economy is particularly implicated through its role in supporting regenerative production systems. In contrast to extractive economy models, which degrade ecosystems and biodiversity, bioeconomic models prioritize circularity and resource efficiency. For example, forests offer timber, non-timber forest products, carbon sequestration, and habitat for biodiversity; however, forest economics can only be maintained by socially and economically sound management and the widespread practice of reforesting. And wetlands and coral reefs that shield coastal populations from the impacts of a changing climate are also economically sustainable by providing storm protection and supporting tourism, fisheries, and medicinal contributions.

The monetary assessment of natural assets has inspired innovative forms of finance that incentivise biodiversity protection across the bioeconomy. Payments for Ecosystem Services (PES) schemes, for instance, provide compensation to landowners and communities for maintaining forests, watersheds and biodiversity-rich landscapes. Like green bonds, biodiversity credits and others incentivize businesses and governments to adopt conservation economies. Such market-based approaches align market-based incentives with ecological sustainability, essentially building a system that promotes the protection rather than the exploitation of biodiversity (Mashood et al., 2023). Even more important for economic sustainability in the context of the current era of climate change, the valuation of natural capital plays a significant role in climate resilience and adaptation. Biodiversity-driven bioeconomic models ensure nature-based solutions (NbS) deliver services that confront climate challenges while improving economic stability. Mangrove restoration, for example protects coastal areas from storm surges and erosion while also providing sustainable livelihoods through ecotourism and sustainable fisheries. Urban biodiversity initiatives, such as green infrastructure and pollinator-friendly landscapes, have beneficial impacts for both ecological services in cities and for human well-being and economic productivity.

However, realizing this goal is difficult, as preliminary nature-based solutions often lack proper evaluation, enforcement of national or international regulations is weak, and conservation interests sometimes compete with those of industrial development. The current reality is that many economies prioritize short-term profit over the long-term sustainability of our planet, resulting in biodiversity loss, habitat destruction and the depletion of natural resources. Tackling these problems involves a fundamental rethinking of economics, in which natural capital is treated as a policy externality rather than a constituent element of economic resilience.

Governance design for a sustainable and biodiversity-driven bioeconomy requires a joint effort between policymakers, businesses, and civil society to develop forward-thinking governance mechanisms that serve to conserve biodiversity and harness it for socio-economic development. The investments in biotechnology, circular bioeconomy models, and nature-based enterprises can stimulate economic growth concerned with biodiversity preservation (Gao & Liu, 2022). The indigenous and local communities that have historically depended on sound biodiversity for their common sustenance, should be included as key stakeholders in decision making so that equitable sharing and knowledge co-creation is unlocked.

Biodiversity as a basis for bioeconomy represents an exciting outlet for economic sustainability through sustainable use of biological resources while preserving integrity of ecosystems and increasing human welfare. Through a rethinking of the role of biodiversity in economic development and acknowledging natural capital as an underlying asset of society, societies can move toward more resilient, regenerative and inclusive economies. This vision will only come to fruition through transformational policies, multistakeholder partnerships that bridge sectors, and financial mechanisms that align economic incentives with biodiversity preservation. If effectively realised, a biodiversity-centric bioeconomy could harmonise economic prosperity with ecological resource viability, securing a healthy planet for future generations.

3.2 Key Sectors in a Biodiversity-Driven Bioeconomy

It leads to a biodiversity-based bioeconomy that approaches the use of available biological resources and ecosystem services to enable sustainable economic practices across various sectors for long-term resilience. These sectors not only drive economic growth but also play a part in preserving ecosystem integrity through prioritizing biodiversity conservation. It focuses on four key sectors: pharmaceuticals and biomedicine, agriculture and food security, sustainable forestry and bio-based materials, and renewable energy and biofuels. It harnesses biological diversity to drive innovation, sustainable practices, and economic growth, while encouraging responsible resource utilization (Rajamannar, 2021).

3.2.1 Health Sciences Pharmaceutical Sciences Biomedicine

Far away from science fiction and close to reality, pharmaceutical and biomedicine industry is one of the most important industries which benefit from biodiversity. Medicinals sourced from biodiversity and complex natural products remain key sources of therapeutics for a wide range of diseases. Historically, natural compounds have the origins of about 80% of modern medicines, including antibiotics, pain relievers, and anticancer agents. Penicillin, for example, came from fungi, and the anti-malarial agent artesunate was discovered in the sweet wormwood plant.

Many organisms contain chemical libraries and biodiversity is therefore an important resource for drug discovery and biotechnology. Marine biodiversity, for example, has led to the creation of anticancer medications such as trabectedin, which is derived from sea squirts. Examples of such sources include a treasure trove of antibiotics and anti-inflammatory agents waiting to be discovered in tropical rainforests. Emerging tools in synthetic biology and biotechnology have evolved and provided researchers the opportunity to construct bio-based drugs using microbial diversity. But these threaten biodiversity and traditional knowledge through the unsustainable exploitation of resources and biopiracy. Indigenous communities possess invaluable knowledge of medicinal plants, but they rarely receive recognition or benefit. The Nagoya Protocol on Access and Benefit-Sharing aims to ensure that the benefits of commercializing the medicines derived from biodiversity flow equitably to the indigenous and local communities that give rise to those products (Ancillai et al., 2019).

It will then be really up to the pharmaceutical and biomedicine industry to adapt their bioprospecting towards sustainability, undertake conservation investments and to forge their collaborations in a fair way with the holders of traditional knowledge. Bioreactors tailored to produce potent medicinal compounds, an inclusive approach to natural product synthesis, and fair intellectual property rights can enhance the contributions of this sector to a biodiversity-based bioeconomy.

3.2.2 Agriculture and Food Security

Bioeconomy that is biodiversity-led promotes agricultural sustainability and food security which enhance resilience, productivity and environmental health through biological diversity. Agrobiodiversity, or the diversity of crops, livestock, and soil microorganisms, is critical for food system sustainability. It promotes climate adaptation, increase of nutritional diversity, and pest resistance, and consequently reduces the dependence on chemical inputs. Industrial agriculture has created monocultures that leach soils' nutrients, boost susceptibility to pests and diseases and lead to biodiversity loss. A move toward agroecological practices like crop rotation, intercropping and permaculture would help support ecosystem health while actually improving yields. Indigenous and traditional farming systems like milpa agriculture in Mesoamerica, for example, involve polycultures that naturally nourish the soil by engaging multiple species.

Diversity conservation of seeds is important for a resilient agricultural system. Organizations such as the Svalbard Global Seed Vault and community-driven seed storage centers preserve genetic diversity for food security in the future. The biotechnological solutions such as precision breeding, gene editing e.g., CRISPR etc. can help develop resilient crops with minimized environmental footprints. In a bioeconomy, sustainable livestock production is part of it, too. Embedding silvopastoral systems, in which livestock graze in treed pastures, you maximally increase biodiversity and minimize emissions of methane. Innovations in alternative proteins, including insect farming and lab-grown meat, offer viable possibilities for alleviating environmental pressures stemming from traditional animal agriculture. To secure food security in a biodiversity-based bioeconomy of the future, expanding regenerative agriculture, nurturing smallholder farmers, and researching soil microbiomes must be prioritized. Agriculture has always been an integral of our ecosystem and aspect of our well-being must act in sync with biodiversity conservation.

3.2.3 Sustainable Forestry and Bio-based Materials

Sustainable forestry is a basic pillar of a biodiversity embracing bioeconomy. We depend on forest ecosystems for timber, non-timber forest products (NTFPs), carbon sequestration, and habitat conservation, also helping industries like construction, packaging, and textiles. But these benefits are threatened by deforestation and land degradation. Utilizing a biodiversity-centered approach creates long-term sustainable practices within forest sector whilst extracting what balances out with the ecology (George, 2023). Sustainable forestry certifications like FSC (Forest Stewardship Council) certification support logging methods that preserve biodiversity. Agroforestry systems, as well as selective logging systems, maintain ecosystem functions and offer sustainable economic utilization of forest environment as opposed to clear-cutting. Through community forestry programs that incorporate the traditional ecological knowledge of indigenous peoples, we can actively manage and ensure the sustainable long-term stewardship of forests.

Beyond timber alternatives, bio based materials provide exciting new options to replace fossil fuel-based products. In a similar vein, biodegradable packaging and insulation are provided by mycelium-based biomaterials made from funghi. Bioplastics are made from biodegradable or renewable resources such as cornstarch or seaweed, accommodating to petroleum-based plastics in a way that helps minimise waste. Reforestation and afforestation projects help build resilient ecosystems while creating more economic opportunities. The continued integration of technological innovations, like the use of remote sensing for monitoring forests, will keep pushing the boundaries of sustainable forestry (e.g. prospective AI-driven conservation strategies) and genomics for identifying tree species to capture resilience. More strict policies of biodiversity conservation in forest-based industries help reduce ecological loss at the benefit of economic profits

3.2.4 Renewable Energy and Biofuels

A bioeconomy based on observations of the varieties of life helps develop clean energy solutions that use biological resources without disruption. In the field of bio-energy, especially biofuels, biomass conversion to bioethanol, biodiesel and biogas is developed for fossil fuel substitutes. Not all biofuels are sustainably produced though; monoculture plantations for biofuel crops on an industrial scale threaten deforestation and biodiversity. Second and third-generation biofuels, made from algae, agricultural residue, and waste biomass, offer more sustainable alternatives. Algae-based biofuels, for example, needless land and water than conventional bio-fuel feedstocks in addition to producing high yields. On the other hand, cellulosic biofuels are produced from lignocellulosic biomass, such as crop residues and waste, and do not compete with food production while eliminating further environmental damage (Kuusinen, 2019).

Beyond liquid fuels, bioenergy solutions also include. Biogas systems use organic waste and manure to produce renewable electricity and the byproduct, fertilizer, aligning with principles of circular bioeconomy. Microbially produced biohydrogen is a promising clean fuel with near-zero emissions. To ensure the sustainable development of bioenergy, biodiversity conservation principles should be integrated into bioenergy policy. Avoid production of bioenergy (biofuels, bioelectricity) in the large-scale land conversion of natural ecosystems; supportive investments in agroforestry-based bioenergy systems (whereby energy crops are intercropped with food crops or trees) can be a win-win solution. Such bioproduct certification schemes include the Roundtable on Sustainable Biomaterials (RSB) and are needed at governments and industries level to create high quality bioenergy without harming biodiversity.

The biodiversity-first renewable energy transition needs broader source portfolios, low-impact biofuels, and enhanced circular bio-economy models. By prioritizing the integration of energy production and biodiversity conservation, the bioeconomy has the potential to help achieve a more resilient and cleaner energy future. These four sectors pharmaceuticals and biomedicine, agriculture and food security, sustainable forestry and bio-based materials, and renewable energy and biofuels illustrate the capacity of a bioeconomy anchored in biodiversity to combine ecological sustainability and economic prosperity. Putting conservation first, encouraging innovation and ensuring good governance can turn biodiversity into an engine for sustainable development (Naji et al., 2024). To realise this vision, multi-stakeholder collaboration, policy incentives, and investment in nature-based solutions that provide the balance between economic prosperity and ecological resilience are needed.

Figure 2. Key sectors in a biodiversity-driven bioeconomy

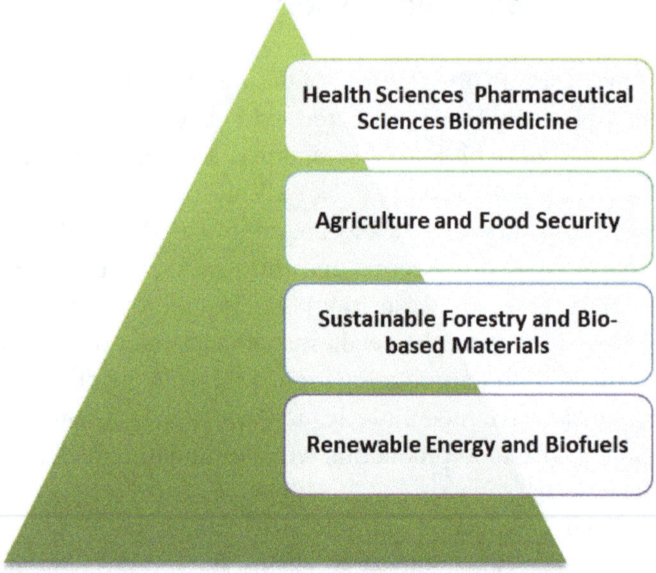

4. SOCIAL DIMENSIONS OF A BIODIVERSITY-DRIVEN BIOECONOMY

The biodiversity-driven bioeconomy is not simply an economic and environmental model; it also has major social dimensions that influence human wellbeing, equity, cultural identity, and community resilience. Such sustainable usage of biodiversity for economic growth has both a direct impact on the livelihoods and social structures of various people in addition to enhancing the local communities connection with the natural environment. Social Dimensions that need to be addressed to integrate biodiversity in a bioeconomy include: employment opportunities, rights of indigenous and local communities, food and nutrition security, public health and social equity (Wheeler, 2023).

The major social benefit of a bioeconomy based on biodiversity is the generation of sustainable livelihoods. Millions of people are working in industries that are rich in biodiversity, such as eco-tourism, sustainable forestry, agroecology and biotechnology. Biodiversity is the economic backbone for small-scale farmers, communities dependent on forests, and practitioners of traditional medicine. It is necessary to turn bioeconomics social sustainable if bioeconomic models support inclusive growth and particularly the inclusion of women, indigenous peoples, and smallholder farmers. Fair trade principles, access to market opportunities, and financial support systems (e.g., microfinance for biodiversity derived sectors/enterprise) can help balance economic benefits (Sedkaoui, 2018). Also, important social dimension is the role of indigenous and local communities. Indigenous knowledge systems have long been critical to conserving biodiversity. Traditional agriculture, herbal medicine, community-managed conservation areas, etc. are some of the ways local communities embody sustainability in their everyday lives.

Biopiracy and exploitation are still major challenges, as corporations and research institutes benefiting from traditional knowledge without proper compensation. One way to address such issues is through an international agreement known as the Nagoya Protocol on Access and Benefit-Sharing (ABS), which ensures that biodiversity-related knowledge and genetic resources are obtained ethically, and that indigenous peoples who provide such knowledge and resources receive fair benefit from their use. Policies for a socially responsible bioeconomy should focus on, among others, strengthening legal systems for protecting traditional knowledge, promotion of community-owned bioeconomic enterprises, and full inclusion of indigenous voices in relevant policy framing.

Food security and nutrition quotient are also at the core of the social impact of a bioeconomy. Diversity in agricultural landscapes is less susceptible to climate shocks, pests, and diseases—this makes biodiversity a key factor of resilient food systems. This breed and crop genetic erosion, fueled by industrial monoculture farming, has compounded food insecurity in many areas. A sustainable bioeconomy based on biodiversity promotes agrobiodiversity and the growth of native and climate-resilient crops, which contribute to dietary diversification. To ensure food security while also nourished diversity (Neuwirth, 2023), It can start local seed banks and participatory breeding programs, and incentivize traditional practice for farming.

When it comes to public health, biodiversity provides traditional medicine, pharmaceutical innovations and supply of ecosystem services, such as clean air, water and disease regulation. But biodiversity loss can lead to exacerbated health crises by disrupting ecosystems and increasing the risk of zoonotic diseases which has spilled over from wildlife). Integrating One Health approaches, where the health of people, animals and the environment are intertwined, is essential for a socially responsible bioeconomy. This also speaks to the need for investment in nature-based healthcare solutions from research of medicinal foods and plants, to the design of green urban spaces that contribute to both wildlife habitat and public health. Biodiversity-based industries are often driven by disparities in wealth and power. Many bioeconomic markets are dominated by large agribusiness and pharmaceutical corps who have the ability to drive small-scale producers out of the markets. To prevent the widening of social inequities, inclusive policies that recognize land tenure rights, equitable resource distribution, and transparent benefit-sharing mechanisms are essential (Schaeffer, 2017). Empowerment of local communities through education and supply chain capacity-building initiatives can also enable them to fully engage in bioeconomic sectors. Socially inclusive knowledge-driven bioeconomy for biodiversity should focus on community empowerment, social justice and governance by morals. Combining biodiversity conservation with equity-driven economic models will help create a more sustainable and equitable world in which nature and people flourish.

5. ETHICAL DIMENSIONS IN A BIODIVERSITY-DRIVEN BIOECONOMY

A bioeconomy underpinned by biodiversity comes with important ethical issues and implications, as it touches on questions of environmental stewardship, social justice and equity, and the rights of indigenous peoples and local communities. It also reveals the relevance of a broader ethical context, as using biodiversity for economic gain can lead to exploitation, inequity and ecological damage, and there are economic and ecotechnological developments that allow for better internalised usage and equilibrate concerns throughout. In a bioeconomy, what it means to govern ethically is to reconcile economic benefits with moral duties to nature and society (Reviglio della Venaria, 2020).

Biopiracy and fair benefit-sharing is one of the most urgent ethical dilemmas. Genetic resources from biodiversity hotspots, often sited on Indigenous lands, are the basis for many industries, including pharmaceuticals, biotechnology, and agriculture. These communities have maintained and expanded biological knowledge for centuries, but they often receive little or no acknowledgment or monetary compensation when their resources are commercialized. The goal of the Nagoya Protocol on Access and Benefit-Sharing (ABS) is to achieve a fair compensation of the communities most affected by the harvesting of their genetic resources. While these practices are now being stopped, several old laws still exist, where companies patent knowledge without the consent of indigenous groups. It is critical that ethical bioeconomic models reinforce regulatory structures that disallow exploitation, and ensure that corporate and research partnerships with local communities are transparent.

An ethical dilemma is seeing how biodiversity is commercialized and privatized. The growth of bioeconomic markets may produce over-commercialisation where biodiversity is viewed exclusively as commodity and not as a common global heritage. This raises concerns about patents on biomolecules and intellectual property rights (IPRs) that limit access to genetic resources. Privatizing biodiversity under the control of a few multinational corporations can block access for small farmers, traditional healers and conservationists who rely on these resources for their survival (Duan, et al., 2024). Teamless Bioresources Downsides of bioeconomiesEco-bioeconomic policies should not only foster policymeasures such as open-access biodiversity databases, that IPRs do not limit the free use of traditionalrural resources and that there are mechanisms for publicinvolvement in biodiversity governance.

5.1 Ethical Concerns in Bioprospecting and Biopiracy: Intellectual Property Rights (IPRs) and Patenting of Genetic Resources

Sustainability and intergenerational equity are likewise pivotal ethical values. A biodiversity based bioeconomy must function within planetary limits, preventing resource extraction from causing irreversible ecological degradation. Overharvesting of medicinal plants, genetic modification of species without ecological assessments, large-scale monoculture farming can erode biodiversity for future generations. They should call for regenerative and restorative practices with a focus on ecosystem integrity rather than short-term economic gain. In light of the uncertainty that surrounds synthetic biology, genetic engineering, and other various biotechnological interventions, the precautionary principle, which emphasizes a precautionary approach to socio-ecological systems, should be paramount in bioeconomic decision-making.

Social justice and equity need to be fundamental aspects of bioeconomic strategies. Many regions with high levels of biodiversity are also home to economically disadvantaged communities, but the economies associated with biodiversity often see the profits move to other, more developed economies. With this comes a moral imperative to ensure that the economic benefits are equitably distributed, and that local communities are afforded opportunities to engage in bioeconomic activities themselves. Ethical bioeconomic policies that respect land tenure rights, prevent forced displacement, and assist locally owned biodiversity-based firms must be placed at the center of any new approaches. Gender justice is also critical; women are central to biodiversity conservation, traditional knowledge systems, etc. and do not often have equal access to finance and decision-making mechanisms because of structural barriers (Tzampazaki et al., 2024).

5.2 The Precautionary Principle in Biotechnology

Among the bioeconomic sectors bioeconomy as biomedicine, aquaculture, wildlife-based there are ethical problems related to animal treatment. Bioeconomic models that are truly ethical must be humane and sustainable, meaning that what we do with biodiversity does not involve animal cruelty, ecosystem disruption, or unsustainable use of wildlife. The nascent notion of ecosystem rights proposes that ecosystems should be considered and treated as legal persons, affording them legal protection against unsustainable anthropogenic practices. Other countries, like Ecuador and New Zealand, have already introduced legal rights for rivers and forests in recent years, paving the way for the inclusion of ethical considerations into biodiversity governance.

Thirdly, ethics of governance in a biodiversity-based bioeconomy should enable inclusiveness, transparency, and accountability. Corruption and poor oversight can result in the misuse of biological variety assets, striking at conservation objectives. Public participation, environmental justice, and corporate responsibility will be essential principles for governments, corporations, and research institutions in all aspects of biodiversity management. Indigenous voices, civil society organizations, and ethical review boards should be incorporated into policies determining biodiversity-based economic activities (Hanafiah et al., 2024).

Therefore, the ethical landscape of a biodiversity-led bioeconomy encompasses a broad spectrum that must negotiate not only human rights, and environmental stewardship, but also intergenerational equity, underscoring the necessity of a holistic viewpoint. Sustainable, ethical bioeconomic models must be underscored by equitable benefit-sharing, transparent governance, sustainability, social equity and the recognition of indigenous knowledge and nature's rights. We can incorporate ethical values into economic systems that are centered around biodiversity, and in that way, counter inequalities, ensuring that both people and the planet benefit from the bioeconomy well into the future.

6. REGULATORY AND GOVERNANCE FRAMEWORKS: SOCIAL AND ETHICAL CONSIDERATIONS INTO POLICY

Strong legislative and governance frameworks will be required to embed social and ethical values into policy to enable the transition to a biodiversity-driven bioeconomy. Biodiversity is governed in a way that it is exploited sustainably, fairly, and responsibly, striking a balance between economic development, protecting the environment, communities rights, and future generations. Through a lack of comprehensive regulations, bioeconomic activities are likely to exacerbate disparities, environmental destruction, violation of ethics (biopiracy, unfair appropriation of indigenous rights and traditional knowledge, and other commitments from companies). It must therefore incorporate the principles of social justice, ethical accountability, and ecological sustainability into bioeconomic governance in national and international policies (Tembrevilla et al., 2024).

The legal protection of biodiversity and genetic resources is one of the key blueprints for governance. International treaties like the Convention on Biological Diversity (CBD) and its supplemental Nagoya Protocol on Access and Benefit-Sharing (ABS) offer regulatory models for ethical biodiversity utilization. The Nagoya Protocol is a treaty aimed at making sure that local communities and indigenous peoples are compensated when genetic resources and traditional knowledge contribute to bioeconomic innovations. However, enforcement remains problematic, with many corporations and research institutions exploiting legal loopholes to commercialize genetic materials without adequate benefit-sharing. An emphasis on local application of the ABS framework and transparent mechanisms for monitoring compliance must be reinforced with a view to enforcing ethical standards in the governance of biodiversity (Bala & Gupta, 2024).

The other significant regulatory question surrounds intellectual property rights (IPRs) and patents over biological resources. International patent regime, based on TRIPS Agreement prefer corporate interests over traditional knowledge and collective ownership. Multinational big business in many cases patents plant species, medicinal compounds and agricultural biodiversity, rendering them inaccessible to indigenous farmers, healers and researchers. Regulators will need to reshape intellectual property systems to develop alternative models protecting more than just big data monopolies, including community-held patents, open-source biodiversity databases, and traditional knowledge registries that will discourage biopiracy and strengthen Indigenous resistance (Hazarika & Rahmati, 2023). The frameworks would safeguard indigenous knowledge and prevent the corporatization of biodiversity. Governance policies also need to integrate social equity and equitable distribution of resources. A bioeconomy based on biodiversity-pool should not exclude rich countries, large agribusinesses, and big biotech cartel against small-scale farmers, harvesters, and local enterprises. This is a general recommendation, and policy interventions can consist of many approaches, including land tenure rights for indigenous populations, equitable access to bioeconomic markets, and financial support for local biodiversity-based industries. Governance models that ensure local communities include participatory governance and contribute to decisions concerning biodiversity at a national and international level must also be made mandatory.

Environmental sustainability regulations however also need to be in accordance to bioeconomic policies to curb overexploitation, habitat and biodiversity loss. Regulations that govern biotechnology, genetic engineering, and synthetic biology should reflect the precautionary principle to ensure that bioeconomic innovations help avoid unintentional ecological harm. All large scale biodiversity based industrial projects must require an environmental impact assessment (EIA) (legally required) before going ahead. In addition, circular economy business models that encourage waste minimization, bioproduct recycling (or upcycling), and ecosystem regeneration need to be integrated into bioeconomic governance schemes.

The frameworks of ethical governance should include accountability of corporations and transparency of biodiversity use. Biotechnology, pharmaceuticals, agribusiness and renewable bioenergy are embarked upon by large corporations that need stringent ethical oversight including mandatory corporate social responsibility (CSR) reporting on biodiversity impact, third-party audit of the environmental consequences, reporting of biodiversity related patents and profits in published and accessible documents (Varela et al., 2024). Governments and the international regulatory body should also create measures for anti-corruption, including preventing illicit biotrade, illegal deforestation and fraudulent biodiversity commercialization. Multilateral institutions like the United Nations Environment Programme (UNEP), the Food and Agriculture Organization (FAO) and the World Trade Organization (WTO) can help harmonise bioeconomic policies between jurisdictions. Such institutions can play a role in fostering global policy coordinated around bioeconomic objectives which are aligned with sustainable development goals (SDGs), climate action plans, and biodiversity conservation strategies. The other essential component for DE2100 is the need for regional and national governance bodies to be able to develop their own regulatory frameworks, according to their ecological goals and socio-economic environments, whether it be the European Green Deal, the African Union's Agenda 2063, or the ASEAN Biodiversity Strategy.

Education and public awareness are additional governance tools that are critical to secure the ethical application of a biodiversity-driven bioeconomy. Governments need to provide community education programs, scientific literacy campaigns, and biodiversity ethics training to policymakers, businesses, and researchers. An engaged public can keep corporations and governments in check, demanding fair biodiversity policy from civil society organisations, grassroots movements and environmental activism. A complete regulatory and governance framework is essential for incorporating social and ethical aspects into biodiversity-based bioeconomic policies. Policy needs to focus on equitable benefit-sharing, establishment of intellectual property reforms, social equity, environmental sustainability, corporate accountability, and participatory governance (Gupta, 2024). This will ensure that the bioeconomy aligns and reinforces biodiversity conservation and ethical practices in economic activity, contributing to both human and environmental well-being for generations to come.'

7. CONCLUSION AND RECOMMENDATIONS

Biodiversity-driven bioeconomies raise important social and ethical challenges that must be addressed through policy and governance frameworks. International legal instruments, including Convention on Biological Diversity (CBD), Nagoya Protocol and the United Nations Sustainable Development Goals (SDGs) establish guided principles for ethical and equitable utilization of biodiversity. However, inconsistencies in national policy implementation have left governance gaps, where countries lack the institutional capacity or political will to implement biodiversity protection and trade agreements. Thus, creating, reinforcing legal frameworks, seeking international cooperation, and supporting multi-stakeholder governance are all crucial in ensuring that bioeconomic efforts do not extend beyond ethical and socially acceptable limits.

The bioeconomy offers promise for sustainable production and consumption but also substantial challenges to integrating social and ethical dimensions into economic choices. Biodiversity markets, without proper management, may lead to the pursuit of profit ahead of conservation and equity. Such power dynamics between multinational companies and local communities or between multinationals and governments can result in exploitative activities at the expense of conserving biodiversity. Thus, ethical considerations need to go beyond compliance with legal frameworks to take into account more broadly defined moral obligations requiring that biodiversity be treated as a global commons rather than a commodity to be exploited for economic benefit.

The extent to which a biodiversity-driven bioeconomy will emerge will rely on collaborations between governments, businesses, scientists, and communities in forging ethical parameters, equitable policies, and sustainable practices. Investment in education and capacity–building as well as interdisciplinary research will be critical in shaping the bioeconomy as a socio–economic system that preserves human rights, ecological integrity and ethical responsibility in the pursuit of scientific and economic goals. This is a work that is simply vast; and all of its dimensions are inter-related more than one might think. Implementing social justice, environmental stewardship, and ethical governance within bioeconomic policies will pave a path towards more resilient, inclusive, and sustainable economies that uphold the respect for nature and human dignity.

REFERENCES

Ahmed, A. (2022). Marketing 4.0: The Unseen Potential of AI in Consumer Relations. *International Journal of New Media Studies: International Peer Reviewed Scholarly Indexed Journal*, *9*(1), 5–12.

Ancillai, C., Terho, H., Cardinali, S., & Pascucci, F. (2019). Advancing social media driven sales research: Establishing conceptual foundations for B-to-B social selling. *Industrial Marketing Management*, *82*, 293–308. DOI: 10.1016/j.indmarman.2019.01.002

Auttri, B., Chaitanya, K., Daida, S., & Jain, S. K. (2023). Digital Transformation in Customer Relationship Management: Enhancing Engagement and Loyalty. [EEL]. *European Economic Letters*, *13*(3), 1140–1149.

. Babet, A. (2020). Utilization of personalization in marketing automation and email marketing.

. Bala, R., & Gupta, P. (2024). Virtual Reality in Education: Benefits, Applications and Challenges. *Transforming Education with Virtual Reality*, 165-180.

. Bhatnagar, S. (2022). Digital Disruptions and Transformation of Bank Marketing.

. Çalışkan, G., Yayla, İ., & Pamukçu, H. (2023). The use of augmented reality technologies in tourism businesses from the perspective of UTAUT2. *European Journal of Innovation Management*.

Duan, W., Khurshid, A., Khan, K., & Calin, A. C. (2024). Transforming industry: Investigating 4.0 technologies for sustainable product evolution in china through a novel fuzzy three-way decision-making process. *Technological Forecasting and Social Change*, *200*, 123125. DOI: 10.1016/j.techfore.2023.123125

. Gao, Y., & Liu, H. (2022). Artificial intelligence-enabled personalization in interactive marketing: a customer journey perspective. *Journal of Research in Interactive Marketing*, (ahead-of-print), 1-18.

George, A. S. (2023). Future Economic Implications of Artificial Intelligence. *Partners Universal International Research Journal*, *2*(3), 20–39.

Gupta, V. (2024). An Empirical Evaluation of a Generative Artificial Intelligence Technology Adoption Model from Entrepreneurs' Perspectives. *Systems*, *12*(3), 103. DOI: 10.3390/systems12030103

Hanafiah, M. H., Asyraff, M. A., Ismail, M. N. I., & Sjukriana, J. (2024). Understanding the key drivers in using mobile payment (M-Payment) among Generation Z travellers. *Young Consumers*, *25*(5), 645–664. DOI: 10.1108/YC-08-2023-1835

Hazarika, A., & Rahmati, M. (2023). Towards an evolved immersive experience: Exploring 5G-and beyond-enabled ultra-low-latency communications for augmented and virtual reality. *Sensors (Basel)*, *23*(7), 3682. DOI: 10.3390/s23073682 PMID: 37050742

. Henke, N., & Jacques Bughin, L. (2016). The age of analytics: Competing in a data-driven world.

Kacprzak, A., & Hensel, P. (2023). Exploring online customer experience: A systematic literature review and research agenda. *International Journal of Consumer Studies*, *47*(6), 2583–2608. DOI: 10.1111/ijcs.12966

Kamal, M., & Himel, A. S. (2023). Redefining Modern Marketing: An Analysis of AI and NLP's Influence on Consumer Engagement, Strategy, and Beyond. *Eigenpub Review of Science and Technology*, *7*(1), 203–223.

Kaushal, V., & Yadav, R. (2023). Learning successful implementation of Chatbots in businesses from B2B customer experience perspective. *Concurrency and Computation*, *35*(1), e7450. DOI: 10.1002/cpe.7450

King, K. (2022). *AI Strategy for Sales and Marketing: Connecting Marketing, Sales and Customer Experience*. Kogan Page Publishers.

. Kuusinen, M. (2019). Scenarios for digital marketing: a Delphi-based analysis for 2028.

Lavuri, R., & Akram, U. (2023). Role of virtual reality authentic experience on affective responses: Moderating role virtual reality attachment. *Journal of Ecotourism*, •••, 1–19.

Mashood, K., Kayani, H. U. R., Malik, A. A., & Tahir, A. (2023). ARTIFICIAL INTELLIGENCE RECENT TRENDS AND APPLICATIONS IN INDUSTRIES. *Pakistan Journal of Science*, *75*(02). Advance online publication. DOI: 10.57041/pjs.v75i02.855

Naji, K. K., Gunduz, M., Alhenzab, F., Al-Hababi, H., & Al-Qahtani, A. (2024). A Systematic Review of the Digital Transformation of the Building Construction Industry. *IEEE Access : Practical Innovations, Open Solutions*, *12*, 31461–31487. DOI: 10.1109/ACCESS.2024.3365934

Neuwirth, R. J. (2023). Prohibited artificial intelligence practices in the proposed EU artificial intelligence act (AIA). *Computer Law & Security Review*, *48*, 105798. DOI: 10.1016/j.clsr.2023.105798

Rajamannar, R. (2021). *Quantum marketing: mastering the new marketing mindset for tomorrow's consumers*. HarperCollins Leadership.

. Reviglio della Venaria, U. (2020). Personalization in Social Media: Challenges and Opportunities for Democratic Societies.

Sadowski, J. (2023). Total life insurance: Logics of anticipatory control and actuarial governance in insurance technology. *Social Studies of Science*, •••, 03063127231186437. PMID: 37427796

Schaeffer, E. (2017). *Industry X. 0: Realizing digital value in industrial sectors*. Kogan Page Publishers.

Sedkaoui, S. (Ed.). (2018). *Big data analytics for entrepreneurial success*. IGI Global.

Soliman, M., & Al Balushi, M. K. (2023). Unveiling destination evangelism through generative AI tools. *ROBONOMICS: The Journal of the Automated Economy*, *4*(54), 1.

Tembrevilla, G., Phillion, A., & Zeadin, M. (2024). Experiential learning in engineering education: A systematic literature review. *Journal of Engineering Education*, *113*(1), 195–218. DOI: 10.1002/jee.20575

Tzampazaki, M., Zografos, C., Vrochidou, E., & Papakostas, G. A. (2024). Machine Vision—Moving from Industry 4.0 to Industry 5.0. *Applied Sciences (Basel, Switzerland)*, *14*(4), 1471. DOI: 10.3390/app14041471

Varela, L., Putnik, G., & Romero, F. (2024). Collaborative manufacturing and management contextualization in the Industry 4.0 based on a systematic literature review. *International Journal of Management Science and Engineering Management*, *19*(1), 78–95. DOI: 10.1080/17509653.2023.2174200

Vinaykarthik, B. C. (2022, October). Design of Artificial Intelligence (AI) based User Experience Websites for E-commerce Application and Future of Digital Marketing. In *2022 3rd International Conference on Smart Electronics and Communication (ICOSEC)* (pp. 1023-1029). IEEE.

Wheeler, J. (2023). *The Digital-First Customer Experience: Seven Design Strategies from the World's Leading Brands*. Kogan Page Publishers.

Zaki, H. O. (2022). The Impact Of Artificial Intelligence On Content Marketing. *Journal of Strategic Digital Transformation In Society*, *2*(3).

Chapter 3
Nexus Between Bio-Resources, Traditional Knowledge, and Biodiversity Laws in Achieving Sustainable Development

Aniket Nihal
https://orcid.org/0009-0008-9261-8436
Independent Researcher, India

ABSTRACT

Bio-resources and Traditional Knowledge associated therewith, have been used since ancient times for a wide range of purposes due to its high applicability across diverse sectors for fulfilment of human own needs and environmental conservation. This creates huge opportunities in the field of modern scientific research and scope for commercial gain. However, its exploitation has its own interlinked repercussions such as biodiversity loss, socio-economic implications and patent revocation. In an attempt to mitigate these repercussions, legal instruments have been explored mainly focusing on sharing of benefit framework under biodiversity laws of international and national regimes majorly focusing on Indian perspectives. It has been found that the legal intervention in access and use of bio-resources, bio-resources-based innovations supported with Intellectual Property Rights protection aid in accomplishing the majority of the total goals of Sustainable Development Goals.

DOI: 10.4018/979-8-3373-2003-8.ch003

INTRODUCTION

Bio-resources and Traditional Knowledge (hereafter referred to as TK) associated therewith, have been used for different purposes such as in folk remedies, traditional cosmetics, conservations, cooking and many other since time immemorial. It has also been utilized in the modern era for the development of drugs, modern medicine practices-based treatment of diseases, agricultural practices, cosmetics, food, biotechnology, and clothing industries, etc. Because of its numerous applications and uses, it has often been utilized for commercial gain, which may lead to the over-exploitation of the biodiversity heritage of the concerned region. Monopolization of bio-resources-based products by using the instrumentality of patents by commercial organizations without getting prior permission and acknowledgement from the government authorities, local tribal communities and without sharing of benefits with its stakeholders led to uncontrolled exploitation of resources, biodiversity loss, and also has socio-economic and heritage loss to countries from where such resources are being exploited for commercial gain (Greiber et al., 2012, p. 6).

Profit-driven commercial players mostly multinational companies obtain highly valued bio-resources and/or TK associated therewith and thereafter secure exclusive ownership on it without obtaining permission from their exclusive owner (such as bio-resource providing country, local and tribal communities) and without sharing of benefits which are arising from the commercialization of these resources (Rabitz, 2015). Such exploitation of resources severely affects the biodiversity and ecological balance. However, it has been seen that repercussions have not only been limited to biodiversity loss but also has socio-economic implications. For instance, the company "Monsanto India Ltd." developed genetically modified "Bt. Cotton seed" which is resistant to Bollworm, thereafter regulated the price, terms and conditions of its use, due to which farmers are not allowed to reuse the same seeds in the next year. In a repercussion, farmers suffered a huge financial loss, and this leads to a suicide crisis among them. According to the claim made by the State Biodiversity Board (hereafter referred to as SBB) of Andhra Pradesh, India, the company has developed the "Bollguard-11" cotton seeds by using the bacterial gene information which is indigenous to the Andhra Pradesh State of India. In most cases, it has been seen that companies and/or research and development organizations exploit these resources, and obtain patents on it for commercial gain. The problem becomes more severe when to reduce cost in drug development, like in search of new molecules or bio-active compounds, conducting a number of scientific research experiments, and high risk of payment in liabilities when an adverse situation occurs during pre- and post-clinical trials; companies use TK based bioprospecting. This surely reduces the investments in drug developments but takes away the livelihood of the indigenous local communities if the patent is granted without recognizing and sharing of benefits between the local communities, who preserve the centuries-old TK and practices for the treatment of diseases. Obtaining patents on bio-resources which were exploited based on the TK of the regional community which has been passed to generation to generation is subject to biopiracy, as it does not fulfil the two essential criteria of newness and inventiveness which is a must for obtaining a patent (Singh et al., 2014).

While discussing the biopiracy, it is worth to revisit the three classical landmark cases *viz.*, the turmeric, the neem and, the basmati case which serve as the best example of revocation of patents due to biopiracy, wherein the TK of Indian regional communities associated with the bio-resources had been patented in the foreign country without sharing the benefits or acknowledging government authority or local communities. The Turmeric (*Curcuma longa* Linn.) which is a rhizome of the plant of the ginger family (Zingiberaceae) is known to have a number of beneficial applications such as used in the treatment of diseases due to its pharmacological properties, cooking ingredient in Indian cousins, applications in cosmetics and, many more (Traditional Knowledge Digital Library [TKDL], n.d.). In India, turmeric has been used for ages, and the knowledge associated with it has been passed down through generations. In 1995, a peculiar and landmark dispute emerged when two United States-based Indian nationals were granted a United States patent by "United States Patent and Trademark Office" (hereafter referred to as US PTO) on turmeric powder for its medicinal property specifically for wound healing. This grants them an exclusive right for selling and distributing turmeric powder. India strongly opposed the granted patent on turmeric powder by US PTO. "The Council of Scientific & Industrial Research, India," contended that wound healing properties of turmeric power have been in practice in India since ages and this is not a novel invention and cited their claim with 32 documentary references which were written in different languages including Sanskrit. US PTO revoked the patent in 1998, on the ground that the claims made by patent holders were evident and predictable, and this was based on TK of India and does not fulfil the necessary criteria of novelty and inventiveness (Lakshmanan, 2015).

Medicinal properties, environmental, agricultural and cosmetics applications of different parts (such as leaves, seeds, flowers, and barks) of the neem tree (*Azadirachta Indica*) and its various extracts have been reported more than 2000 years ago in Indian literature, and their applications are in practice since ages in India (Singla, 2020). Conflict arose when a patent on the method of inhibiting the growth of fungus on the plants by the use of a certain formulation of neem oil to "W.R. Grace Company and US Department of Agriculture" granted by European Patent Office (hereafter referred to as EPO) in 1994 (Lakshmanan, 2015). The activists and Indian farmers opposed the granted patent and strongly contended that anti-fungal properties of neem seeds had already existed and in practice since centuries in

India. The activists also produced ancient ayurvedic literature that described the anti-fungal properties of hydrophobic extracts of neem seeds in controlling fungal skin diseases in humans and fungal infections in plants (Balasubramanian, 2017, pp. 3-6) With the evidence submitted by the activists, in 2000, EPO cancelled the granted patent on the ground that it lacks novelty and inventiveness. An appeal against the revocation was filed by the patent holder, however, it was again finally revoked in 2005 by EPO (Lakshmanan, 2015).

Basmati Rice (*Oryza sativa* Linn.) is a legendary traditional rice of India which is mostly known for its unique aroma and grown since thousands of years in the north-western part of India, mostly in the Punjab state. A United States-based company "RiceTech Inc." (hereafter referred to as RT Inc.) developed rice similar to Indian basmati rice by crossing it with long grain and semi-dwarf varieties and obtained a patent on it by the US PTO in 1997 (Lakshmanan, 2015). This has affected the Indian export market of Basmati rice, loss in Indian biodiversity heritage, and the restriction of the use of the name "Basmati" by Indian farmers who are the sole traditional owners for its development, conservation and trade (Singla, 2020, p. 8). India opposed this patent at US PTO. After opposition, RT Inc., out of 20 claims which were made in the original patent application withdrew 15 claims. Out of the remaining 5 claims, 3 claims were retained the patentability by US PTO and the remaining 2 claims were amended (Lakshmanan, 2015). Thus, due to the United States patent protection of plant varieties, RT Inc. was allowed to use the label on its strain as "Superior Basmati Rice" (Balasubramanian, 2017, p. 5).

As we have observed from the above cases wherein the patent in a third country, if granted, was revoked, if challenged by the parties whose interest was conflicting with the patent holder. This type of patent based on the bioprospecting on the basis of TK of regional communities is subject to biopiracy. This chapter analyzes sharing of benefit framework under biodiversity laws of international and national regimes along with equitable principle and constitutional provision as a legal obligation to share benefits to the conserver and keeper of bio-resources and TK in the Indian context to establish linkage between environmental components, legal instruments and technology in achieving sustainable development.

LAW AND BIO-RESOURCES

The world community came up with the "Convention of Biological Diversity, 1992" (hereafter referred to as CBD, 1992) to prevent biopiracy and to protect bio-diversity from over-commercial exploitation. It also addresses the socio-economic-cultural and aesthetic aspect of biodiversity. The CBD, 1992 came into force on 29 December 1993 which introduced a concept of controlled and regulated access of genetic resources; and fair and equitable sharing of benefits (hereafter referred to as FES benefits) which aims to protect the interest of TK holders like tribes or local communities. Like any other legal framework, the preamble of CBD, 1992, acts as a heart of the convention as well. The preamble expressively says about the global concern on conservation and sustainability of bio-resources for maintaining life on earth. Additionally, it emphasizes on the *in-situ* and *ex-situ* conservation of ecosystems and the protection of natural habitats. According to the Article 1 of CBD, 1992, it establishes three primary objectives: first, the conservation of biodiversity; second, the utilization of biodiversity in a sustainable manner; and third, benefit sharing arising from the use of genetic resources, which includes access to these resources and the transfer of technologies. The third objective has also meant to deal with the requirements to share the cost and the benefit of biodiversity conservation among developing and developed countries. Article 15 of CBD, 1992, attempted to balance the interest of stakeholders in genetic resources. This protects the interest of users of such resources by getting access to the biodiversity as well as the suppliers of resources by getting FES benefits. However, it is worth to mention that under CBD, 1992, it does not protect other bio-resources, rather protects only genetic resources (Greiber et al., 2012, pp. 4-12). Article 15(4) of CBD talks about the grant of access shall be done on "Mutually Agreed Terms" (hereafter referred to as MAT), and Article 15(5) talks about the access to such genetic resources are subject to "Prior Informed Consent" (hereafter referred to as PIC) of parties providing such resources. So, through these two clauses, it grants, subject to MAT, authorized access to genetic resources, control over the use of such resources, and, set the FES benefits arising from their utilization. The idea behind PIC is to allow those parties who are authorized to control such genetic resources, or has sovereign right on such resources, or are affected by the use of such resources to take a decision prior to access by potential users. However, Article 15(5) also read with Article 15(1) of CBD, 1992, refers that by exercising their sovereign right parties may or may not require PIC. Thus, it totally depends on their government and local laws. In the context of TK, Article 8(j) of CBD, 1992, talks about the preservation and maintenance of TK holders and/or indigenous communities and promotes FES benefit arising out from the utilization of their genetic resources and associated TK therewith.

Eighteen years after the adoption of the CBD in 1992, the Nagoya Protocol was adopted in 2010 by the Conference of the Parties to the CBD, held in Nagoya, Japan. This landmark treaty addresses the proper implementation of access to genetic resources and the sharing of FES benefits. This was the significant achievement of the global community in conserving biodiversity, particularly concerning the framework for FES benefits (Secretariat of the Convention on Biological Diversity, 2011). It is legally binding, and serves as a supplementary agreement to the CBD, 1992. Article 1 of the Nagoya Protocol addresses its objective which is mainly based on the third objective of the CBD, 1992, which refers to "the fair and equitable sharing of the benefits arising from the utilization of genetic resource." It recognizes the rights of indigenous/local/regional people and communities over those resources and associated TK, and further clarifies what may be included in the benefits sharing, such as proper access to genetic resources, technology transfer, and funding (Nagoya Protocol, art. 5, para. 4 read with Annexure).

India's Stand on Access and Use of Bio-resources

The Indian government became a party to the CBD, 1992 in 1994 and thereafter in response to its commitment toward principles and objectives enacted the "Biological Diversity Act, 2002" (hereafter referred to as BD Act, 2002) to conserve and regulate responsible use of bio-resources, and to establish FES benefits framework in a country (De Penning & De Penning, 2025). Additionally, previously enacted legislation "The Protection of Plant Varieties and Farmers' Rights Act, 2001" also provides an FES benefits framework, and acknowledges the conservation of bio-resources particularly to plant genetic materials. To meet the procedural requirement under the BD Act, 2002, and to set the guidelines and formula to determine methods for FES benefits, the central government first framed "The Biological Diversity Rules, 2004" and thereafter replace it with new "the Biological Diversity Rules, 2024 (hereafter referred to as BD Rules, 2024); and "National Biodiversity Authority" (hereafter referred to as NBA) first made "Guidelines on Access to Biological Resources and Associated Knowledge and Benefits Sharing Regulations, 2014" and thereafter replace it with new "the Biological Diversity (Access to Biological Resources and Knowledge Associated thereto and Fair and Equitable Sharing of Benefits) Regulations, 2025 (hereafter referred to as BD Regulations, 2025). The said BD Rules, 2024 and BD Regulations, 2025 have been made to operationalize the significant amendment made in the BD Act, 2002 through "The Biological Diversity (Amendment) Act, 2023".

Section 21(2) of the BD Act, 2002, specifies what should be termed as "Benefit Sharing." Under this section, FES benefits have been determined as: joint ownership of Intellectual Property Rights (hereafter referred to as IPR) to the NBA or to identified benefit claimers; transfer of technology; providing monetary as well as non-monetary compensation; providing better living standard to the benefit claimer by setting up of production and research unit in the local area; involving benefit claimers, Indian researchers and regional people in research and development; set-up of venture capital fund to help the cause of benefit claimers.

The Rule 14 of the BD Rules, 2024, allow NBA to restrict or prohibit the access of bio-resources and TK associated therewith, in case of access of threatened or endemic species or those resources as notified under the BD Act, 2002; if such access may cause adverse impact on the environment, ecosystem, livelihood and/ or socio-cultural aspects of the locals; and also in case if it affects to the national interest and international agreements which India is part of it. However NBA may permit in case of cultivated species even if it falls under above mentioned cases based on merit case to case basis for specified period.

Regulation 3, 4, 5, 7, 8, and 9 of the BD Regulations, 2025, provides the modes of FES benefits which may include upfront payment as determined by NBA, which may also be waived in case of conducting academic research. The BD Regulations, 2025, provides standardized FES benefits ranges from Nil to 0.6% on annual gross ex-factory sale price of product excluding government taxes for different category of person based on their annual turnover that ranges from up to INR 5 crore to above INR 250 crore and fixed percentage for higher economic valued bio-resources such as red sanders, sandalwood or such other as decided by NBA. Regulation 11 read with Form F, provides the criteria for the determination of FES benefit. According to which, the FES benefits may be done in two ways by monetary and/or non-monetary mode, which is subject to the agreement between the applicant and NBA on a case-by-case basis. This regulation also specifies the options available under both modes of FES benefits. Options under monetary benefits include: up-front, one-time, and milestone payment; share of license charges, and royalty; funding for research and contribution in different biodiversity funds at various levels (national, state, or regional); joint venture and joint ownership with Indian organizations of the intellectual property developed. On the other hand, non-monetary benefits include: technology transfer; institutional capacity building; infrastructure development; sustainable use of resources; regional economic development and establishment of the production unit in the local area from where bioresources were collected; sharing of research outcomes; conducting research focusing priority needs in India including food, health and livelihood; providing financial assistance and educational training related to conservation and sustainability to locals, setting up of venture capital fund, joint ownership of IPR, and any many other as NBA deem fit. However determination of FES benefits depends on sector and various other factors *inter alia* environmental impact, quality of bio-resources, sustainability, potential commercial use, sustainability and other relevant considerations. Further it also provides that in case of contribution toward non-monetary benefits, monetary benefits may be relaxed or reduced. Comparative dissection of provisions of the CBD, 1992 and the BD Act, 2002 on regulating access and use of bio-resources has been presented in Table 1.

Equitable Principle and Constitutional Provision in Protection to Benefit Claimers

Besides rights and recognitions given to owner, conserver, or holder of genetic and bio-resources and/or associated TK therewith under biodiversity laws regarding FES benefits, the equitable principle *inter alia* doctrine of "unjust enrichment," and constitutional provision under Article 21 of the Constitution of India provide protection to them as well. According to the said doctrine, it restricts any person from enriching at the cost of any other person (Takwani, 2021, pp. 452-453). That means, it prevents others from enjoying benefits that have been received from a person who has suffered a loss. In the present context, using bio-resources and associated TK and gaining profit therefrom, without sharing benefits among its stakeholders and without incurring liabilities and responsibilities towards environment and local/tribal/indigenous people and communities may also amount to unjust enrichment. Thus, that should be prevented from enriching at the cost of others. Here, "others" under this context, may be referred to TK keepers, conservers of bio-resources and the governmental, local authority, who, since ages made their efforts to safeguard the resources and developed the traditional remedial practices from ancient research on bio-resources for the benefits of mankind and environment, and whose livelihood also depends on those resources. Although in absence or in presence of international and national regulatory guidelines and other legal framework, or in any expressed or implied agreement between the user and provider of resources, the equitable principle of doctrine of unjust enrichment should also be applied under these circumstances. The profit bearer shall also bear liabilities and responsibilities towards bio-resources keepers and associated TK developers who may be regarded as the ancient innovators.

Moreover, the rights and recognition have also been protected under "right to livelihood" guaranteed under Article 21 according to which "No person shall be deprived of his life or personal liberty except according to procedure established by law (Government of India, 2024). The honorable Supreme Court of India has recognized the right to livelihood under Article 21 in the famous "pavement dwellers" case (*Olga Tellis v. Bombay Municipal Corporation*, 1985; Pandey, 2021, p. 287). In generality, in developing countries, the livelihood of people at large, especially those who are local/tribal/indigenous, depends on bio-resources and/or associated TK therewith. Therefore, any activities which have an impact on such resources especially in terms of commercial exploitation, and which cause harm to the livelihood of people, such activities must be restricted, and the right must be protected. The examples which state about the effects of commercial exploitation on socio-economic aspects of local people have already been discussed. The Constitution of India under Article 21 itself provides protection of right to livelihood. So, directly, it provides protection to these people, and indirectly it recognizes FES benefits. On dissection of this concept, it could be inferred that, accessor, or user of bio-resources whose activities in relation to use of such resources if adversely impacted on livelihood of its provider, the same may be compensated in terms of FES benefits.

Initiatives and Legal Instruments in Safeguarding the Rights by Filling the Loopholes

According to the World Intellectual Property Organization (n.d.), TK-based innovations benefit from various instruments of IPR. However, in practice, most of the TK keepers pass their intellectual traditional old practices and knowledge from generation to generation verbally and very few of them are preserved as in written form. In general, this verbal transfer of knowledge is either not or difficult to be protected by the instrument of IPR. These loopholes create a scope of biopiracy for bio-pirates. To encounter the problem of recognizing indigenous TK and practices which are either not documented or not in the public domain or remain hidden within a regional community, the Indian government through its various organizations started documentation of such heritage to get the benefits in the globalized market. This initiative led to the set-up of the "Traditional Knowledge Digital Library" in the country (Traditional Knowledge Digital Library [TKDL], n.d.). This pioneering initiative of the Indian government restricted third-party organizations from obtaining patents due to their availability in the public domain. In addition, the legislative framework, with regards to the access and use of bio-resources and associated TK, particularly FES benefits framework under national and international laws, promotes the "collaborative culture" in research and trade as well, to bring its stakeholders under one roof.

So, TK in bioprospecting and obtaining patents has a brighter side but only if user companies/organizations collaborate with the provider local communities, indigenous tribes and government agencies or regional regulating authorities, then this could become an important source of revenue for tribal people of developing countries. This collaboration led to the sharing of benefits among companies, researchers, government regulating authority and local tribal communities and; recognizing the source of bio-resources may establish mutual beneficial relationships among its different stakeholders and promote sustainable growth of developing countries (Singh et al., 2014). Using the instrument of FES benefits framework not only benefits the resource providing countries and its locals but also protects the image of private companies whose reputation may get damaged due to public protests on misuse or exploitation of their resources (Greiber et al., 2012, p. 12).

BIO-RESOURCES AND SUSTAINABLE DEVELOPMENT ALIGNMENT

Bio-resources have long played a vital role in supporting human life and driving social advancement. Owing to the dependence on bio-resources in the life of humans, and other living creatures, inclusion of importance of bio-resources is an inseparable part of any initiatives made on sustainable development. United Nations member States in the United Nations Sustainable Development Summit in 2015 set out 17 goals of sustainable development that are necessary to safeguard the sustainable future (Department of Economic and Social Affairs, United Nations, n.d.). The interconnection between bio-resources and SDGs are mutually beneficial with each complementing the other. Effort to achieve clean water, sanitation, and healthy eco-system (goal 6), responsible consumption of resources (goal 12), mitigating climate change (goal 13), sustainable use of marine and terrestrial resources (goal 14 and 15) and along with other goals that promote societal advancement are interlinked to contribute in the conservation of bio-resources (Secretariat of the Convention on Biological Diversity, 2020, p. 27). It has been seen that bio-resources and/or their byproducts and/or associated services thereto, positively impacts and advances the Sustainable Development Goals (hereafter referred to as SDGs) in various ways. The direct impact on SDGs come from the trade of bio-resources when local and tribal communities use such resources for their own livelihood and fulfilment of basic needs without external intrusion. There have been catenae of case studies documented so far which demonstrate how human perspectives on bio-resources significantly turned into bio-economy for a better future. With respect to the direct contribution of bio-resources to SDGs, it plays a fundamental role in achieving Goal: 1 (no poverty), 2 (zero hunger), 3 (good health and well-being), 8 (decent work and economic growth), 14 (life below water), and 15 (life on land).

Bio-resources are essential for economic development and ensuring food and nutritional security. These resources are used by people of all classes to meet their nutritional and health needs which contribute to goal 2 and 3. Rural and tribal people heavily rely on bio-resources for earning to support their livelihood and basic needs which contributes to goal 1 and 8 in eradicating poverty and economic growth of rural population. Considering the larger perspective, all living creatures on the planet depend on bio-resources as their main foundational source of food which can be said to be its contribution to life below water and on land as well. Besides that, the value of bio-resources extend far beyond trade, food and nutritional importance. It also contributes in other ways to human well-being. This is clearly seen in Liberia, wherein aquatic and coastal resources protect a large portion of the population from climatic disasters like storms and floods, while also supporting their livelihood (The Economics of Ecosystems & Biodiversity, n.d.).

Next is the indirect impact on SDGs which come from the trade of bio-resources regulated under legal instruments driven by technology for the purpose of commercialization which consequentially contributes to the majority of SDGs. Numerous case studies on FES benefits as documented so far as *inter alia* shown below and interlinked with its SDGs dimension as shown in Table 2 demonstrate how this intervention fosters collaboration between corporate bodies, research organizations, and TK keepers for ultimately working toward a sustainable future.

Benefit Sharing Case Studies on Access to Bio-resources

The landmark example on importance of FES benefits framework on access of bio-resources is the "PepsiCo seaweed case," which is a numerously cited case by NBA itself in their reports. The seaweed (*Kappaphycus alvarezii* or *Euchemia cottonii*) is cultivated by fishermen of Tamil Nadu coastal village. The "PepsiCo India Holdings Pvt. Ltd." exported these marine macroalgae for commercial purposes to other countries like Malaysia, Philippines, and Indonesia. As per the agreement with NBA, the company paid a royalty of 5% of free on board of cost of seaweed, amounting to Rupees of 39 lakhs for exporting 2000 metric tons. This fund was used for the conservation of biodiversity and socio-economic development of the region; and benefits were distributed to 754 benefit claimers with the help of the SBB of Tamil Nadu (National Biodiversity Authority, n.d.; National Law School of India University, 2019).

Another most popular illustration is the "Neem leaves case," wherein, "Bio India Biologicals" company enter into an agreement with a Japanese firm to export neem (*Azadirachta indica*) leaves from India (Amarchinta village of district Mahboobnagar of Andhra Pradesh State) to Japan. The exporter company paid a royalty of 5% of free on-board which amounted to more than 55,000 Rupees to "Amarchinta Biodiversity Monitoring Committee." The said amount was used for planting neem saplings in the region and creating biodiversity conservation awareness among locals (National Biodiversity Authority, n.d.).

The Gram Moongligai case involved a direct exchange between local communities and corporations. Collectors and cultivators from several Indian States including Odisha, Madhya Pradesh, Chhattisgarh, and Tamil Nadu provided medicinal plants to commercial organizations through "Gram Moongligai Co. Ltd." and in return, were compensated with both financial and non-financial benefits (Ministry of Environment, Forest and Climate Change, 2018, p. 15).

Benefit Sharing Case Studies on Integration of Biotechnology and IPR

There are several cases which illustrate how the FES benefits framework as an instrument of IPR, plays a vital role in the development of tribal/local community, the growth of the country's economy, scientific development as well as in conservation of biodiversity. The most historic example is of the "Kani case of Arogyapacha," on the benefit sharing principle concerning with the use of TK associated bio-resources with their native owners before the set-up of the national legal framework. In 1987, a research group from the "All India Coordinate Research Project on Ethnobiology" with the help of Kani tribes of the Agasthyamalai hills located in Kerala State, discovered a fruit of the plant "Arogyapacha" (*Trichopus zeylanicus*) which is consumed by the tribal people for health purpose (mostly for improving fatigue and energy). After that, team of researchers conducted a series of biochemical and pharmacological studies on the plant and finally come up with a formulation of herbal drug called "Jeevani" which has an immune-stimulating and anti-fatigue effect (National Law School of India University, 2019, pp. 42-47). The "Jawaharlal Nehru Tropical Botanical Garden and Research Institute" (hereafter referred to as JNTBGRI) got a patent on the technology for making this herbal drug. In 1996, JNTBGRI transferred the license to "Arya Vaidya Pharmacy Ltd, Coimbatore" for commercialization of "Jeevani" drug for a fee of 50,000 USD and 2% royalties at ex-factory sale. This fee and royalties were shared with a ratio of 1:1 with Kani tribal communities through trust (National Law School of India University, 2019). So, this fund was used for the development of Kani tribes, documentation of TK of Kani tribes, and conservation of Trichopus plant of Agasthyamalai hills of Kerala (World Intellectual Property Organization, 2010).

Another landmark case that is first of its own kind after enactment of the BD Act, 2002, is "The Pinak tablet case," wherein an herbal composition termed as "Pinak" tablet was formulated using a combination of medicinal plants including Pangara mool (*Erythrina indica*); Mogra mool (*Jusminum sambac*); Jambhul chhal (*Eugenia jambolana*); and Amba chhal (*Magnefera indica*) (Raut, et al., 2013). As a result, the NBA set a 2% royalty on the product's gross sales or revenue (National Biodiversity Authority, n.d.).

Indigenous TK on medicines, folk remedies and biodiversity plays a vital role in medical and biotechnological research, which results in the scientific as well as socio-economic development of the whole world community. The development of anti-cancer drugs such as vincristine and vinblastine are one of the examples of TK-based drug development. These drugs are developed from a plant commonly known as "bright eyes" (*Catharanthus roseus*) which is native and indigenous to Madagascar (Bency Baby & Suriyaprakash, 2021).

Hoodia cactus (*Hoodia gordonii*) have been eaten by native San tribal people of Africa for its appetite suppressant properties. The P57 (an appetite-suppressing element found in the Hoodia plant) was patented by the "South African Council of Scientific and Industrial Research" (hereafter referred to as SA CSIR) for a dietary product. In 1997, the license was transferred to Phytopharm (a UK-based Biotech company), and then to Pfizer in 1998 (TKDL, n.d.). As the patent was based on the bio-resources and associated TK, the San tribal people move a motion against the SA CSIR and the company. After a negotiation, both parties entered into an agreement on benefit sharing, whereby the San tribal people would get a royalty of 6-8% of future commercial profits (Rabitz, 2015).

In "The Castor oilseed case," ICAR-IIOR, Hyderabad, collaborated with "Netherlands Biotechnology Programme" for research on Castor (*Ricinus communis*) and formulated a pesticide with the use of microbes collected from villages of Andhra Pradesh, India. The license was transferred to entrepreneurs for commercialization of research outcomes. A license fee and about 0.1 to 0.5% of royalty on the product's yearly gross sales were set. A percentage of license fee was directly transferred to Biodiversity Management Committees (hereafter referred to as BMC) of the village and 95% of royalty through SBB (Ministry of Environment, Forest and Climate Change, 2018).

Another example is the development of prostratin (12-deoxyphorbol), a compound that has immense value in HIV therapy. This phorbol ester was extracted from the plant *Homalanthus nutans* from Western Samoa, and the basis of selection of this plant for research was based on the TK of Samoan indigenous people who use this plant for the treatment of liver diseases (later identified as hepatitis). Due to the use of genetic material and TK of Falealupo village, over 480,000 USD had been given to the regional community for the development of health facilities, education and other infrastructural development of the region (Gupta et al., 2005). So, this TK-based research not only benefits the world medical fraternity but also to the development of Indigenous communities due to the FES benefits framework.

BIO-RESOURCES MANAGEMENT: LEGAL PERSPECTIVES AND CHALLENGES

As discussed, the bio-resources contribute to SDGs, either directly or indirectly or both. However, due to its vast diversity and abundance, it becomes a huge challenge for institutions to manage such diversified resources. In India, these resources are managed and monitored under the three-tiered institutional system established and constituted under the BD Act, 2002, namely, NBA, SBB, and "Biodiversity Management Committees" (hereafter referred to as BMC). Each of them has its own function and power granted under the BD Act, 2002. There are certain issues concerning the management of bio-resources, *inter alia* documentation of bio-resources, sustainable and responsible utilization of bio-resources, and issues with the influx of FES benefits to the benefits claimers are pointed out below:

Documentation of Bio-resources

Section 41 under Chapter X of the BD Act, 2002 provides provision for the constitution of BMC and also states about the purpose for which it shall be constituted which is construed as a function that they have to perform within their local jurisdiction. According to this Section, BMC shall be constituted by a local body for its territorial jurisdiction for three main purposes, namely, "promoting conservation, sustainable use and documentation of biological diversity." Sub-section (2) of this Section, puts NBA and SBB under obligation to consult BMC while taking any decision pertaining to the use of bio-resources occurring within its territorial jurisdiction. Documentation of bio-resources is important in the sense that unless and until we know about what the bio-resources are available in that particular area, we would not be able to manage it. Moreover, it also helps institutional machinery to identify the potentially tradable bio-resources, as not all resources are very useful for human needs. Identification of potentially tradable bio-resources would help to better utilize such resources for fulfilling human needs. According to Sub-section (1) of Section 41, BMC shall be responsible for the documentation of bio-resources. "Documentation" of known and unknown diversified bio-resources of biodiversity rich countries is a challenging task that requires a huge institutional effort. In this respect, Rule 11(h) of the BD Rules, 2024, requires that NBA shall provide guidance and technical support to BMC through the SBB or Union Territory Biodiversity Council or directly in preparing, validating and maintaining Biodiversity Registers. The preparation of People's Biodiversity Registers (hereafter referred to as PBR) is a tedious process and the same requires cooperation of a large number of people to share their TK (National Biodiversity Authority, 2013, p. 5). As of January 2025, more than 2.77 lakhs of BMC have been constituted; and the total number of PBR is 2,73,145 consisting of data from a total of 28 States and 8 Union Territories of India (National Biodiversity Authority, 2025). Owing to the fact that India is one of the megadiverse countries of this planet, this much of documentation is not sufficient and major milestones are still ahead to achieve.

Sustainability Challenges

The world faces a tough challenge in balancing the needs of a growing human population with the health of the environment. Achieving SDGs such as goal 12, 14, and 15 that deal with responsible and sustainable utilization of resources concurrently while achieving other goals is a challenge for us to balance our own needs with the future necessity. A major conflict exists between human needs and environmental health. The "Living Planet Report, 2020," states that humanity is over-exploiting Earth's bio-resources by at least 56% (Almond et al., 2020, p. 9). Although bio-trade provides vital economic support to locals, it raises serious concerns and causes harm to the environment. The decline in species is the result of over-utilization, whether for survival or for pure profit, puts bio-resources in jeopardy. In the "Kani case of Arogyapacha," the forest department banned the commercial harvesting of the Arogyapacha plant to prevent damage to the forest's long-term health. However, cultivation and harvesting of the said plant continued after employing sustainable practices with the help from JN TBGRI, and Integrated Tribal Development Program (World Intellectual Property Organization, 2010).

Lack of Proper System for Distribution of Benefits

Beyond the issues of conservation and sustainability, a significant concern is the failure of legal machinery for providing benefits to the resource providers. Flaws in these machinery often deprive providers of benefits they are entitled to. In the "PepsiCo Seaweed" case, a large royalty payment made to the NBA struck in a procedural backlog and hence reportedly remains unutilized (Jishnu, 2012). Similarly, in the Kani case, because the Kani community people are scattered and not unified at one place, only few who were present at the time of agreement had received benefits leaving others deprived (National Law School of India University, 2019, pp. 44-45). Thus this also brought to light a problem with benefit distribution.

CONCLUSION

The biopiracy creates a conflict of interest between the user, exclusive owner and the provider of bio-resources and associated TK. Serious repercussions of biopiracy could be seen which is not only restricted to revocation of patents and biodiversity losses but also affects the livelihood of the local communities. The Monsanto "Bt. Cotton" seed case is a clear-cut example of how the commercialization of such resources affects the socio-economic life of the farmers and locals. However, on the other side, TK-based bioprospecting under the instrumentality of FES benefits framework could be very beneficial not only in research and business fraternity, but also for the development of local communities, and socio-economic growth of undeveloped countries that hold the heritage of bio-resources associated TK. Indigenous TK on the cure of diseases is a result of thousands of years of ancient practices, which could be termed as "Ancient Research." There are numerous examples, some of them also stated in this chapter, that show the development of drugs based on the TK of tribal people. The FES benefits framework is an instrument that joins ancient and modern research. Hence, bridging the gap between research and commercialization, which also aids in achieving sustainable development (Nihal, 2022).

This chapter further suggests how biotechnological and legal intervention in the use of bio-resources aid in achieving SDGs. The environmental components supported by scientific innovations contribute in accomplishing the majority of the SDGs. As per the yearly report of NBA, it has received nearly INR 8.4 crores as benefit sharing during 2020-2021 (National Biodiversity Authority, 2021). This shows high revenue generation potential of legally regulated access and use of bio-resources. Bio-trade solely contributes to the achievement of goals on reducing poverty, zero hunger, good health, sustainable economic growth, maintaining life below water and on land. While, legal intervention aided with technology and IPR in the access and use of bio-resources helps to strengthen such goals and additionally help in achieving goal on ensuring inclusive and equitable quality education, clean water and sanitation, sustainable industrialization and foster innovation, making cities and human settlement sustainable, responsible use of natural resources, and to mitigate climate change. However, there is also a socio-legal challenge with regard to the responsible influx of benefits to society, that arises from the utilization of bio-resources, which needs to be explored in-ground practicality.

REFERENCES

Almond, R. E. A., Grooten, M., & Petersen, T. (2020). *Living planet report 2020 - Bending the curve of biodiversity loss.* World Wildlife Fund. https://files.worldwildlife.org/wwfcmsprod/files/Publication/file/279c656a32_ENGLISH_FULL.pdf

Balasubramanian, S. (2017). *Traditional knowledge and patent issues: An overview of Turmeric, Basmati, Neem cases*. Intellectual Property and Technology Law updates, S&A IP-Tech. https://www.manupatrafast.in/NewsletterArchives/listing/SAIPTech%20Singh%20Associates/2017/Mar/IPTech%20March17.pdf

Bency Baby, T., & Suriyaprakash, T. N. K. (2021). Intellectual property rights: Bioprospecting, biopiracy and protection of traditional knowledge - An Indian perspective. In Prabu, S. L., & Suriyaprakash, T. N. K. (Eds.), *Intellectual Property* (p. 6). IntechOpen.

De Penning & De Penning. (2025). The National Biodiversity Authority: New Rules and Their Implications. https://depenning.com/blog/the-national-biodiversity-authority-new-rules-and-their-implications/

Department of Economic and Social Affairs. United Nations. (n.d.). The 17 Goals. https://sdgs.un.org/goals

Government of India. (2024). *The Constitution of India.* Government of India. Art. 21. https://cdnbbsr.s3waas.gov.in/s380537a945c7aaa788ccfcdf1b99b5d8f/uploads/2024/07/20240716890312078.pdf

Greiber, T., Moreno, S. P., Ahren, M., Carrasco, J. N., Kamau, E. C., Medaglia, J. C., Oliva, M. J., & Perron-Welch, F. (2012). *An explanatory guide to the Nagoya Protocol on access and benefit-sharing*. International Union for Conservation of Nature.

Gupta, R., Gabrielsen, B., & Ferguson, S. M. (2005). Nature's medicines: Traditional knowledge and intellectual property management. Case studies from the National Institutes of Health, USA. *Current Drug Discovery Technologies*, *2*(4), 16–17. DOI: 10.2174/157016305775202937 PMID: 16475917

India code: Biological diversity act, 2002. (2002). https://www.indiacode.nic.in/handle/123456789/2046?sam_handle=123456789/1362

Ingle, A. P., Philippini, R. R., Martiniano, S., Marcelino, P. R. F., Gupta, I., Prasad, S., & Silva, S. S. (2020). Bioresources and their significance: Prospects and obstacles. In Kataki, R., Pandey, A., Khanal, S. K., & Pant, D. (Eds.), *Current Developments in Biotechnology and Bioengineering* (pp. 3–4). Elsevier., DOI: 10.1016/B978-0-444-64309-4.00001-5

Jishnu, L. (2012, October 15). Curious case of seaweed. *Down to Earth.* https://www.downtoearth.org.in/coverage/curious-case-of-seaweed-39207

Lakshmanan, P. K. (2015). An inquiry into the Biological Diversity Act, 2002 in the light of the Nagoya Protocol on access and benefit Sharing. *SSRN*, *6*, 35–55. DOI: 10.2139/ssrn.3048709

Ministry of Environment. Forest and Climate Change. (2018). *Implementation of Nagoya Protocol on access and benefit sharing India's experience.* Ministry of Environment, Forest and Climate Change. http://nbaindia.org/uploaded/pdf/Implementation%20of%20Nagoya%20Protocol%20in%20India.pdf

National Biodiversity Authority. (2013). Revised People's Biodiversity Register Guidelines. http://nbaindia.org/uploaded/pdf/PBR%20Format%202013.pdf

National Biodiversity Authority. (2021). *Annual Report 2020-21.* National Biodiversity Authority. http://nbaindia.org/uploaded/pdf/AR_English_2020-21.pdf

National Biodiversity Authority. (2025, January 23). People's Biodiversity Register. http://nbaindia.org/content/105/30/1/pbr.html

National Biodiversity Authority. (n.d.) *Access and benefit sharing in India a handbook for researchers.* National Biodiversity Authority. http://nbaindia.org/uploaded/pdf/IDB_ABS.pdf

National Biodiversity Authority. (n.d.). *Access and benefits sharing experiences from India.* http://nbaindia.org/uploaded/pdf/ABS_Factsheets_1.pdf

National Law School of India University. (2019). *Handbook on biodiversity laws, access and benefit sharing.* National Law School of India University. https://ceerapub.nls.ac.in/wp-content/uploads/2019/05/Handbook-on-Biodiversity-law-Access-and-Benifit-Sharing.pdf

Nihal, A. (2022). *Deciphering the role of environmental law in bio-pharmaceutical research: In the context of IPR* [Paper presentation]. Proceedings of International Conference on Recent Advances in Biotechnology, Abstract Book, Dr. B.R. Ambedkar National Institute of Technology, Jalandhar, 152.

Olga Tellis v. Bombay Municipal Corporation (1985) 3 SCC 545.

Pandey, J. N. (2021). *Constitutional Law of India* (58th ed.). Central Law Agency.

Rabitz, F. (2015). Biopiracy after the Nagoya Protocol: Problem structure, regime design and implementation challenges. *Brazilian Political Science Review*, *9*(2), 30–53. DOI: 10.1590/1981-38212014000200010

Raut, R. G., Pawar, P. S., & Chavan, H. H. (2013). A community trial of Pinak® medicine in the management of snakebite cases in a rural setting of Western Maharashtra, India. *International Journal of Medicine and Medical Sciences*, *5*(4), 164. Retrieved December 07, 2023, from https://academicjournals.org/journal/IJMMS/article-full-text-pdf/293184488

Secretariat of the Convention on Biological Diversity. (2011). *Nagoya Protocol on access to genetic resources and the fair and equitable sharing of benefits arising from their utilization to the Convention on Biological Diversity*. Secretariat of the Convention on Biological Diversity. https://www.cbd.int/abs/doc/protocol/nagoya-protocol-en.pdf

Secretariat of the Convention on Biological Diversity. (2020). *Global Biodiversity Outlook 5*. Secretariat of the Convention on Biological Diversity. https://www.cbd.int/gbo/gbo5/publication/gbo-5-en.pdf

Singh, R. D., Mody, S. K., Patel, H. B., Devi, S., Modi, C. M., & Kamani, D. R. (2014). Pharmaceutical biopiracy and protection of traditional knowledge. *International Journal of Research and Development in Pharmacy and Life Sciences*, *3*(2), 833–838. https://ijrdpl.com/index.php/ijrdpl/article/download/372/369

Singla, A. (2020). *Protection of traditional knowledge in India with reference to Neem, Turmeric, Basmati Rice*. https://www.researchgate.net/publication/340977589_'Protection_of_Traditional_Knowledge_in_India_with_reference_to_Neem_Turmeric_Basmati_Rice'_Name_-Aditya_SinglaProtection_of_Traditional_Knowledge_in_India_with_reference_to_Neem_Turmeric_Basmati_Rice

Takwani, C. K. (2021). *Lectures on Administrative Law* (7th ed.). Eastern Book Company.

The Economics of Ecosystems & Biodiversity. (n.d.). *TEEB Liberia*. https://teebweb.org/our-work/country-studies/enrtp2013/liberia/

Traditional Knowledge Digital Library. (n.d.). *Bio-piracy of traditional knowledge*. https://www.tkdl.res.in/tkdl/langdefault/common/Biopiracy.asp?GL=Eng

World Intellectual Property Organization. (2010). *Using traditional knowledge to revive the body and a community*. https://www.wipo.int/ipadvantage/en/details.jsp?id=2599

World Intellectual Property Organization. (n.d.). *Traditional knowledge.* https://www.wipo.int/tk/en/tk/

ADDITIONAL READING

Secretariat of the Convention on Biological Diversity. (2011). *The Convention on Biological Diversity.* Secretariat of the Convention on Biological Diversity. https://www.cbd.int/doc/legal/cbd-en.pdf

Secretariat of the Convention on Biological Diversity. (2011). *Nagoya Protocol on access to genetic resources and the fair and equitable sharing of benefits arising from their utilization to the Convention on Biological Diversity*. Secretariat of the Convention on Biological Diversity. https://www.cbd.int/abs/doc/protocol/nagoya-protocol-en.pdf

The Biological Diversity (Access to Biological Resources and Knowledge Associated thereto and Fair and Equitable Sharing of Benefits) Regulations, 2025. https://cdnbbsr.s3waas.gov.in/s3fcdb3b4550e745d29a64a696047067b7/uploads/2025/05/202505151889481287.pdf

The Biological Diversity Act, 2002, No. 18, Acts of Parliament, 2003 (India). https://www.indiacode.nic.in/bitstream/123456789/21545/1/the_biological_diversity_act%2C_2002.pdf

The Biological Diversity Rules. 2024. https://keralabiodiversity.org/wp-content/uploads/2024/11/Biological-Diversity-Rules-2024.pdf

The Protection of Plant Varieties and Farmers' Rights Act. 2001, No. 53, Acts of Parliament, 2001 (India). https://www.indiacode.nic.in/bitstream/123456789/1909/1/A2001-53.pdf

KEY TERMS AND DEFINITIONS

Access: In the present context, it means collecting, procuring, or processing of bio-resources or associated TK in or obtained in India for research, bio-survey, and commercialization purposes (BD Act, 2002, Section 2(a)).

Benefit claimers: It includes conservers, creators or holders of bio-resources, their by-products and associated TK (BD Act, 2002, Section 2(aa)).

Biological resources: Any natural resources having a biological origin, which may include all micro as well as macro-living forms like plants, animals, bacteria, fungi etc., as a whole and/or their parts, or by-products (Ingle et al., 2020).

Biopiracy: It is the mining of bio-resources and/or TK associated therewith, and thereafter getting exclusive ownership, and utilizing it for commercial gain without obtaining permission from their exclusive owner (such as bio-resource providing country, local and tribal communities) and without sharing the benefits which are arising from the commercialization of these resources (Rabitz, 2015).

Codified traditional knowledge: A knowledge/information obtained from books (BD Act, 2002, Section 2(ea)).

Traditional knowledge: A knowledge and practice that originated in ancient times, and has subsequently evolved, and being passed down from generation to generation within a community (WIPO, n.d.).

Value-added products: In the biodiversity context, a product containing the unrecognizable and physically inseparable form of plant and animal's parts or extracts (BD Act, 2002, Section 2(p)).

APPENDIX

Table 1. Comparative dissection of provisions of CBD, 1992 in relation to BD Act, 2002 as amended to date

Matters	CBD, 1992	BD Act, 2002
Objectives		
Purpose	To provide provisions for: conservation, and sustainable use of biodiversity; and FES benefits arising out of use, and access of genetic resources and transfer of technologies.	To give effect to the objectives of CBD, 1992. However, it relates to "biological resources" instead of "genetic resources."
Access		
Who requires approval	According to Article 15(1) and (2), it recognizes the sovereign rights of State, and further it recognizes authority of State to determine and put conditions for access of genetic resources. Thus, the contracting party, who obtained and uses such resources shall get approval from the provider State which is subject to their national legislation.	Under section 3, following persons requires approval of NBA: not a citizen of India; a citizen of India but "non-resident" defined under section 2(30) of Act No. 43 of 1961; corporate body, association or organization not incorporated or registered in India, if so, controlled by "foreigner" defined section 2(27) of Act No. 18 of 2013.
Exception to approval	In general, no such exception. However, with respect to applicability, for the purpose of CBD, 1992, only those genetic resources that are of country of origin or such resources which are acquired by contracting party in accordance with CBD, 1992 are only considered under the scope of Article 15, 16 and 19.	Under following conditions as mentioned under section 5(3) requirement of approval under section 3 shall not apply: when in consonance with the policy guidelines of Central government; or approval by Central government.
Who required to give/ take prior intimation/ consent, in the case may be	According to Article 15(1), the same shall be determined under national legislation of State, and according to Article 15(5) access is subject to PIC.	Under section 7(1), the person other than those categories of person referred under section 3(2), shall give prior intimation to SBB, to access bio-resources and associated knowledge, for the purpose of commercial utilization.

continued on following page

Table 1. Continued

Matters	CBD, 1992	BD Act, 2002
Who and When, not required to give/ take prior intimation/ consent, in the case may be	The same shall be determined under national legislation of State, as per Article 15(1) and (5).	According to Proviso to section 7, requirement of giving prior intimation shall not applied to: codified TK; medicinal plants which are cultivated, and its products; and regional people and communities, which also include growers and cultivators, and person (such as *vaids*, *hakims,* and registered AYUSH practitioners) who, as profession for livelihood, practices indigenous medicines that includes Indian system of medicine. It is worth to mention here that, for giving prior intimation under section 7, it does not include access for research purpose. That means, for research purpose, person other than as referred under section 3(2), no such prior intimation is required. However, categories of persons as mentioned under section 3(2) approval shall be required even for research purpose.
Access to/transfer of research result	Article 15(6) says about full participation of provider of genetic resources in scientific research. Article 19(1) says about participation in biotechnology-based research especially in case of developing countries which has provided such resources. Article 19(2) says, each contracting party shall take measures to promote access of research result, especially to developing countries on MAT based on fair and equitable basis.	Section 4 put restriction on transfer of research result to certain persons. According to which, no person shall transfer result of research to person referred under section 3(2) without prior approval of NBA. Section 20 provides that if research result on bio-resources or associated TK has to be transferred to person referred under section 3(2) by any person, then an application has to be made to NBA.
Access to and transfer of technology (hereafter referred to as ATT)	Article 16(1) recognizes ATT as essential elements for achieving objectives of CBD, 1992. Article 16(3) provides that, developing countries shall be provided with the ATT including technology which is subject to IPR on MAT. Article 16(2) specifically states about the nature of MAT, that in case of developing countries the ATT shall be made under fair and most favorable terms including concessional and preferential terms.	It provides expressed provision related to transfer of research result under section 4 and section 20. However, it remains silent specifically on ATT. Moreover, in terms of benefit arises out of use of bio-resources, under section 21 which deals with determination of FES benefits, the ATT with respect to or matters connected to those resources shall be decided on condition of approval and in consonance with MAT. Section 21(2)(b) include transfer of technology as one of the methods to share benefits.
Conservation and sustainability		

continued on following page

Table 1. Continued

<table>
<tr><th>Matters</th><th>CBD, 1992</th><th>BD Act, 2002</th></tr>
<tr><td>Cooperation</td><td>Article 5 says about cooperation between contracting party as far as its possible, practicable and appropriate.</td><td rowspan="2">It remains silent on providing expressed provision on conservation and sustainable use of bio-resources. However, it provides function to NBA to advise the Central government relating to conservation and sustainability under section 18(3)(a); and under section 23(a) to State government by SBB. Further, section 24 gives power to SBB to restrict certain activities which violates objectives of this Act.</td></tr>
<tr><td>Efforts for conservation and sustainability</td><td>Article 6 states about the general measure, whereas Article 11 talks about incentive measures that could be taken for conservation and sustainable use of biodiversity, for instances: - development and adoption of strategies, policies, and plans. Article 7 talks about the identification of biodiversity necessary for the said purpose, identification of such activities which may have adverse effect on the biodiversity; monitoring of the techniques and; maintenance and organization of data derived under said identification and monitoring. Article 8 talks about in-situ and Article 9 about ex-situ conservation of biodiversity. Article 10 provides the ways in which biodiversity could be used in sustainable manner, which also include to support local population to develop remedial action and joint cooperation. Promoting conservation and sustainability directly or indirectly, through research and training provided under Article 12 and through public education and awareness under Article 13. Provision as to environment impact assessment to avoid adverse effect on bio-resources has been provided under Article 14(a). Article 16(1) states about providing of ATT to contacting party for conservation and sustainability. Further, Article 17 states about exchange of information and Article 18 about technical and scientific cooperation for conservation and sustainable use of biodiversity. Article 19(3) says about the need for setting up of protocol for safe transfer, handling and use of any modified living organism which has been resulted from biotechnology-based research and that may have negative impact on biodiversity.</td></tr>
<tr><td colspan="3">Benefit sharing</td></tr>
</table>

continued on following page

Table 1. Continued

Matters	CBD, 1992	BD Act, 2002
FES benefits	Article 16 (3) provides that, the ATT shall be provided to contracting party, specially to developing country, who provided genetic resources. Article 19(1): participation of resource provider contacting party specifically developing country in biotechnological research. Article 16(1) provides FES benefits in terms of active participation in biotechnological research, and Article 15(6) for full participation in scientific research. Article 19(2) deals on access of research result and FES benefits arises from the biotechnological research based on genetic resources. Apart from this expressed provision, the CBD, 1992, failed to provide expressed provision as to ways in which benefits shall be shared, it rather emphasizes more on conservation and sustainability.	Section 21 provide provision related to determination of FES benefits by NBA. Under section 21(1) NBA shall determine as per the condition of approval and MAT. Section 21(2) in list the manners in which benefits shall be determined, which shall be subject to the regulation made in this behalf, for instance: - benefits may include grant of joint ownership of IPR to NBA, and benefits claimers if identified; establishment of production and research units in the areas of benefit claimers; associations of Indian researchers, local people and benefit claimers; set up of venture capital funds; monetary and non-monetary compensation to benefit claimers.

continued on following page

Table 1. Continued

Matters	CBD, 1992	BD Act, 2002
Provision with regards to patents and other IPR	Article 16 deals with the ATT including technology protected by patents and other IPR.	According to section 6(1) it put obligation to certain persons as refereed under section 3(2), when applying for IPR, obtain prior approval from NBA, before grant of such IPR. Section 6(1A) put obligation to any other person other than those referred under section 3(2), when applying for IPR, to register with NBA, before grant of such IPR. According to section 6(1B), any person referred under section 7, who has obtained IPR, has to take prior approval from NBA at the time of commercialization. For the purpose of section 6(1), (1A), and (1B), it is applicable when any invention is based on research or information on a bio-resource or associated TK which is accessed from India. Section 18(4) give power to NBA to take necessary measures to oppose the grant of IPR in country outside territory of India on any bio-resources or associated TK found or brought from India.
Provisions to make protocols, regulations, or guidelines	Article 28 talks about adoption of protocols. Under which, according to paragraph 1, the contracting parties shall formulate and adopt protocols with regards to this convention.	Section 64 which gives power to NBA to make regulations for the purpose of this Act. Section 21(4) gives power to NBA to frame guidelines by regulations for the purpose of Section 21 which states about determination of FES benefits; and section 18(1) which states that NBA shall make regulations for access to bio-resources and associated TK, and determination of FES benefits.

Table 2. Trade of bio-resources under legal guiding framework contributes to SDGs

Case	Resources utilized	Matters involved	Contribution to concerned SDGs
The PepsiCo seaweed case	Marine-based bio-resources	Access to bio-resources	The fund was used for conservation of bio-resources and socio-economic development of areas rich in biodiversity (National Biodiversity Authority, n.d.; NLSIU, 2019). This contributes to SDGs: 8, 11, 12, 13, 14, and 15.
Neem leaves case	Plant-based bio-resources	Access to bio-resources	Monetary benefits were shared and utilized for planting neem saplings and creating awareness on biodiversity conservation among locals (National Biodiversity Authority, n.d.). This contributes to SDGs: 12, 13, and 15.
The Gram Moongligai case	Plant-based bio-resources	Access to bio-resources	The company provided training on sustainable cultivation and harvesting practices. Monetary benefits support livelihoods and non-monetary benefits support capacity building, research, and conservation of biodiversity (Ministry of Environment, Forest and Climate Change, 2018, p. 15). This contributes to SDGs: 1, 2, 8, 11, 12, 13, and 15.
The Kani case of Arogyapacha	Plant-based bio-resources and associated TK	Ethical practice, biotechnological intervention, technology transfer, IPR protection (bioprospecting-based patents)	Money and training provided to tribes for cultivation, as result, secured their stable income and promoted sustainable cultivation. Funds from the Trust were utilized for socio-economic development of Kani community, such as providing health, medical, educational facilities, and awareness on conservation of bio-resources (WIPO, 2010; NLSIU, 2019). This contributes to SDGs: 1, 2, 3, 4, 6, 8, 9, 11, 12, 13, and 15.
The Pinak tablet case	Plant-based bio-resources	Medical research, innovation, and IPR protection through Patent law	Monetary benefits arose due to BD Act, 2002 and IPR protection promote sustainable bio-economy, medical research and developments, and responsible use of resources (National Biodiversity Authority, n.d.), and hence, this contributes to SDGs: 3, 8, 9, 12, and 15.
San tribal people case	Plant-based bio-resources and associated TK	Biotechnological intervention, and IPR protection through Patent law	Monetary benefits given to tribal people which supported their livelihood (Rabitz, 2015), and hence, this contributes to SDGs: 1, 2, 8, 9 and 15.

continued on following page

Table 2. Continued

Case	Resources utilized	Matters involved	Contribution to concerned SDGs
The Castor oilseed case	Plant and microbes-based bio-resources	Biotechnological intervention, research collaboration, technology transfer and IPR protection	Monetary benefits from license fee and royalty utilized by BMC in conservation, protection of bio-resources, conducting awareness programs on ABS mechanism, and regeneration of endangered species (Ministry of Environment, Forest and Climate Change, 2018). This contributes to SDGs: 8, 9, 11, 12, and 15.
Western Samoa case	Plant-based bio-resources and associated TK	Medical research, innovation, and IPR protection through Patent law	Monetary benefits were given to indigenous people for development of health facilities, education, and other infrastructural development (Gupta et al., 2005). This contributes to SDGs: 1, 2, 3, 4, 6, 8, 9, and 15.

Chapter 4
Strategies for Improving Ecosystem Resilience:
The Promotion of Biodiversity

Shanmuga Sundari M.
https://orcid.org/0000-0001-5755-474X
BVRIT HYDERABAD College of Engineering for Women, India

Rishitha Reddy E.
BVRIT HYDERABAD College of Engineering for Women, India

ABSTRACT

Ecosystem resilience is the ability of natural systems to recover from disturbances while maintaining essential functions. Biodiversity enhances this resilience by supporting ecological balance, redundancy, and adaptability. However, habitat loss, climate change, and pollution threaten biodiversity and ecosystem stability. This chapter explores how biodiversity supports resilience through species interactions, genetic and functional diversity, and habitat connectivity. It discusses conservation strategies like restoration, sustainable land use, and policy support. Real-world examples from forests, wetlands, and farms show biodiversity's role in resilience. The chapter concludes with recommendations to integrate science, policy, and community efforts for conserving biodiversity and strengthening ecosystems.

DOI: 10.4018/979-8-3373-2003-8.ch004

1. INTRODUCTION

The Importance of Ecosystem Resilience

Ecosystem resilience (Vasiliev, 2022) is the ability of ecosystems to endure disturbances, reorganize, and maintain their essential functions. This concept has become increasingly significant as humanity faces growing environmental challenges such as climate change, habitat destruction, and pollution. Resilient ecosystems are fundamental to sustaining life on Earth and ensuring human well-being.

These ecosystems provide critical services essential for survival, including food production, clean water, carbon storage, and climate regulation. For instance, forests act as major carbon sinks, storing approximately 289 gigatons of carbon in their biomass and soils, thereby mitigating climate change impacts. Wetlands, on the other hand, filter pollutants from water and support biodiversity by providing habitats for numerous species. For instance, the Ramsar Convention estimates that wetlands provide ecosystem services worth over $47 trillion annually, including flood regulation and water purification.

Biodiversity plays a vital role in enhancing ecosystem resilience. A diverse range of species ensures functional redundancy—where multiple species can perform similar ecological roles—allowing ecosystems to adapt even when certain species are lost. For example, in grasslands, various plant species contribute to soil stabilization and nutrient cycling. Similarly, diverse pollinator populations ensure effective reproduction in flowering plants, which is crucial for food security.

Resilient ecosystems also possess adaptive capabilities to cope with changing environmental conditions. As climate change intensifies, ecosystems face challenges such as altered rainfall patterns and extreme weather events. Systems with high genetic diversity—like coral reefs—are better equipped to withstand temperature fluctuations and maintain their integrity.

The socio-economic benefits of resilient ecosystems are immense. Healthy ecosystems support billions of livelihoods globally. Forests directly sustain approximately 1 billion people through activities like logging and ecotourism, while fisheries provide food and income for millions worldwide. Additionally, ecosystems such as coastal mangroves protect communities from natural disasters like storm surges and flooding while fostering marine biodiversity.

In summary, ecosystem resilience is a cornerstone of environmental sustainability and human prosperity. Promoting resilience through conservation efforts, sustainable land management practices, and policies that prioritize ecological health is crucial for securing a stable future for both nature and humanity.

Figure 1. Ecosystems planting

Key Drivers of Ecosystem Degradation

Ecosystem degradation represents a major threat to biodiversity and human welfare (Sathvara et al., 2024) globally. Various human-induced factors contribute to this degradation, undermining ecosystem resilience and functionality.

Land-use change is one of the most significant drivers of degradation. Urbanization and agricultural expansion have altered approximately 75% of Earth's land surface since the mid-20th century. Deforestation—largely driven by agriculture—accounts for about 80% of global deforestation rates. Monoculture plantations like soybeans or palm oil significantly reduce biodiversity and disrupt essential services such as carbon storage and water regulation.

Climate change further exacerbates these issues by altering temperature and precipitation patterns. Rising temperatures cause heat stress on sensitive species and increase the frequency of extreme weather events like hurricanes and droughts. Coral reefs are particularly vulnerable; warming oceans lead to widespread coral bleaching that devastates reef ecosystems.

Pollution, especially agricultural runoff containing fertilizers, results in nutrient pollution in aquatic systems. This leads to harmful algal blooms that deplete oxygen levels (hypoxia) in water bodies, creating "dead zones" that severely impact marine life. Over 400 such zones have been identified globally.

Invasive species pose another challenge by outcompeting native species or irreversibly altering habitats. Zebra mussels in North American freshwater systems exemplify this issue—they disrupt local ecosystems while causing economic damage by clogging water intake systems.

Overexploitation, particularly in marine environments, has led to the collapse of numerous fish stocks due to unsustainable fishing practices that prevent population recovery.

These drivers often interact in complex ways, creating feedback loops that amplify degradation further—for example, deforestation leading to soil erosion that reduces agricultural productivity, prompting more land conversion for farming.

In conclusion, addressing ecosystem degradation requires a multifaceted approach encompassing sustainable land management practices, pollution control measures, invasive species management strategies, and biodiversity conservation policies aimed at promoting ecological integrity alongside sustainable development.

2. THE ROLE OF BIODIVERSITY IN ECOSYSTEM RESILIENCE

Biodiversity plays a pivotal role in maintaining ecosystem resilience, enabling ecosystems to adapt to disturbances, sustain essential functions, and recover from environmental stressors. It serves as the foundation for ecological stability (Levis et al., 2024) and ensures the continuity of ecosystem services that are vital for human survival. Biodiversity contributes to resilience through mechanisms such as ecological stability, functional redundancy, and the implementation of conservation strategies.

Biodiversity and Ecological Stability

Figure 2. Worldwide ecosystems

Ecological stability is one of the most significant benefits of biodiversity. Diverse ecosystems are better equipped to resist disturbances and recover from them due to the presence of varied species that fulfill complementary roles. For instance, grasslands with a mix of deep-rooted and shallow-rooted plants can stabilize soil during droughts while preventing erosion during heavy rains. Similarly, coral reefs with genetically diverse algae symbionts exhibit greater resilience to marine heatwaves compared to less diverse reefs. These examples highlight how biodiversity enhances an ecosystem's ability to withstand external pressures.

Trophic complexity is another mechanism through which biodiversity fosters stability. In multi-layered food webs, species interactions create buffers that prevent cascading collapses when one species is affected. The reintroduction of wolves in Yellowstone National Park exemplifies this concept; wolves controlled elk populations, allowing riparian vegetation and beaver colonies to recover, which restored balance across the ecosystem. Genetic diversity further strengthens resilience by enhancing adaptive capacity. Boreal forests with diverse tree genotypes have been shown to tolerate temperature fluctuations significantly better than monocultures.

However, biodiversity's stabilizing effects are not limitless. Studies have revealed trade-offs between temporal stability and resistance to acute stressors like warming.

For example, while diverse aquatic ecosystems recover faster from disturbances, they may struggle during sudden shocks. The Amazon rainforest provides a stark example of how fragmented habitats can undermine ecological stability; deforestation has reduced its resilience by 77% since 2000, disrupting moisture recycling and threatening its ability to recover from environmental stressors.

Functional Redundancy and Ecosystem Services

Functional redundancy is another critical aspect of biodiversity that ensures the continuity of ecosystem services despite disturbances. It refers to the presence of multiple species that perform similar ecological roles, providing a safety net for ecosystems when certain species decline or disappear. For example, diverse pollinator networks—including bees, butterflies, and beetles—maintain crop yields even when honeybee populations face declines due to disease or habitat loss.

Redundancy supports a range of ecosystem services essential for human survival. Wetlands with varied plant species effectively remove nitrogen pollutants through complementary root structures, ensuring clean water supplies. Multi-species fisheries sustain catches even when key stocks decline due to overfishing or environmental changes. Soil microbes with redundant nutrient-cycling roles maintain fertility despite drought-induced species loss, supporting agricultural productivity.

From an economic standpoint, the ecosystem services supported by biodiversity represent immense financial value. Studies estimate that global ecosystem services—such as pollination, water purification, and soil formation—contribute over $125 trillion annually to the world economy. The loss of biodiversity can therefore impose enormous hidden costs. For instance, the decline in pollinators could reduce global crop output by up to $500 billion per year. Quantifying such economic values underscores that investing in biodiversity is not only an environmental necessity but also an economically sound strategy within the bioeconomy framework.

Nevertheless, redundancy has its limits. Coral reefs demonstrate this vulnerability; while they boast high fish diversity, 58% of critical functions such as algal grazing depend on single species. This leaves reefs prone to collapse if those species are lost due to warming oceans or pollution. Similarly, the collapse of Atlantic cod stocks in the 1990s revealed how overexploitation can exceed compensatory capacities within marine food webs, leading to widespread disruptions.

Strategies for Promoting Biodiversity

Given its importance for resilience, promoting biodiversity requires targeted conservation efforts at multiple scales. Habitat protection is one of the most effective strategies for safeguarding biodiversity. Expanding protected areas to cover 30% of terrestrial and marine environments by 2030—a goal outlined in the Kunming-Montreal Framework—could preserve habitats for 80% of endangered species worldwide. Costa Rica's network of ecological corridors demonstrates the success of this approach; it has increased forest cover by 25% and bird diversity by 18% since 2000.

Sustainable land-use practices also play a vital role in promoting biodiversity. Agroecological methods such as crop diversification enhance farmland biodiversity while maintaining agricultural yields. For instance, California ranchers have successfully conserved native grasslands by integrating sustainable grazing practices into their operations.

Policy innovation (Kattel, 2022) is another crucial avenue for promoting biodiversity conservation. Redirecting harmful agricultural subsidies—estimated at $500 billion annually—toward agroforestry programs and soil health initiatives could significantly enhance resilience across agricultural landscapes. Community-led conservation efforts further amplify these strategies; Indigenous fire management practices in Australian savannas have reduced wildfires by 75% while simultaneously enhancing biodiversity.

Technological advancements offer additional tools for conserving biodiversity. Bioacoustic monitoring enables researchers to assess ecosystem health through soundscapes, while AI-driven genetic rescue programs introduce heat-tolerant coral symbionts to reefs threatened by warming oceans.

Biodiversity is fundamental to ecosystem resilience through its contributions to ecological stability and functional redundancy while offering pathways for conservation through innovative strategies. By integrating habitat protection measures with sustainable practices and adaptive governance frameworks—as demonstrated in successful restoration projects like those in Costa Rica—ecosystems can better withstand disturbances and continue providing essential services for both nature and humanity. However, addressing systemic drivers such as climate change and land-use pressures remains critical for ensuring that ecosystems retain the diversity needed to adapt to an uncertain future.

Biodiversity and the Bioeconomy

Biodiversity is not only vital for ecological stability but also a cornerstone of the emerging bioeconomy, which aims to use biological resources sustainably for economic development. Integrating biodiversity into bioeconomic planning offers opportunities to generate income, promote innovation, and support livelihoods while maintaining ecological integrity.

Sustainable Agriculture: Integrating biodiversity-based practices such as intercropping, organic farming, reduced tillage, and conservation agriculture enhances ecosystem health and economic resilience. For instance, intercropping systems—where multiple crops are cultivated together—maximize resource use efficiency, improve soil fertility, and reduce dependence on synthetic fertilizers and pesticides. These biodiversity-friendly systems increase total yields and ensure long-term food security and profitability for farmers.

Biomimicry: Nature-inspired innovation, or biomimicry, translates biological principles into sustainable technologies and industrial solutions. Examples include energy-efficient building designs modeled after termite mounds or water-harvesting technologies inspired by desert beetles—demonstrating how biodiversity drives innovation and contributes to the bioeconomy.

Ecotourism: Biodiversity-rich ecosystems provide significant economic opportunities through ecotourism. Protected areas, coral reefs, and wildlife reserves attract millions of visitors annually, generating revenue for local communities and promoting conservation-based livelihoods. For example, ecotourism in Costa Rica contributes over 5% of its GDP while financing habitat protection and biodiversity monitoring programs.

Bioprospecting: The exploration of biological resources for pharmaceutical, agricultural, and industrial applications—when conducted ethically—can create sustainable economic value. For example, bioactive compounds derived from rainforest plants have led to life-saving medicines, emphasizing the economic potential of conserving natural habitats.

Integrating biodiversity into agricultural and industrial systems creates economic opportunities for local communities while promoting ecological balance. Sustainable use of biodiversity ensures that the bioeconomy remains regenerative rather than extractive.

3. RESTORATION OF DEGRADED HABITATS

Figure 3. Roadmap for expected ecosystem

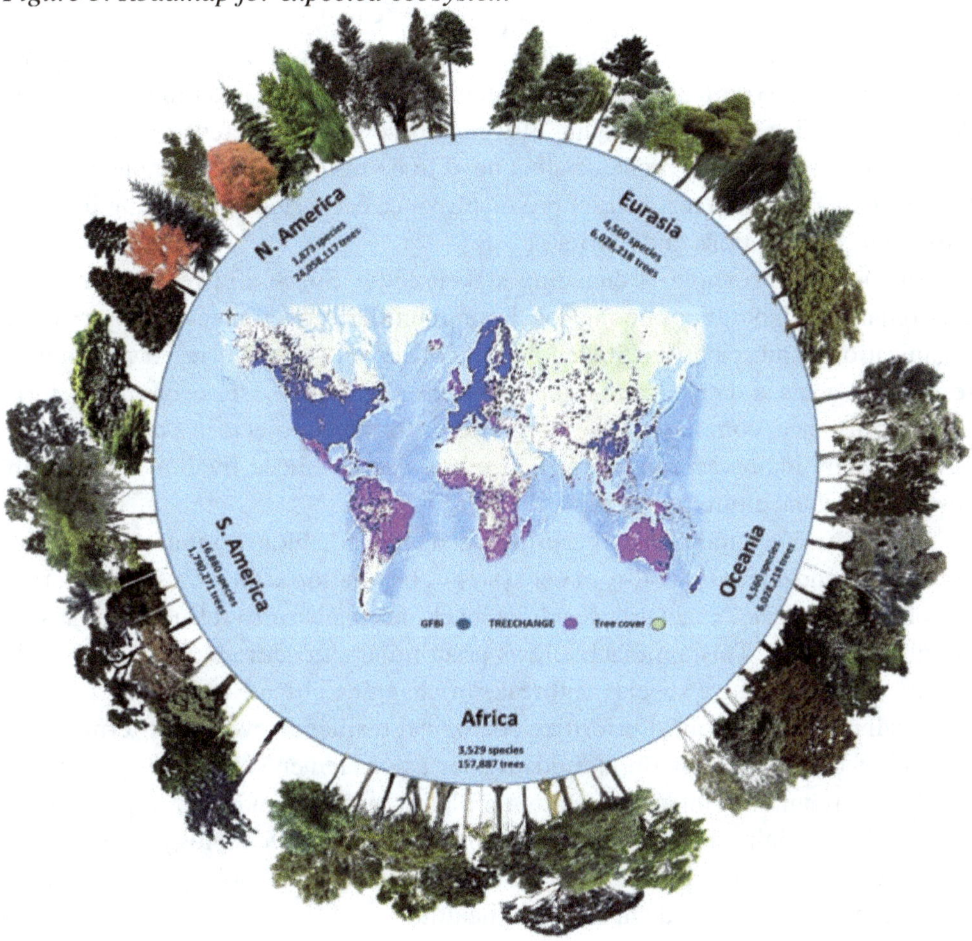

Restoration of degraded habitats is a vital process aimed at rehabilitating ecosystems that have been damaged by human activities such as deforestation, pollution, mining, and urbanization. It seeks to reestablish ecological functions, enhance biodiversity, and ensure the long-term sustainability of ecosystems. With habitat loss being one of the leading causes of species extinction and ecosystem service decline, restoration has emerged as an essential tool for reversing environmental degradation and combating climate change. This section explores the principles guiding habitat restoration, techniques employed to restore ecosystem functionality, and success stories that highlight lessons learned from restoration projects worldwide.

Principles of Habitat Restoration

Effective habitat restoration is guided by several foundational principles that ensure sustainable outcomes. The first principle is understanding baseline conditions. Restoration (Kennedy et al., 2023) efforts must begin with a comprehensive assessment of the original ecosystem structure, composition, and function prior to degradation. This involves studying historical baselines or reference conditions to determine the ecological processes that need to be restored. For example, paleontological data can help reconstruct pre-disturbance ecosystems, providing insights into species interactions and nutrient cycles.

Another key principle is engaging stakeholders. Successful restoration projects require collaboration among local communities, governments, conservation organizations, and scientists. Stakeholder involvement fosters shared ownership of restoration goals and ensures that interventions align with socio-economic priorities. For instance, community-led mangrove restoration projects in Southeast Asia have not only improved coastal resilience but also enhanced livelihoods through sustainable aquaculture practices.

Adaptive management is another critical principle in habitat restoration. Given the complexity and unpredictability of ecosystems, restoration efforts must incorporate ongoing monitoring, evaluation, and feedback mechanisms to refine strategies as conditions evolve. This approach allows practitioners to address unforeseen challenges, such as invasive species outbreaks or changing climate patterns.

Finally, restoration must prioritize ecological resilience over short-term gains. Instead of focusing solely on individual species or elements, modern approaches emphasize rebuilding interactions among plants, animals, and microorganisms to create self-sustaining ecosystems. For example, microbial technology has emerged as a core component of new restoration techniques due to its ability to reconstruct biogeochemical cycles and enhance soil health.

Techniques for Restoring Ecosystem Functionality

Figure 4. City ecosystem

A wide range of techniques is employed to restore ecosystem functionality based on the type of degradation and the specific goals of the project. Reforestation and afforestation are among the most common methods used to rebuild forest ecosystems. Planting native trees and shrubs not only restores vegetation cover but also enhances carbon sequestration and stabilizes soils against erosion.

Wetland rewetting is another effective technique for restoring aquatic ecosystems that have been drained or polluted by human activities. By reestablishing hydrological regimes and planting native wetland vegetation, these projects improve water filtration capacity while providing habitats for diverse species.

Soil remediation is crucial in areas affected by industrial pollution or mining activities. Techniques such as phytoremediation—using plants to absorb contaminants—

and bioengineering approaches like microbial inoculation help detoxify soils and restore their fertility.

Marine habitat restoration focuses on rebuilding coral reefs and mangroves that have been damaged by warming oceans or coastal development. Coral transplantation involves relocating healthy coral fragments to degraded reefs to accelerate recovery, while mangrove planting stabilizes coastlines and supports marine biodiversity.

Passive recovery is also gaining recognition as a viable technique for restoring ecosystem functionality. This approach involves halting harmful activities (e.g., logging or grazing) and allowing natural processes to facilitate recovery over time. Studies have shown that passive recovery can be as effective as active interventions in certain contexts, particularly when propagules (seeds or organismal dispersers) are present in surrounding landscapes.

Emerging technologies are revolutionizing restoration practices. For instance, genomic tools enable practitioners to select resilient species for reforestation projects based on their genetic traits, while AI-powered models predict climate impacts on restored habitats to optimize interventions.

Success Stories and Lessons Learned

Numerous success stories from around the world demonstrate the potential of habitat restoration to achieve ecological recovery while benefiting local communities. One notable example is the Loess Plateau restoration project in China. Once a severely degraded landscape plagued by soil erosion and poverty, targeted interventions transformed the region into a thriving ecosystem with improved agricultural productivity and biodiversity. Key strategies included terracing slopes to reduce erosion, planting native vegetation to stabilize soils, and engaging local communities in sustainable land-use practices.

Another success story comes from Costa Rica's efforts to restore its tropical forests through ecological corridors connecting fragmented habitats. These corridors have increased forest cover by 25% over two decades while boosting bird diversity by 18%. The project highlights the importance of landscape-level planning in overcoming habitat fragmentation—a major driver of biodiversity loss.

In marine environments, coral reef restoration projects in Australia's Great Barrier Reef have shown promising results despite challenges posed by climate change. Techniques such as coral transplantation combined with selective breeding of heat-tolerant coral strains have accelerated reef recovery following mass bleaching events.

These projects offer valuable lessons for future restoration efforts. First, long-term monitoring is essential for assessing progress and adapting strategies as conditions change. Second, integrating socio-economic benefits into restoration goals ensures sustained local support—whether through ecotourism initiatives or sustainable agriculture practices linked to restored habitats.

Finally, collaboration across sectors—science, policy-making, community engagement (Yadav et al., 2023—enhances the scalability of successful interventions. For example, international frameworks like the UN Decade on Ecosystem Restoration provide platforms for sharing best practices globally while mobilizing resources for large-scale projects.

Restoration of degraded habitats is an indispensable strategy for reversing environmental damage and enhancing ecosystem resilience in the face of global challenges such as climate change and biodiversity loss. By adhering to principles such as stakeholder engagement and adaptive management while employing diverse techniques tailored to specific contexts—from reforestation to marine habitat restoration—practitioners can achieve meaningful ecological recovery. Success stories from around the world underscore the transformative potential of habitat restoration when implemented thoughtfully and collaboratively.

4. ADAPTIVE MANAGEMENT: A RESPONSIVE APPROACH

Figure 5. Adaptive management for ecosystem

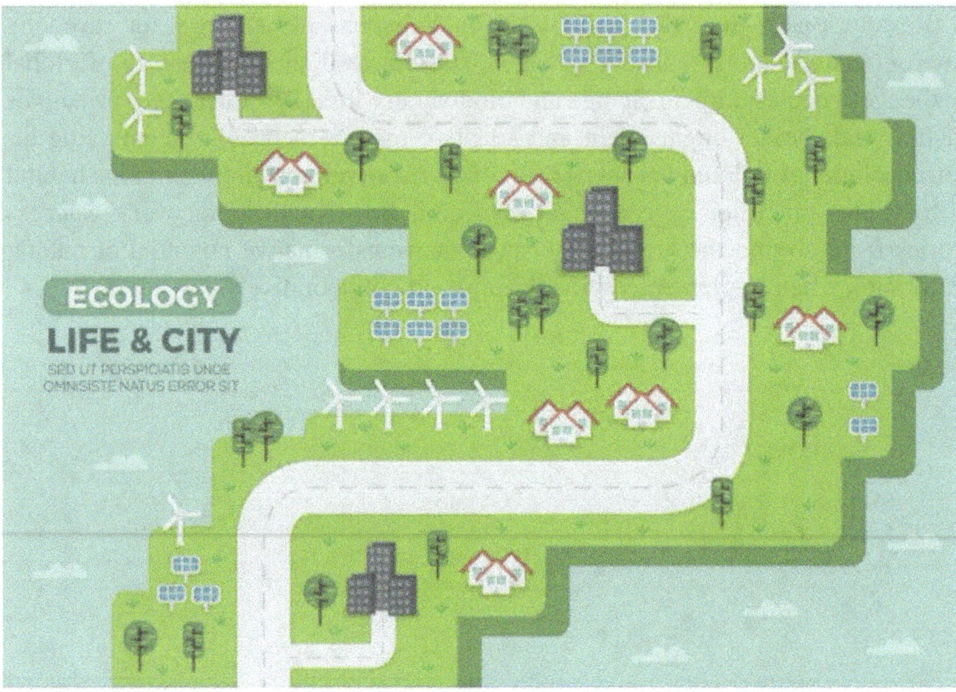

Adaptive management is a well-organized, step-by-step process designed to improve how we make decisions when things are uncertain. It's all about learning by doing, where we constantly tweak our management plans based on what we observe and learn. Because it's flexible and responsive, it's great for dealing with complicated systems like ecosystems, where things are naturally unpredictable. This section will explain what adaptive management is, its main ideas, the tools we use to keep track of things and get feedback, and real-world examples of how it's used.

Definition and Core Principles

Adaptive management is like a dynamic way of making decisions that brings learning into the way we manage things. It works as a never-ending loop of planning, doing, watching, figuring out what happened, and making adjustments. This lets managers react to new info or changing situations while also getting better at predicting what will happen over time. Unlike old-fashioned, set-in-stone approaches, adaptive management knows that ecosystems are complex and unpredictable, so we need flexible plans that can change as we learn more.

At its heart, adaptive management is based on a few key ideas. First, there's iterative learning, where every choice we make is both a step toward our goals and a test to see how the system works. For example, if we're trying to bring back a wetland, we might try different ways of managing the water flow to see which one helps the most plants and animals while preventing floods. This means that even when things don't go as planned, we still learn something valuable for the future.

Another important idea is flexibility in decision-making. Adaptive management understands that our initial plans might need to change as things change or unexpected problems pop up. For instance, if a harmful species suddenly appears or there's a big storm, we might need to shift our focus during habitat restoration.

Getting everyone involved is also key to adaptive management (Wang et al., 2024). By including local communities, scientists, politicians, and other stakeholders in making decisions, we encourage teamwork and make sure that different viewpoints are considered. This makes our actions more fair and effective.

Finally, adaptive management emphasizes keeping a close eye on things over the long term as a basis for making informed adjustments. By collecting data regularly, managers can see how well their actions are working compared to what they wanted to achieve, and then adjust their plans accordingly.

There are two main types of adaptive management: active and passive. Active adaptive management involves intentionally testing different ideas through experimental actions to get a better understanding of how the system works. On the other hand, passive adaptive management focuses on tweaking plans based on what we see happening, without specifically testing different options.

Tools for Monitoring and Feedback Integration

Keeping track of things and getting feedback are super important parts of adaptive management. These tools give us the info we need to see how things are going and make better decisions down the road. And, thanks to new technology, we can now monitor things in real-time and analyze data more effectively than ever before.

One of the most popular tools is remote sensing, which uses satellite images and drones to keep an eye on changes in land cover, plant health, and water quality over large areas. For instance, remote sensing has been a game-changer for tracking deforestation in the Amazon rainforest and seeing how well reforestation efforts are working.

Ecological indicators are another key tool for checking in on ecosystem health. These include things like how many different species there are, how healthy the soil is, how clean the water is, and how much carbon is being stored. For example, we can check on coral reefs by looking at things like coral cover and fish populations. Citizen science is also becoming more and more important in adaptive management. It involves getting local communities to help collect data. Platforms like iNaturalist let people share their observations of where different species are, which can then be used to spot trends or new threats.

The data we collect with these tools needs to be used to make decisions through feedback loops. Adaptive management uses systems like Bayesian networks, which use probability models to update our predictions based on new evidence. These networks help managers weigh the pros and cons of different goals—like balancing water use for farming with keeping wetlands healthy—and figure out the best strategies when things are uncertain.

Besides these ecological tools, social feedback is also becoming recognized as a key part of adaptive management. Surveys and interviews with stakeholders give us valuable insights into how people feel about our actions and highlight areas where we might need to make changes to better align with community needs.

Case Studies in Adaptive Management

There are many examples from around the world that show how adaptive management has been successfully used to tackle complex environmental problems while also learning and coming up with new ideas.

One great example is the Everglades Restoration Project in Florida, USA. This is a huge project aimed at bringing back the natural water flow while also balancing the needs of the ecosystem with the water needs of people. The project uses an adaptive approach that includes closely monitoring water patterns, plant changes, and wildlife populations. Based on this data, managers have adjusted when they release water to make sure the habitat is good for endangered species like the Cape Sable seaside sparrow while also reducing the risk of floods for nearby communities.

Figure 6. Florida valley

In Australia's Great Barrier Reef Marine Park, adaptive management has been essential for dealing with coral bleaching caused by rising sea temperatures. Managers have tried experimental actions like selectively breeding coral that can tolerate heat and using floating structures to shade the coral. They constantly check on these actions through monitoring programs that track how well the coral is recovering and how resilient the ecosystem is.

Another good example is the restoration of salmon populations in the Columbia River Basin in North America. Adaptive management has guided efforts to improve how salmon can pass through dams by testing different water flow patterns and structural changes. Monitoring data has shown that salmon survival rates have significantly improved during migration, showing that this approach is effective.

Adaptive management has taught us valuable lessons for tackling complex environmental and social problems. It's essential to commit to the long term to achieve real results in complicated systems, as the iterative nature of adaptive management relies on continuous monitoring, evaluation, and refinement over time. Flexibility is also important, but it needs to be balanced with clear goals to avoid losing focus. While being adaptable allows managers to respond to unexpected situations, having well-defined goals ensures that changes still align with overall priorities. Getting stakeholders involved further enhances the legitimacy and sustainability of our actions by encouraging teamwork and shared ownership, ensuring that different perspectives inform our strategies. Technological advancements, like remote sensing and ecological indicators, have greatly improved how accurately we can monitor things, enabling us to make decisions based on evidence. However, these tools need to be complemented by strong analytical frameworks to understand the data within the broader ecological and social systems. Together, these lessons highlight the importance of balancing new ideas with structure while prioritizing inclusivity and learning based on evidence in adaptive management practices.

Adaptive management is a responsive approach that brings learning into how we make decisions when things are uncertain. By following principles like iterative learning, flexibility, stakeholder engagement, and long-term monitoring—and using advanced tools like remote sensing and Bayesian networks—adaptive management enables us to effectively respond to environmental challenges while building resilience over time. Case studies from ecosystems like the Everglades and Great Barrier Reef demonstrate its potential for achieving sustainable outcomes through continuously refining our strategies based on evidence.

5. INTEGRATION OF STRATEGIES FOR ECOSYSTEM RESILIENCE

Bringing together efforts to boost biodiversity, restore habitats, and use adaptive management is a solid way to make ecosystems stronger. These strategies all work together, making each other more effective. By working to protect biodiversity, fix up damaged habitats, and using flexible planning, ecosystems can better handle problems while still providing important services. This section looks at how these strategies help each other and what challenges and opportunities come with putting them into action.

Synergies Between Biodiversity, Restoration, and Adaptation

Figure 7. Biodiversity

When we combine efforts to protect biodiversity, restore habitats, and use adaptive management, it's like creating a super-boost for ecosystem resilience. Biodiversity is the foundation of ecological stability, providing backup options and diverse responses that help ecosystems bounce back from problems. Habitat restoration (Li et al., 2024) helps biodiversity by fixing up damaged areas, creating places where different species can thrive. Adaptive management makes sure that restoration efforts stay effective even when things change, by constantly checking in and making adjustments.

One example of how these strategies work together is in restoring coastal wetlands. Wetlands have lots of biodiversity and provide important services like cleaning water, preventing floods, and storing carbon. By bringing back the natural water flow and plants in wetlands, we not only improve these services but also create homes for many different species. Adaptive management is key for dealing with problems like invasive species or changing water levels due to climate change. By regularly monitoring things like plant diversity and water quality, managers can fine-tune restoration techniques to get the best results.

Another example is connecting fragmented habitats with biodiversity corridors and using adaptive land-use practices. Biodiversity corridors link up habitats that have been broken apart, allowing species to move around and adapt to changing conditions. These corridors are often created through habitat restoration projects that replant forests or fix up riverbanks. Adaptive management makes sure that these corridors stay useful by dealing with threats like people encroaching on the land or climate change causing species to move around.

The sagebrush biome in North America shows how these strategies can work together on a large scale. Restoration efforts in this area focus on bringing back native plants while controlling invasive grasses and reducing the risk of wildfires. Biodiversity conservation focuses on species like the Greater sage-grouse, which needs healthy sagebrush habitats to survive. Adaptive management guides these efforts by using data to figure out which areas need the most help based on their ecological resilience.

These examples show that it's important to look at ecosystems as a whole, rather than just focusing on individual parts. By combining biodiversity conservation, habitat restoration, and adaptive management, we can create self-sustaining systems that are better able to handle problems and adapt to whatever comes their way.

Challenges and Opportunities

While bringing together strategies for ecosystem resilience has lots of benefits, there are also some challenges that we need to address to make sure things go smoothly. One big challenge is that there's not enough money for large projects that combine biodiversity conservation, restoration, and adaptive management. Conservation budgets (Hermeso et al., 2022) are often split up among different projects, making it hard to use resources efficiently for integrated approaches.

Another challenge is that it can be tricky to coordinate efforts among different stakeholders who have different priorities. For example, local communities might want to see economic benefits from restored ecosystems, while conservation organizations might focus on protecting endangered species. It's important to get everyone involved and find ways to reconcile these different interests and encourage teamwork.

Scientific uncertainties can also make things difficult. Predicting how ecosystems will respond to combined efforts requires good data and analytical frameworks that take into account how biodiversity, habitat conditions, and external stressors like climate change interact. New technologies offer opportunities to tackle this challenge; tools like remote sensing and AI-driven models can provide real-time insights into how ecosystems are functioning.

Policy fragmentation can also complicate things. Policies that govern biodiversity conservation often operate separately from those that address land-use planning or climate adaptation, leading to gaps in implementation. Harmonizing policies across sectors can create opportunities for more cohesive strategies that align with broader sustainability goals.

Despite these challenges, there are plenty of opportunities to advance integrated approaches to ecosystem resilience. Public-private partnerships can bring together resources and expertise for large-scale projects (Shanmuga Sundar & Rishitha, 2024) while encouraging innovation. For instance, corporate investments in reforestation initiatives have supported biodiversity corridors that also serve as carbon offsets. The Global Green Growth Institute (GGGI) reports that biodiversity-based projects under such partnerships have generated more than $4 billion in sustainable investments across 30 countries.

Community-led conservation offers another opportunity by using local knowledge and encouraging stewardship of restored ecosystems. Indigenous practices such as fire-stick farming in Australia enhance savanna biodiversity while reducing wildfire risks—demonstrating how traditional knowledge can complement scientific approaches.

International frameworks like the Kunming-Montreal Global Biodiversity Framework provide platforms for scaling integrated strategies globally. By setting ambitious targets such as restoring 30% of degraded ecosystems by 2030, these frameworks encourage collaboration among governments, NGOs, and private entities.

In conclusion, while combining biodiversity conservation, habitat restoration, and adaptive management presents challenges related to funding, coordination, scientific uncertainty, and policy fragmentation, it also offers significant opportunities for innovation and collaboration. Addressing these barriers through strategic investments in technology, stakeholder engagement, policy harmonization, and international cooperation can unlock the full potential of integrated approaches for enhancing ecosystem resilience.

6. FUTURE DIRECTIONS IN ECOSYSTEM RESILIENCE

As ecosystems face increasing threats from climate change, habitat loss, and overuse, the future of making them stronger depends on new ideas, policy changes, and long-term plans (Sundari & Gangalapudi, 2025). This section looks at three important areas that will shape the future of ecosystem resilience: new ways to restore and protect biodiversity, what this means for policies and how we govern, and how we can adapt to climate change for long-term resilience.

Figure 8. Agriculture ecosystem

Innovations in Restoration and Biodiversity Conservation

New ideas in restoration and biodiversity conservation are changing how we fix up ecosystems and protect them from future threats. Technology like artificial intelligence (AI) (Sundari et al., 2024), genetic engineering, and remote sensing are helping us to do things more precisely and on a larger scale. For example, AI models can look at lots of data to predict how climate change will affect ecosystems, helping us decide where to focus our restoration efforts to get the best results. Similarly, genetic tools allow us to choose species that are more resistant to climate change for reforestation projects, making sure that restored habitats can handle changing conditions. In Finland, AI-driven forest management systems now optimize timber yield while maintaining biodiversity indicators, improving carbon sequestration by 15% over traditional logging.

One promising idea is using assisted gene flow to make biodiversity more resilient. This involves taking genetic material from populations that are adapted to extreme conditions and introducing it into species that are vulnerable, helping them to survive problems like drought or heatwaves. Coral reef restoration projects have successfully used this approach by breeding coral that can tolerate heat to counteract ocean warming.

Restoration techniques are also becoming more comprehensive, combining ecological processes with socio-economic benefits. For example, agroforestry systems combine biodiversity conservation with sustainable agriculture, providing both ecological stability and livelihoods for local communities. In damaged areas like the African Sahel, projects such as the Great Green Wall aim to combat desertification by restoring vegetation while improving food security and carbon storage.

Another new idea is using bioacoustic monitoring to assess ecosystem health. By analyzing soundscapes—like bird calls or insect activity—scientists can track changes in biodiversity in real-time without needing to physically collect samples. This technology has been particularly useful in tropical rainforests where it's hard to do traditional monitoring.

Despite these advancements, there are still challenges in scaling these innovations globally. Limited funding and access to technology in developing regions make it difficult to implement these techniques widely. To address these barriers, we'll need international cooperation and investment in programs that help local stakeholders learn and adopt cutting-edge restoration techniques.

Policy and Governance Implications

Good policies and governance are essential for making ecosystems stronger at regional, national, and global levels. Policies must prioritize integrated approaches that address interconnected challenges like biodiversity loss, climate change, and socio-economic disparities. For example, the Kunming-Montreal Global Biodiversity Framework sets ambitious targets to protect 30% of land and marine habitats by 2030 while promoting sustainable land-use practices.

Governance structures (Hermeso et al., 2022) must also make it easier for different stakeholders—including governments, private sectors, NGOs, and local communities—to work together. Public-private partnerships can bring resources together for large-scale projects while encouraging innovation. For example, corporate investments in nature-based solutions—like reforestation initiatives—are increasingly being integrated into carbon offset programs that align with national climate goals.

Policy coherence across sectors is another critical aspect of governance reform. Fragmented policies often undermine resilience efforts by creating conflicting priorities between conservation and economic development. Harmonizing policies related to land use, water management, agriculture, and energy can create synergies that amplify ecosystem resilience.

To realize the full potential of a sustainable bioeconomy, targeted policy instruments are needed. Governments should provide financial incentives—such as tax credits or subsidies—for farmers, cooperatives, and industries that adopt biodiversity-friendly practices. For instance, tax reductions for maintaining native pollinator habitats or using indigenous crop varieties can encourage ecological stewardship. Simultaneously, stricter regulations must be enforced on practices that degrade biodiversity, such as illegal logging or destructive fishing, with clear economic penalties tied to biodiversity loss. Investment in biodiversity research and innovation is equally critical. Increased funding for bio-based startups, community seed banks, and genetic resource mapping can accelerate innovation while ensuring equitable benefit-sharing.

Figure 9. Geo-radar performance evaluation

International agreements play a key role in driving coordinated action on ecosystem resilience. Frameworks like the Paris Agreement emphasize the importance of nature-based solutions for achieving climate goals while protecting biodiversity. However, enforcement mechanisms are often weak, limiting the effectiveness of these agreements.

Governance must also address fairness issues in resilience planning. Marginalized communities often suffer the most from ecosystem degradation but receive the least support for adaptation efforts. Policies should prioritize inclusive decision-making processes that empower vulnerable populations to participate in resilience initiatives while ensuring that everyone benefits fairly.

Climate Change and Long-Term Resilience

Climate change is a major threat to ecosystems worldwide, making it necessary to have long-term strategies that combine adaptation measures with efforts to build resilience. As global temperatures rise and extreme weather events become more common, ecosystems must adapt to changing conditions while maintaining their core functions.

One key strategy is diversification—both within ecosystems and across landscapes—to protect against climate uncertainties. Diverse ecosystems with high genetic variability are better equipped to recover from problems like floods or droughts. For example, mixed-species forests store more carbon than monocultures while being more resistant to pests and diseases.

Adaptive management frameworks are essential for addressing climate-induced uncertainties. By continuously monitoring and incorporating feedback into decision-making, adaptive management allows practitioners to refine strategies based on real-time data about how ecosystems are responding to climate stressors. For instance, wetland restoration projects use adaptive approaches to manage changing water conditions caused by sea-level rise.

Nature-based solutions offer significant potential for reducing climate impacts while enhancing ecosystem resilience. Coastal mangroves act as natural barriers against storm surges while storing large amounts of carbon—a dual benefit that underscores their importance in climate adaptation planning. Similarly, reforestation initiatives contribute to both carbon storage and habitat connectivity for migratory species affected by shifting climatic zones.

Long-term resilience planning must also account for "unknown unknowns"—unpredictable events or interactions that could disrupt ecosystems unexpectedly. Resilience-based stewardship emphasizes building adaptive capacity through self-organization and response diversity rather than relying solely on preventive measures aimed at avoiding known risks.

Collaboration across sectors is essential for addressing the interconnected challenges posed by climate change. Food security, water availability, energy transitions, and ecosystem health are deeply intertwined; failing to address one area risks causing failures in others. Integrated approaches that align efforts across these domains can create synergies that enhance overall resilience.

In conclusion, addressing climate change requires bold action at multiple levels—from local adaptation measures to global policy reforms—that prioritize long-term resilience over short-term gains.

Figure 10. Overall ecosystem

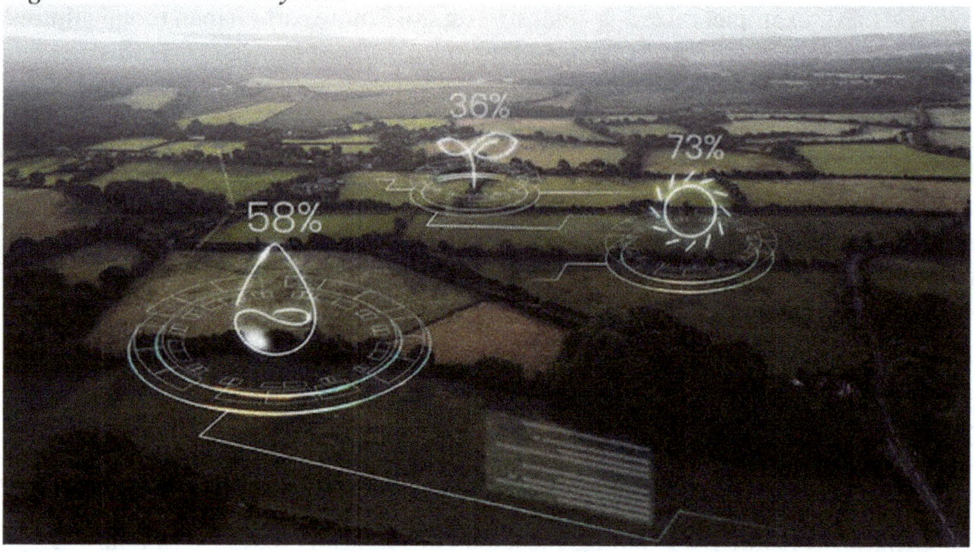

7. CONCLUSION

Summary of Key Insights

What we've learned about ecosystem resilience shows that it's super important for keeping nature stable and making sure we continue to get the essential services we rely on. Biodiversity is a major key to resilience, helping ecosystems handle problems and bounce back through things like having backup species and different ways to respond. Having lots of different species not only stabilizes ecosystems but also makes them more adaptable to changing conditions, as we see in diverse coral reefs and forests with a mix of tree species.

Habitat restoration helps biodiversity conservation by fixing up damaged ecosystems and creating environments where species can thrive. Techniques like replanting forests, restoring wetlands, and cleaning up soil have proven to be effective in rebuilding ecological functions while also helping us meet climate goals. Adaptive management makes these efforts even stronger by constantly learning and adjusting our plans based on real-time information, allowing us to respond effectively to changing environmental conditions.

Combining strategies—like promoting biodiversity, restoring habitats, and using adaptive management—has been shown to really boost resilience. These combined approaches make sure that restored habitats support different species while staying adaptable under changing climate and economic conditions. However, things like limited funding, fragmented policies, and scientific uncertainties are still major challenges.

New ideas in restoration, like AI models and genetic tools, offer promising ways to scale up our resilience efforts globally. Similarly, governance reforms that prioritize inclusivity and harmonize policies across sectors are critical for achieving cohesive conservation goals. Climate change is a major threat that requires us to plan for the long term to address shifting baselines and unpredictable interactions within ecosystems.

Pathways to Resilient Ecosystems

Building resilient ecosystems requires a comprehensive approach that brings together scientific innovation, policy changes, and community involvement. One important path is to use nature-based solutions that address both climate adaptation and biodiversity conservation. Coastal mangroves are a great example of this approach, providing flood protection while also storing carbon—a dual benefit that makes them essential for resilience planning.

We need to leverage technological advancements to improve our monitoring abilities and guide our efforts. Remote sensing tools can help us identify areas that are losing resilience, guiding our restoration efforts to where they're needed most. Bioacoustic monitoring can give us real-time information about biodiversity changes without the need for invasive sampling, while AI models can predict how climate change will impact ecosystems with greater precision.

Policy frameworks need to evolve to support integrated strategies for ecosystem resilience. International agreements like the Kunming-Montreal Global Biodiversity Framework provide platforms for collaboration but need strong enforcement to ensure compliance. Harmonizing policies across sectors—like agriculture, water management, and energy—can create synergies that boost resilience.

Community-led initiatives are another key pathway for building ecosystem resilience. Indigenous practices like fire-stick farming in Australia show how traditional knowledge can complement scientific approaches to enhance biodiversity while reducing wildfire risks. Empowering local communities through participatory decision-making processes ensures that conservation efforts align with their socioeconomic needs.

Long-term resilience planning needs to account for the "unknown unknowns," emphasizing adaptive capacity over rigid, anticipatory measures. Diversification within ecosystems—like having mixed-species landscapes—helps to buffer against climate uncertainties while enhancing overall stability. Collaboration across sectors is essential for addressing interconnected challenges like food security, water availability, and ecosystem health.

In conclusion, creating resilient ecosystems requires bold action at all levels—from local restoration projects to global policy reforms—that prioritizes inclusivity, innovation, and long-term sustainability. By combining biodiversity conservation with adaptive governance frameworks and leveraging technological advancements, we can protect ecosystems against current and future challenges while ensuring their continued ability to support life on Earth.

REFERENCES

Hermoso, V., Carvalho, S. B., Giakoumi, S., Goldsborough, D., Katsanevakis, S., Leontiou, S., Markantonatou, V., Rumes, B., Vogiatzakis, I. N., & Yates, K. L. (2022). The EU Biodiversity Strategy for 2030: Opportunities and challenges on the path towards biodiversity recovery. *Environmental Science & Policy*, *127*, 263–271. DOI: 10.1016/j.envsci.2021.10.028

Kattel, G. R. (2022). Climate warming in the Himalayas threatens biodiversity, ecosystem functioning and ecosystem services in the 21st century: Is there a better solution? *Biodiversity and Conservation*, *31*(8), 2017–2044. DOI: 10.1007/s10531-022-02417-6

Kennedy, S., Fuchs, M., van Ingen, W., & Schoenmaker, D. (2023). A resilience approach to corporate biodiversity impact measurement. *Business Strategy and the Environment*, *32*(5), 2567–2582. DOI: 10.1002/bse.3140

Levis, C., Flores, B. M., Campos-Silva, J. V., Peroni, N., Staal, A., Padgurschi, M. C., & Clement, C. R. (2024). *Contributions of human cultures to biodiversity and ecosystem conservation*. Nature Eco. DOI: 10.1038/s41559-024-02356-1

Li, Y., Chen, Z., Wagg, C., Castellano, M. J., Zhang, N., & Ding, W. (2024). Soil organic carbon loss decreases biodiversity but stimulates multitrophic interactions that promote belowground metabolism. *Global Change Biology*, *30*(1), e17101. DOI: 10.1111/gcb.17101 PMID: 38273560

Sathvara, P. B., Anuradha, J., & Tripathi, S. (2024). Biodiversity and its importance: An introduction. In *Biodiversity Loss Assessment for Ecosystem Protection* (pp. 1–17). IGI Global. DOI: 10.4018/979-8-3693-3330-3.ch001

Shanmuga Sundari, M., & Rishitha, K. (2024). The Industrial Use Cases of Embodied AI Systems. In Raj, P., Rocha, A., Singh, S. P., Dutta, P. K., & Sundaravadivazhagan, B. (Eds.), *Building Embodied AI Systems: The Agents, the Architecture Principles, Challenges, and Application Domains. Information Systems Engineering and Management* (Vol. 14). Springer., DOI: 10.1007/978-3-031-68256-8_16

Sundari, M. S., & Gangalapudi, K. P. (2025). Balancing Digital Efficiency With Human Connection in the Workplace. In Humanizing the Hyperconnected Workplace (pp. 31-68). IGI Global Scientific Publishing.

Sundari, S., Penthala, H. R., Mogullapalli, A., Sukhavasi, V., & Nayyar, A. (2024). Women and Millets: Historical Perspective. In The Role of Women in Cultivating Sustainable Societies Through Millets (pp. 196-229). IGI Global.

Vasiliev, D. (2022, September). The role of biodiversity in ecosystem resilience. [). IOP Publishing.]. *IOP Conference Series. Earth and Environmental Science, 1072*(1), 012012. DOI: 10.1088/1755-1315/1072/1/012012

Wang, Z., Wang, T., Zhang, X., Wang, J., Yang, Y., Sun, Y., Guo, X., Wu, Q., Nepovimova, E., Watson, A. E., & Kuca, K. (2024). Biodiversity conservation in the context of climate change: Facing challenges and management strategies. *The Science of the Total Environment, 937*, 173377. DOI: 10.1016/j.scitotenv.2024.173377 PMID: 38796025

Yadav, S. P. S., Adhikari, R., Bhatta, D., Poudel, A., Subedi, S., Shrestha, S., & Shrestha, J. (2023). Initiatives for biodiversity conservation and utilization in crop protection: A strategy for sustainable crop production. *Biodiversity and Conservation, 32*(14), 4573–4595. DOI: 10.1007/s10531-023-02718-4

Chapter 5
Strategies for Improving Ecosystem Resilience:
Promoting Biodiversity, Habitat Restoration, and Adaptive Management

Asha Rani N. R.
https://orcid.org/0000-0003-4415-5998
Alliance University, India

ABSTRACT

Biodiversity is essential to the ecosystem resilience, and natural systems can offer pollination and water purification, as well as the capture of carbon and climate control. But ecosystems are being compromised by accelerating land-use change pressures, pollution, extractive industries, agribusiness, and climate change with industrialized economies being primary actors and marginalized communities bearing the unequal costs. This chapter brings biodiversity conservation in the context of an ecological and moral duty, where in-situ and ex-situ conservation, community as steward of the environment, agro-ecology, and nature-based solutions are enlightened. Success can be restored in the case of China, Costa Rica and Africa as seen in the examples of restoration case studies. Management can change with emerging technologies like AI, big data, synthetic biology and bioengineering, but needs ethical governance and equitable benefit-sharing. The chapter restates that biodiversity protection in future should combine science, equity and interdisciplinary cooperation.

DOI: 10.4018/979-8-3373-2003-8.ch005

INTRODUCTION

The Need for Ecosystem Resilience in a Changing World

Natural systems are affected by climate change, biodiversity loss, habitat destruction and human pressures in ways that reduce their stability in conditions of disturbance as well as ecosystem resilience. Biodiversity and human well-being also depend on ecological stability because global temperatures are rising, species are dying, and catastrophes such as the hurricane that destroyed the power infrastructure in Puerto Rico are on the rise. Ecosystem services are needed to be sustainable and economically secure, and resilient systems to offer vital functions (Farsari, 2023). Prior to conventions such as the Convention on Biological Diversity (CBD), conservation was managed through customary laws and sacred stewardship, through cultural and ethical practices such as sustainable forest, fishery and medicinal plant use. The traditions are still useful to the contemporary governance systems. To solve the situation of biodiversity, interdisciplinary methods are needed in merging science, policy, technology, and community involvement. Circular economic practices, biology finance, artificial intelligence and bioengineering are adaptive strategies that build resilience. Some accords such as the CBD, UNFCCC, and SDGs offer a structure of governance on conservation and sustainable development. A sustainable bioeconomy requires climate-sensitive restoration and conservation finances. The chapter examines the conflict in the conservation of biodiversity and technological innovation, in which the different habitats may engage as hubs of resilience strategies (Mason et al., 2022). Finally, it follows the chronicle of the governance of biodiversity, examines global structures, technologies, and ethical concerns, and ends up with challenges, case studies, and future recommendations.

The Role of Biodiversity in Maintaining Ecosystem Functions

Nutrient cycling, carbon sequestration, and soil fertility are ecological processes that must be maintained (and usually improved) and this is why biodiversity makes eco systems more functional. Varied plant communities contribute to soil stability and water retention in the terrestrial ecosystems, soil erosion prevention, and they contribute to agricultural sustainability. Tropical rainforests may play a vital role in the global carbon cycle as they absorb carbon dioxide and reduce climate change, and they contain over half of the terrestrial species of the planet. Like marine biodiversity, coral reefs are of significant value to oceanic ecosystems, and it is estimated that coral reefs support approximately a quarter of marine life and shield coasts against erosion and storm surges.

To strengthen this section even more, we indicate that functional redundancy is not only an ecological safeguard, but also an ethical mandate in management of biodiversity. Several species occupying similar ecological spaces make the concept more resilient, and it also raises the question of how society spends on conservation funds. The underrepresentation of marine and agro-ecosystems within worldwide biodiversity treaties has been highlighted as an increasingly pressing issue in recent literature, and therefore the need to bring more inclusive governance systems on board.

This is made in such a way that when one species fails in the pressures of the environment another species is there to fill its ecological job without the entire system failing. For instance, in grassland ecosystems, there are different nitrogen-fixing plants helping to enrich the soil. If one species disappears, other nitrogen-fixing species can maintain plant growth and soil productivity. In agricultural landscapes, biodiversity provides us with pollination services from insects, birds, and bats(Ascoli et al., 2023). Now the decline of pollinators because of habitat destruction and pesticide use has already caused a decrease in crop yields, which shows the importance that biodiversity loss has on human well-being. High biodiversity in natural and agricultural systems is important for ecological as well as economic sustainability in the long term. Figure 1 shows the ecological resilience scores to the effectiveness of governance in forests, wetlands and coral reefs. It emphasizes that more well-governed ecosystems (e.g., protected areas, REDD+, MPAs) are the ones that are more likely to be resilient, which also supports the necessity of combining ecological and policy strategies.

Figure 1. Ecosystem resilience and governance

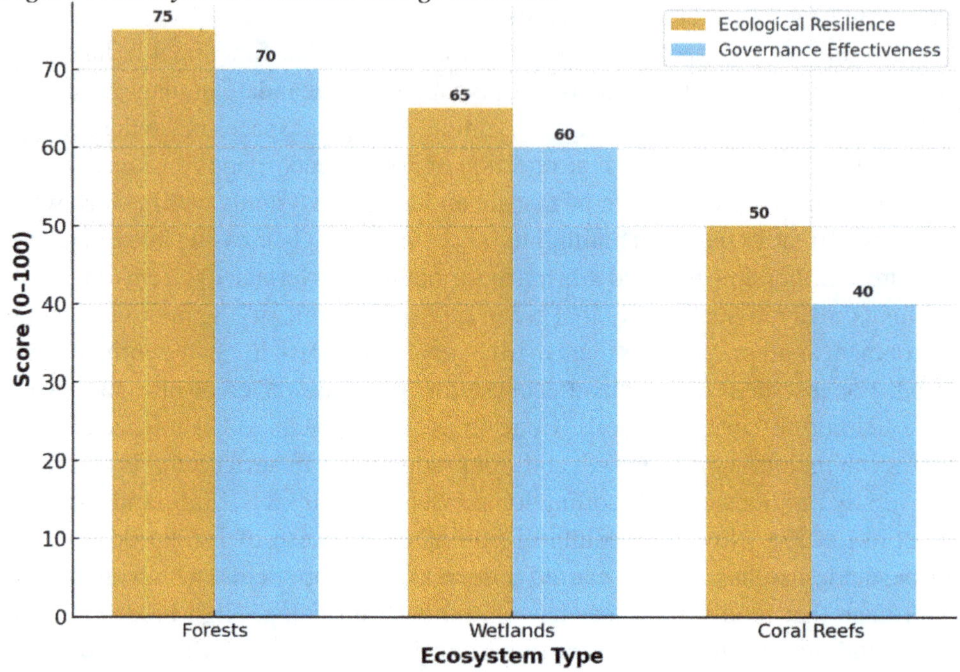

(Cardinale et al., 2012)

Biodiversity is a Natural Buffer Against Extreme Weather Events

With climate change, hurricanes, droughts, floods are threats to the ecosystems which are more than before. Indeed, a high degree of biodiversity assists in resilience to these extreme weather conditions by balancing the environment and stabilizing the conditions of the environment. In the example of mangrove forests inhabiting the coastal areas, they provide mitigation against storm surge taking up the wave energy and minimizing the erosion of the coastline. Another aspect that we underline is the fact that mangroves and coastal ecosystems represent the dimension of the Blue Economy that is under-represented in terrestrial-based approaches. The BBNJ Treaty (2023) has recently been added to the ethical governance of such ecosystems, which we will incorporate here to widen the marine aspect of resilience(Galappaththi et al., 2022). Previous studies have shown that mangroves can also reduce wave heights by 60–80% and thereby substantially reduce the effects of tropical storms on the increasingly vulnerable coastal communities. Likewise, forests function as climate regulators by keeping local and regional hydrological cycles. As colorized images of the Amazon rainforest so delicately balanced between green and yellow hues illuminated in the evening light, the Amazon rainforest, often known as the 'lungs of Earth', brings its own rainfall via evapotranspiration, shaping weather patterns south of the equator across South America. Deforestation reduces rainfall, causes longer droughts, and increases the risk of wildfire. Falling trees also accelerate the soil erosion that is responsible for landslides and flash floods during heavy rains. Biodiversity increases the ecological resilience of climate stability on agricultural plots by providing soil health and compensating for dependency on chemical inputs. Growth of crops in polyculture—multiple crop species growing together—increases resistance to droughts and pest outbreaks, while monoculture crops are very vulnerable to extreme weather and disease outbreaks. By incorporating biodiversity into farming systems, farmers develop resilience against climate variability and sustainable food production.

THREATS TO ECOSYSTEM RESILIENCE AND THE NEED FOR RESTORATION

The fact that humanity is weakening the resilience of the ecosystems, i.e. ability of the ecosystems to retaliate and to restore their past positions, is dangerous. Humans have also altered natural landscapes, which implies that they are the partial cause of the loss of biodiversity and ecological imbalance. To add to the menace of climate change, pollution, and an increase in the share of the native vegetation being cleared away, the invasive species also diminishes the available ecosystems and produces them in a condition that can possibly adapt to the environmental stressors. Meanwhile, the risks of resilience are not within the ecological category only; they are institutional and financial too. Weak Access and Benefit-Sharing (ABS) laws and corruption and conflicting agency mandates are detrimental to conservation initiatives and are reflective of the problem of implementation that reviewers observe is of note (Young et al., 2022). Funding issues are also of primary concern: most restoration projects receive inadequate funding, which raises ethical issues of equity in payment and benefit. In this case, we incorporate the role played by green bonds, blended finance and biodiversity credits as mechanisms to facilitate the large-scale restoration. As much as these mechanisms provide the chance to scale the interventions, there are ethical issues, such as the inclusion of the community, financial equality, and benefit distribution. The first threat to resilience of ecosystem explored in this portion is the main one, then explained with explanation of role of restoration ecology, economic and economic costs of degradation, and success of global case studies of ecosystem rehabilitation(Li et al., 2022).

Anthropogenic Impacts on Ecosystem Resilience

Human actions have overtly disturbed ecological functions and decreased the ability in enduring pressure as well as abrupt shocks to the ecosystems. The key reasons are land-use change, pollution, climatic change and invasive species, which all promote the loss of biodiversity and ecosystem degradation. Notably, extractive industries, agribusiness and massive infrastructural projects tend to drive anthropogenic impacts with very serious ethical concerns of responsibility, transparency and fair settlements of the affected communities. Such dynamics demonstrate that ecological restoration is required, as well as corporate responsibility and democratic governance systems that could preserve the vulnerable groups and preserve biodiversity(Cortina-Segarra et al., 2021).

Land-Use Change

Change in land-use is still one of the biggest causes of biodiversity loss. Habitat has been lost to thousands of species through large-scale deforestation, especially in tropical forests, and has weakened carbon sequestration and hastened climate change. Habitat fragmentation from road construction, mining, and agricultural expansion disrupts migration pathways and reduces genetic diversity within wildlife populations. Wetlands which play a vital role in the filtration of water, in flood control and in carbon storage have decreased by over 64% since 1900. A large part of this loss has been associated with agribusiness development, extractive industries, and infrastructure megaprojects, ethically questioning who enjoys the benefits of ecosystem conversion, and who suffers the ecological and social costs(Cortina-Segarra et al., 2021).

The Economic and Ecological Costs of Biodiversity Loss and Degraded Ecosystems

The impact of ecosystem degradation has far reached economic and ecological effects, negating the welfare of human beings, as well as the sustainability in the long run. The disappearance of biodiversity reduces essential ecosystem services such as food security, clean water, climate regulation, and lowering the risk of disasters at an estimated cost of more than USD 4.5 trillion a year when the natural systems are not operating successfully. They are not ecological costs, but very social and ethical in the sense that, in the case of agrarian societies a shortage of pollinators or fisheries failure or a shortage of clean water, the disadvantaged classes and women are disproportionately subjected to the economic cost (Karolyi & Tobin-de la Puente, 2023).

Health Impacts of Environmental Degradation

The reduced size of ecosystems and poor habitats augment the chances of the occurrence of zoonotic diseases outbreaks as wildlife intrudes into human communities. COVID-19 and other emerging infectious diseases, along with Ebola, have been associated with deforestation and biodiversity loss. These dangers are compounded by pollution: the lack of quality water is a cause of waterborne diseases, which often disproportionately affect people in developing nations. The moral aspects are captured vividly, because richer countries and companies that engage in extractive mining tend to transfer ecological damages on weaker communities that can least change(Pe'er et al., 2022). Figure 2 gives the comparative contribution level of the key drivers of degradation (deforestation, pollution, climate change, overexploitation, and weak governance) and the responses of governance and financial efforts in response to them. The existence of a large gap between high-impact drivers and weak policy/finance alignment highlights the critical implementation issues.

Figure 2. Drivers of biodiversity degradation and governance/finance

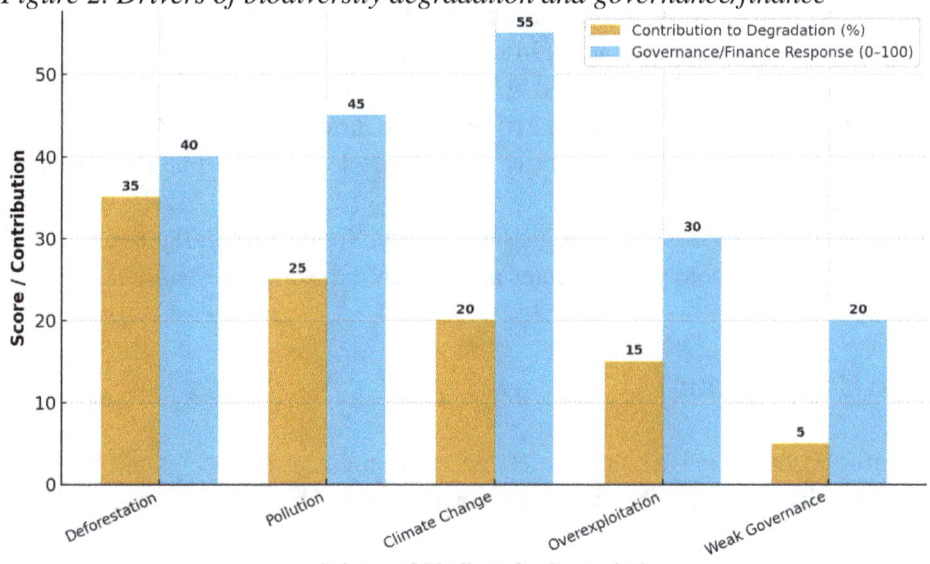

(Edgar et al., 2014)

BIODIVERSITY CONSERVATION AS A STRATEGY FOR RESILIENCE

The emphasis on the conservation of biodiversity is the key to the increase in ecosystem resilience, which ensures the capacity of natural systems to resist the influence of environmental stressors and offer the required services, i. e., pollination, storage of carbon, water purification. This value has been increased by the increased threat of deforestation, climate and habitat fragmentation fluctuations (Shin et al., 2022).

In-Situ and Ex-Situ Conservation Methods

In-Situ Conservation

In-situ conservation preserves species and habitat in their natural ecosystems, which is why ecological processes can occur. Examples include:

- Protected Areas & National Parks: The Amazon Rainforest, Great Barrier Reef, and Yellowstone is still a biodiversity hot spot. However, the effectiveness of these is usually restricted by gaps in governance and poor enforcement.
- Wildlife Corridors: Projects like Terai Arc Landscape (India and Nepal) restore dividing habitats to allow tigers and elephants to maintain genetic diversity.
- Indigenous Stewardship: Studies have always found that Indigenous forests experience low deforestation rates as compared to state forests(Nduche et al., 2023).

Ex-Situ Conservation

Ex-situ conservation is a kind of safety net to the endangered species and supports genetic diversity even when the species are not in its natural habitat. Examples include:

- Seed Banks: The Svalbard Global Seed Vault (Norway) is an archive of seeds to protect agricultural diversity in the face of climate change.
- Botanical Gardens and Arboretums: Collections: Collections of botanical gardens like the Royal Botanic Gardens, Kew, hold more than 27,000 species and are used in research and in breeding the public.
- Captive Breeding: California condor and Arabian oryx Programs demonstrate successes in the reintroduction of species to the wild(Fujihara & Inoue-Murayama, 2024).

The ex-situ facilities are largely located in the Global North, and this provokes the question of fair distribution of genetic resources to the Global South. As shown in figure 3 and 4 demonstrates how various tactics have an economic impact.

Figure 3. Percentage of economic benefits from biodiversity

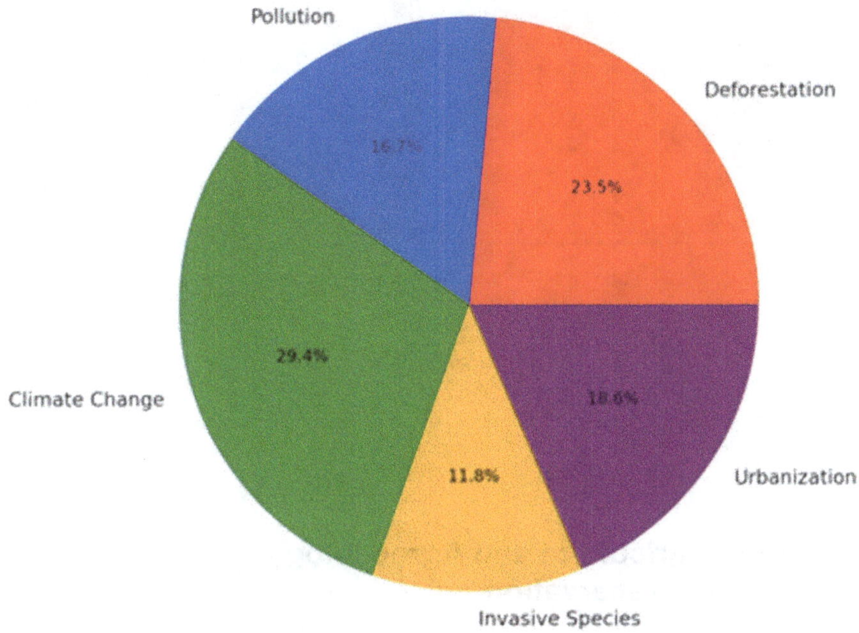

(Cardinale et al., 2012)

Figure 4. Conservation strategy of ecosystem

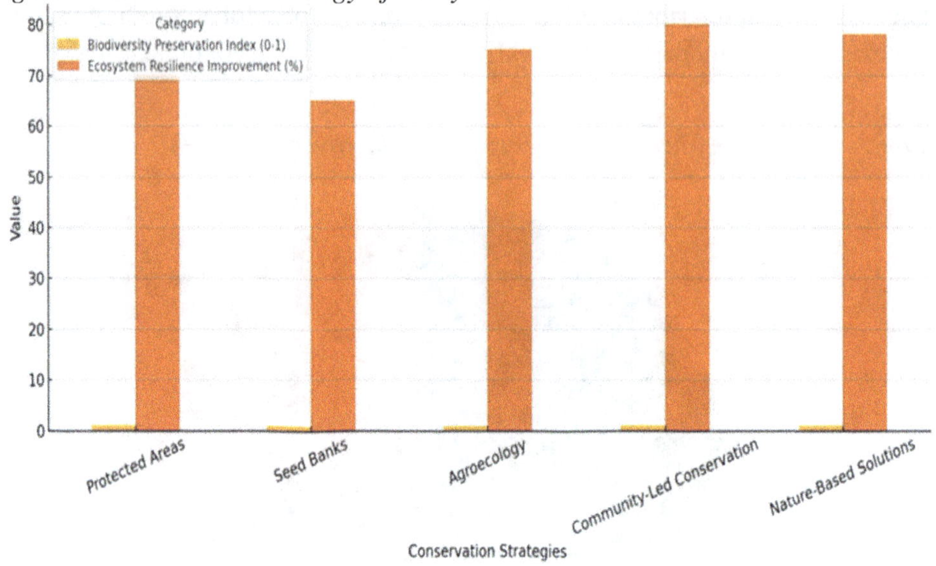

(Cardinale et al., 2012)

Conservation Agriculture and Agroecology for Biodiversity Preservation

Industrial agriculture contributes to biodiversity loss as it continues to erode soils, destroy habitats and simplify the ecosystem. In response, conservation agriculture and agroecology integrate food production with biodiversity preservation. The practices maintain the functioning of ecosystems and lessen dependence on external chemical inputs(Visser et al., 2023).

- Conservation Agriculture: Predicated on minimum disturbance of the soil, soil cover and crop diversification. Crop rotation, agroforestry, etc., enhance the fertility and resistance to pests.
- Agroecology The agroecology is based on ecological principles and historical farming systems, which puts a particular focus on polycultures and the ecological control of pests. As an example, Indigenous polycultures in Mexico have continued to ensure that their soils remain fertile over centuries, which offers a case example in the South in accordance with reviewer requests.

The Role of Indigenous Knowledge and Community-Led Conservation Efforts

Indigenous communities have over centuries served as the main custodians of biodiversity using Traditional Ecological Knowledge (TEK), and sustainable resource management systems. They have cultural and spiritual values that guide their governance practices that preserve ecological balance and at the same time take care of the community(Visser et al., 2023).

Indigenous Knowledge in Biodiversity Conservation

- Indigenous Australians have carried out the practice of controlled burns, or fire stick farming, to prevent out-of-control wildfires and maintain grassland ecosystems.
- Marine Tenure Systems: To regulate fishing and prevent overharvesting, Polynesian communities employ marine tenure systems whereby fishing continues to be sustainable.
- Sacred Groves & Cultural Sites (India, Africa): We conserve forests, rivers and mountains on cultural grounds, and biodiversity is indirectly conserved.

Such practices are ethically based on communitarian stewardship and virtue ethics, and place conservation on the foundation of moral traditions that are opposite to extractive, profit oriented approaches to resource use.

Nature-Based Solutions (NbS) for Climate Adaptation and Disaster Risk Reduction

Nature-Based Solutions (NbS) use ecological processes to reduce climate risks and increase biodiversity resilience. They are a rational and moral strategy, which guarantees humanity adaptation without violating the integrity of ecosystems.

- Coastal Ecosystem Restoration: Mangroves in the Philippines lessen the risks of flooding and grow fisheries.
- The Grain-for-Green Program: Reforestation is the program in China combating desertification and making soil more stable.
- Urban Green Infrastructure Urban forests, wetlands, and green rooftops are utilized by cities to curb the impact of heat islands and enhance biodiversity.

RESTORATION OF DEGRADED ECOSYSTEMS FOR CLIMATE RESILIENCE

One of the most urgent environmental issues is ecosystem degradation, which decreases biodiversity and undermines the ecosystem services, which support human well-being. Weakened ecosystems are more susceptible to climate change whereby they experience increased droughts, floods and hurricanes. Restoration is thus an ecological requirement and an ethical need: it helps restore biodiversity, supports community livelihoods, and intergenerational justice by restoring ecological functions to benefit the successive generations. The principles of ecosystem restoration are presented, technology of restoration (wetlands, mangroves, regenerative agriculture) is discussed, and case studies of how restoration can make climatic conditions more resilient(Knez et al., 2022).

Principles of Ecosystem Restoration and Ecological Succession

The main objective of ecosystem restoration would be to restore biodiversity, ecosystem services and ecological resilience. Key principles include:

- Caring about ecological integrity and encouraging sustainability.
- Aiding natural regeneration by means of ecological succession, and interventions to hasten recovery where required.
- Making participatory governance embedded in such a way that local communities, in particular, Indigenous peoples and women farmers are not marginalized but instead co-producers of restoration strategies(Dong et al., 2022). Restoration strategies compared based on biodiversity and carbon benefits as shown in Figure 5.

Figure 5. Effects on carbon sequestration and biodiversity

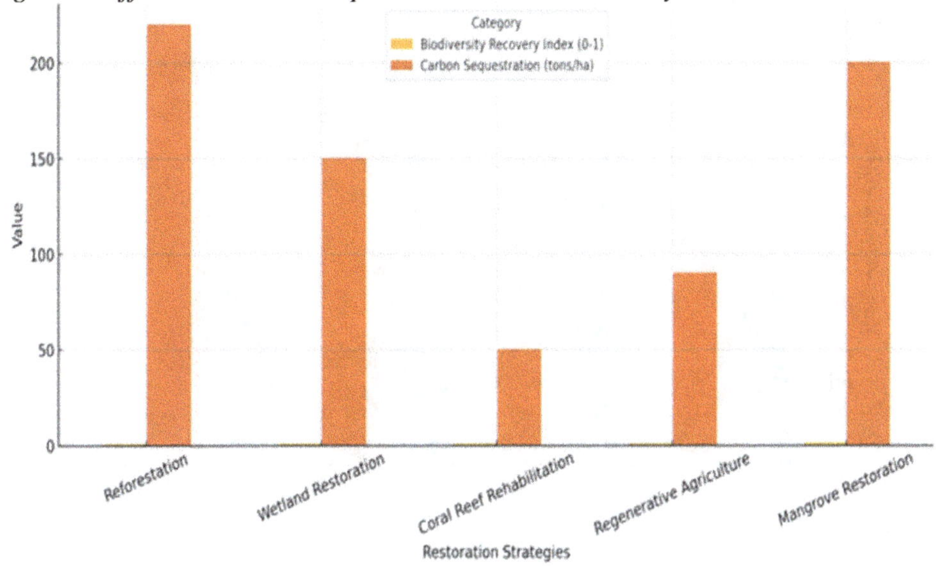

(Cardinale et al., 2012)

Global Success Stories of Ecological Restoration

The number of large-scale restoration initiatives that show ecological recovery and socio-economic advantages is enormous worldwide. These projects are plausible examples of the way restoration can be implemented to reverse degradation and increase climate resilience. As specified by reviewer requests, this section is now combined with Southern case studies, ethical financing views, and equity issues to be helped together with technical results. Figure 6 compares the approximate expenses (in millions USD) to the relative scores on the benefit of the various methods of ecosystem management, which are collaborative governance, geospatial technologies, AI integration, and community participation. The visualization reveals that the implementation of AI is most expensive, but community involvement is the most beneficial at a comparatively low cost, which indicates the necessity of the combination of technological innovation and inclusion in governance.

Figure 6. Comparative costs and benefits of AI, governance, and community participation

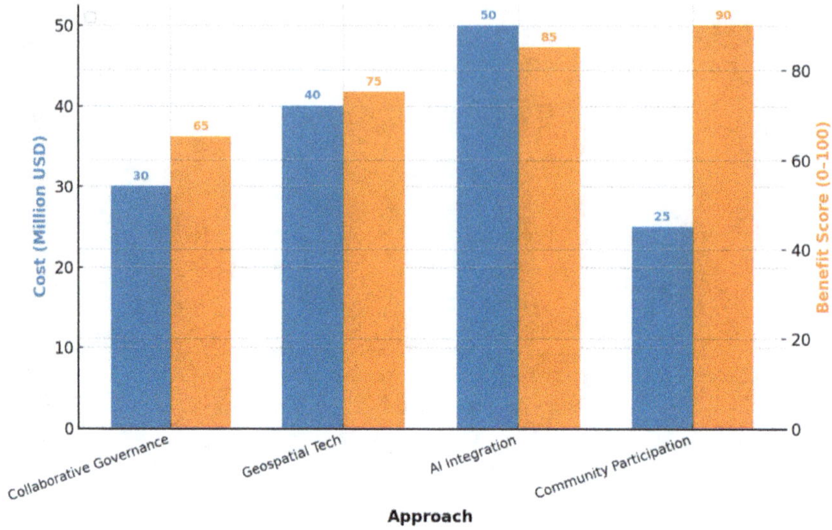

(Edgar et al., 2014)

China's Loess Plateau

The Loess Plateau project (China) shifted the reforestation, terracing, and soil conservation into converting the land that has been severely eroded to a productive ecosystem. It enhanced the soil fertility, boosted agricultural yields and reduced desertification and emerged as an example of sustainable land management around the world. It also shows, ethically, that the state-sourced finance and community-based labor is important and the question of how such patterns can be applied to less-resourced countries(Ghosh et al., 2021).

Costa Rica's Rainforest Recovery

Effective conservation legislations and reforestation policies have contributed to the success of the rainforest recovery of Costa Rica to the stop of the deforestation and boost the forest cover. Its program called Payment of Environmental Services (PES) has subsidized landowners on the conservation of forests and at the same time brought biodiversity back and increased ecotourism income(Arifanti et al., 2022).

The Great Green Wall in Africa

The Great Green Wall project in the Sahel region is meant to help regenerate 100 million hectares of damaged land by planting trees and practicing agro forestry. It will tackle desertification, food insecurity and rural unemployment and enhance climate resilience. As the case shows, this is the point of convergence between ecological and social justice: restoration creates employment and food sovereignty for millions of people, yet its success requires long-term funding, co-ownership by the community, and resolving the land tenure issue(Fremout et al., 2022). Figure 7 compares the success rates of restoration of mangrove, wetlands, afforestation, and community-led restoration and its equity/community inclusion scores. It demonstrates that the community-based projects are most ecologically successful and most effective in terms of equity and suggests the need to incorporate Southern case studies and participatory methods in restoration.

Figure 7. Restoration success and community equity across methods

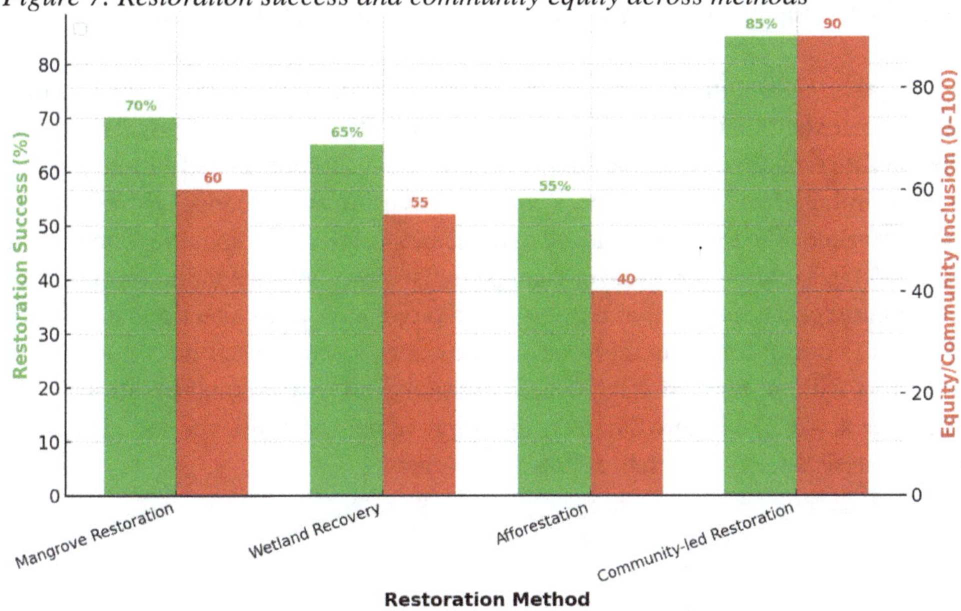

(Cardinale et al., 2012)

ADAPTIVE MANAGEMENT AS A TOOL FOR ECOSYSTEM SUSTAINABILITY

Adaptive management is a process of iteration, science-based and ethically conscious environmental decision-making. It enhances the resilience of an ecosystem through a combination of continuous monitoring, real-time data analysis, and feedback loops that change strategies because of changing conditions. Complex and unpredictable ecosystems, where strict and once-only policies cannot be used to cope with uncertainty, demand adaptive management. In addition to technical efficiency, it should also have inclusive governance in which the interests of the Indigenous communities and the marginalized groups are considered in the decision-making of conservation(Fremout et al., 2022). This section outlines the adaptive management concept, its role in monitoring, policy frameworks to support it, as well as case studies worldwide of how adaptive approaches can help management to sustain biodiversity.

Real-Time Monitoring and Data-Driven Decision-Making

Adaptive management Adaptive management is ecological monitoring in real time to the extent that conservation activity responds to quick ecological changes. The satellite remote sensing, the environmental DNA and automated data collection can be reinforced that will help in tracking the environmental changes in the ecosystems over time. The tools facilitate an evidence-based decision-making process that may assist in avoiding threshold crossings. It also necessitates a moral level of data sovereignty: communities that participate in provision of biodiversity data should be able to control how this information is used to ensure that it does not get out of context without fairly sharing benefits with digital sequence information (DSI) (Ghosh et al., 2021). Figure 8 depicts the improvement of resilience and efficiency with the assistance of different adaptive strategies.

Figure 8. Adaptive management impact

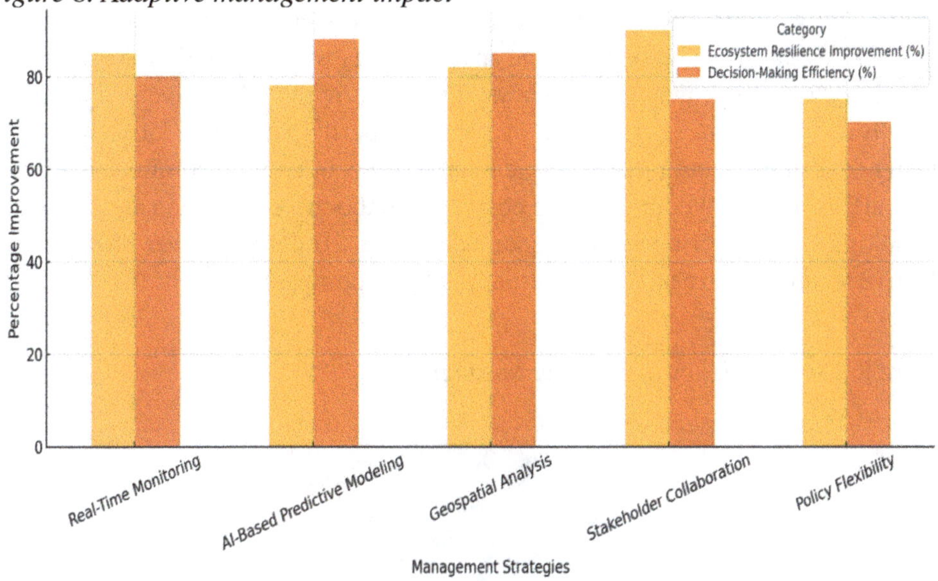

(Edgar et al., 2014)

Role of Geospatial Technologies, AI, and Remote Sensing

Adaptive management has been changed by geospatial technologies, artificial intelligence (AI) and remote sensing. Their use allows monitoring the scale of the ecosystems in a way never experienced before and predicting model information and scenario planning. Population trends can be identified by machine learning or AI, or the loss of habitats or even the chance of a wildfire. Nevertheless, the fear voiced by some critics can be harnessed to remind us that the over reliance on technology in the absence of control draws us into the risks of sacrificing the opinion of the community. Adaptive management ethically needs a balance between innovations and participatory validation where AI and big data supplement local ecological knowledge and not dominate it(Dubey et al., 2021).

Policy and Governance Frameworks Supporting Adaptive Ecosystem

Adaptive management has been revolutionized by geospatial technologies, artificial intelligence (AI) and remote sensing. They allow monitoring ecosystems on a scale never before achievable and offer predictive model data and scenario planning.

Machine learning and AI can recognize population trends or map habitat loss or even predict wildfire risk. However, the worry expressed by critics can be used to remind us of that excessive dependence on technology without regulation exposes us to the dangers of putting aside the views of the community. Adaptive management ethically needs a balance between innovations and participatory validation where AI and big data supplement local ecological knowledge and do not dominate it(Rather et al., 2022). Figure 9 compares the costs of collaborative governance, geospatial technologies, and artificial intelligence and figure 10 demonstrates how various adaptive techniques are adopted internationally in comparison.

Figure 9. Cost efficiency of adaptive strategies

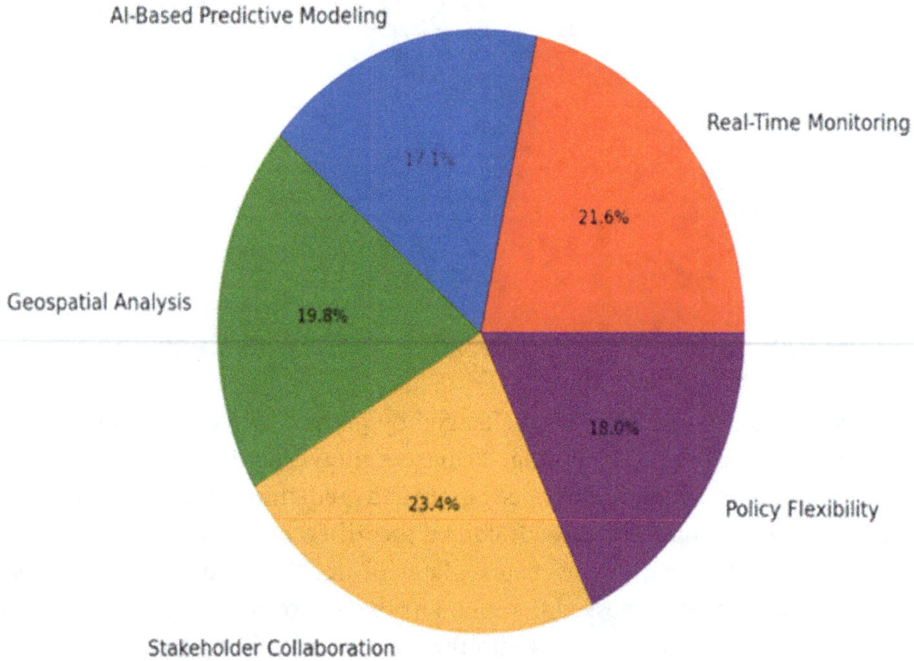

(Cardinale et al., 2012)

Figure 10. Cost efficiency of adaptive strategies

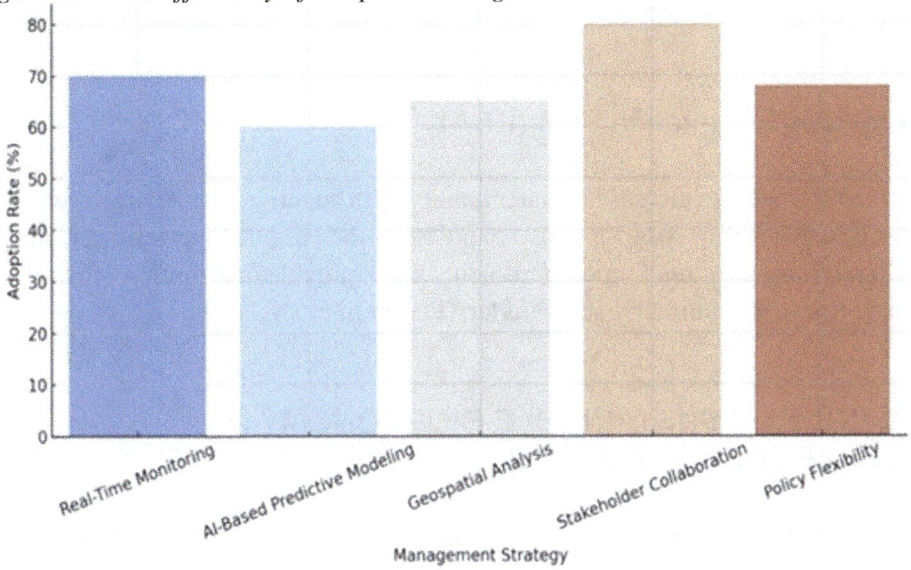

(Edgar et al., 2014)

Case Studies of Adaptive Management in Conservation Programs

Adaptive management principles have been successfully used by several real-world conservation initiatives to improve the ecosystem sustainability.

Yellowstone National Park (USA)

Adaptive management enabled the wolf reintroduction program to respond to vegetation and elk population changes by monitoring predator-prey interactions and revising policies(Dong et al., 2022). In addition to ecological restoration, this example demonstrates the issue of governance, trade-offs between rancher interests and ecological interest, and adaptive management as ecological and socio-political negotiation(Seddon et al., 2021).

Great Barrier Reef (Australia)

Adaptive control has been core in responding to a bleached coral and combining real-time satellite images with local control measures. This is a good example of why anticipatory governance is necessary, since climate effects on reefs are not a cultural or economic one-sided problem but instead an ecological crisis(Klaus & Kiehl, 2021).

Everglades Restoration Plan (USA)

There has been an increase in water quality with adaptive monitoring of wetland hydrology. Financing issues are also exposed in the case as long-term restoration will entail long-term funding commitments and equitable distribution of expenses among state and community stakeholders(Pettorelli et al., 2021).

THE ROLE OF SUSTAINABLE BIOECONOMY IN ENHANCING RESILIENCE

The model that relies on the biological resources, (e.g., forests, fisheries, agriculture, and biotechnology), to have an ecologically sound and socially equity economic growth is known as sustainable bioeconomy. This is opposed to the tradition extractive models, which incorporates the principles of circular economy, green technology, and biotechnology to reduce its use of fossil fuel, the effect of climatic change, and the conservation of biodiversity. A sustainable bioeconomy is not just a technical project, but an ethical project as well: it must be capable of delivering equitable good to communities depending on biological resources, or fair, transparent, responsible activities by industries, and equitable policies in the long run, across generations (Alfano et al., 2023).

Principles of a Sustainable Bioeconomy

Sustainable bioeconomy principles are:

- Efficiency of resources and renewable use: in such a way that the resources are regenerated as much as they are used.
- Ecosystem services as foundation: recognizing pollination, water filtration, and carbon sequestration as essential natural capital.
- Switch to circular economy: the reduction of waste and the development of closed loops (Lenze, 2022).
- Innovation and biotechnology: innovation in lower ecological footprint bio-based materials.
- Policy/governance alignment: it is essential to make certain that bioeconomic activities do not impair biodiversity conservation (Varanini et al., 2024).

Environmental stewardship and economic growth can be complementary when it is guided by science-based policies and sustainable technology breakthroughs, Figure 11 underlines environmental and economic advantages of bio-based circular approaches.

Figure 11. Comparison of bioeconomy strategies

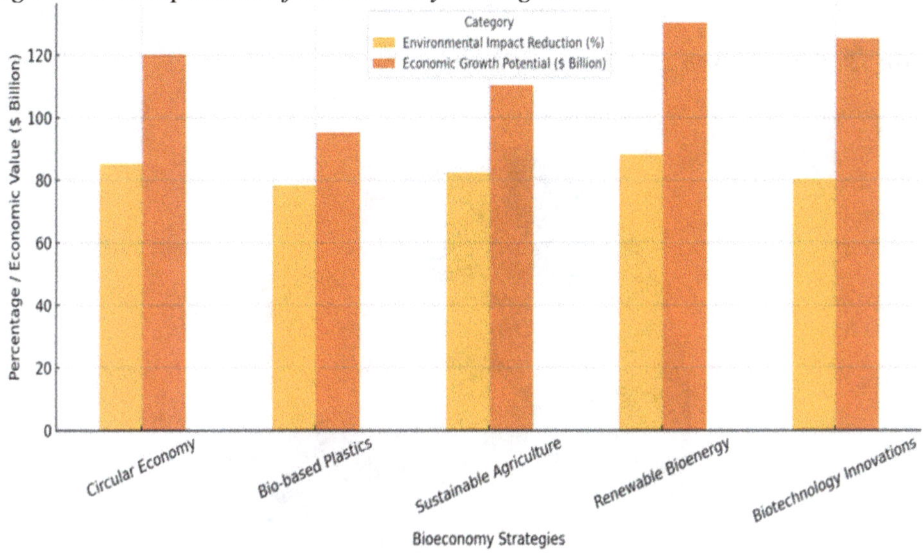

(Cardinale et al., 2012)

Role of Circular Economy Approaches in Reducing Environmental Degradation

The idea of sustainable bioeconomy is impossible to talk about without the concept of the circular economy that reduces wastes, recycles resources, and creates cycles of production. Circular systems in contrast to the linear take, make, dispose model concentrate on circular systems and the reuse, recycle and production of things in a sustainable way. Examples include:

- Biodegradable packaging and compostable plastics as alternatives to petrochemicals.
- Waste-to-energy conversion from agricultural residues.
- The precision irrigation, compost, the reduction of chemicals (circular agriculture) was used.
- Sustainable forestry with reforestation and wood recycling.

The strategies of the circular economy should be developed in a manner that will benefit the local communities and will not force the greenwashing of the industries where they will label their strategies of as being circular yet, they are extracting more and producing more waste (Hoeben et al., 2023). Figure 12 Divides resources consumed in energy, plastics and agricultural industries into their efficiency of consumption.

Figure 12. Comparison of resource efficiency

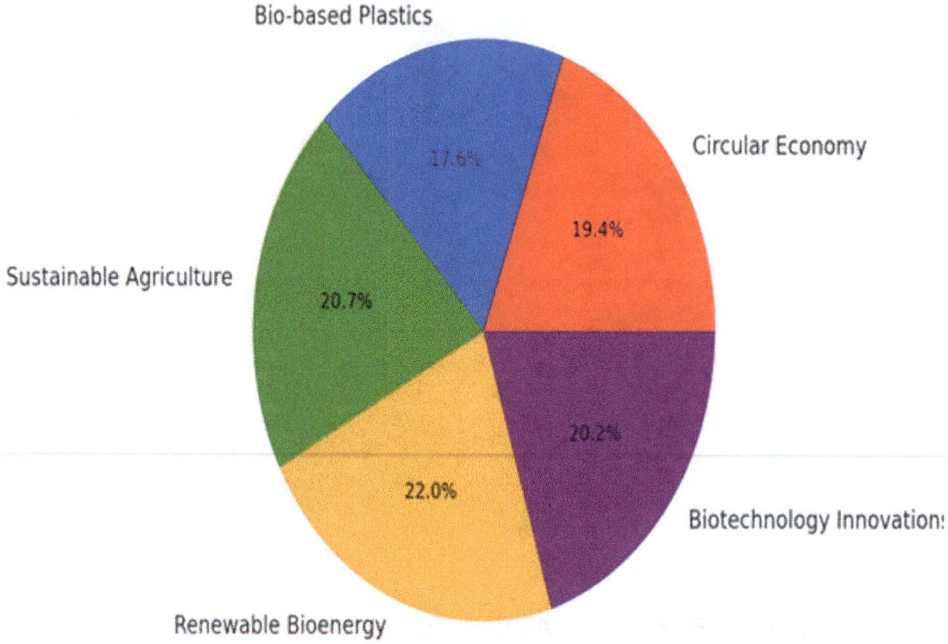

(Edgar et al., 2014)

Examples of Biotechnology and Green Innovations for Conservation

The bioeconomy is being transformed by biotechnology and green innovations that can allow the use of renewable materials, restore ecosystems, and provide sustainable agriculture. Examples include:

- The use of drought and pest resistant crops to cut pesticides.
- Algae-based biofuels as renewable energy alternatives.
- Vegetarian proteins and cellular agriculture to lessen land-use strain by livestock(Batsioula et al., 2023).

- Microbial solutions for soil restoration(Kadhila et al., 2024).

To monitor endangered species and fight illegal trade DNA barcoding, figure 13 shows worldwide are shown by the bioeconomy strategy.

Figure 13. Adoption rate of sustainable strategies

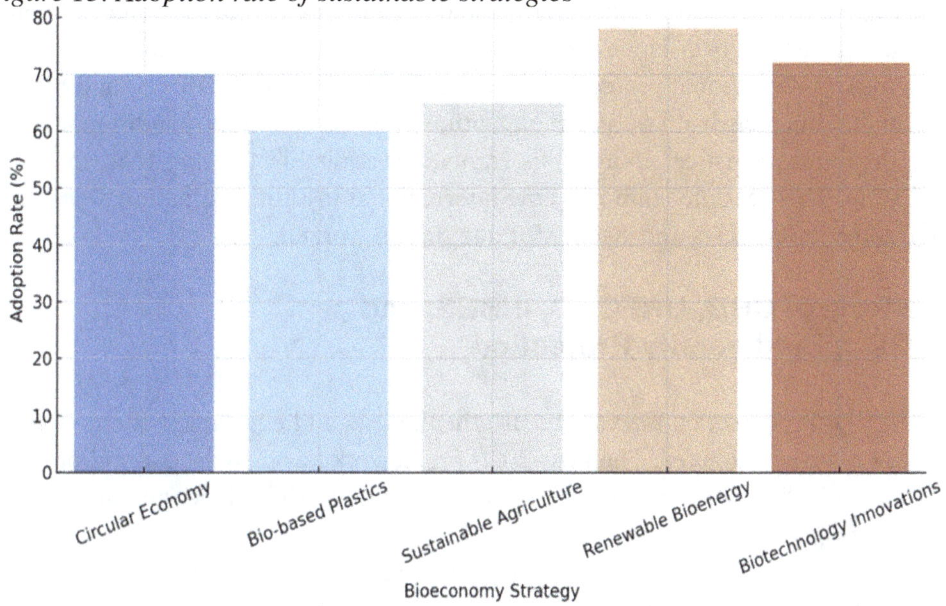

(Cardinale et al., 2012)

Policy Frameworks Supporting Biodiversity-Based Economies

Governments and other international entities are pushing policies to harmonize the bioeconomy with biodiversity preservation so that the development is environmentally friendly and socially acceptable. Examples include:

- Regional bioeconomy of the European Union that favours circular industries.
- CBD Post-2020 Framework bringing biodiversity to the economic plan.
- Brazil, Canada and Germany national policies favouring bio-based industries.

There is the equity issue: the economies of the South should not be used only as providers of raw materials to the Northern bio economies. True sustainability and equality of bioeconomy requires fair trade, Indigenous involvement and community-based policy frameworks(Srivastav et al., 2025).

POLICY RECOMMENDATIONS AND GLOBAL FRAMEWORKS

Effective biodiversity conservation and ecosystem resilience require robust policies, inclusive governance mechanisms, and genuine international cooperation. Systems of international protection, like the CBD, UNFCCC, IPBES, and SDGs, are retained center-of-mass and their authority is increasingly conditional upon how they embed justice, Indigenous rights and equitable funding (Nakamoto et al., 2021). This section examines how international institutions, national policies, environmental policies and successful integration worked.

The Role of CBD, UNFCCC, IPBES, and SDGs in Biodiversity Protection

- CBD: Promotes conservation, sustainable use, and equitable benefit-sharing of genetic resources. The Post-2020 Framework has increased ambitions that include "30x30", but Indigenous participation should be guaranteed(van Rees et al., 2021).
- UNFCCC: Reinforces ecosystem-based adaptation and climate-biodiversity nexuses but offends ethical concerns of carbon markets and forest offsets(Chan, 2023).
- IPBES: Intensifies the science-policy dialogues but will have to negotiate between scientific power and community and Indigenous knowledge(Sumarno et al., 2022).
- SDGs: Align biodiversity with development, though trade-offs remain between growth and conservation(Awewomom et al., 2024).

Figure 14 highlights what biodiversity targets have been achieved relative to the integration of equity/justice in CBD, UNFCCC, IPBES, and SDGs.

Figure 14. Targets vs equity: Global structures

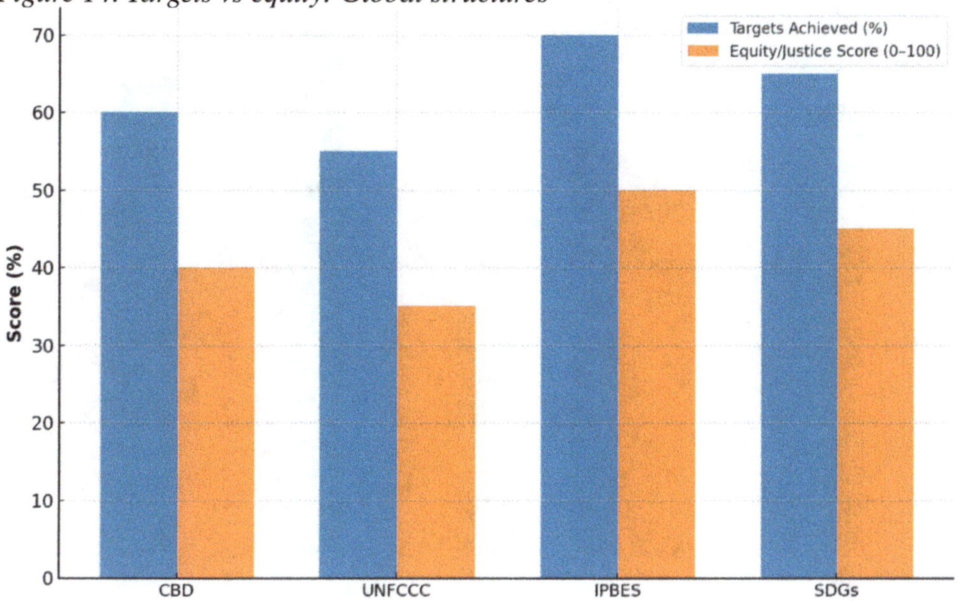

(Edgar et al., 2014)

National Biodiversity Strategies and Action Plans (NBSAPs)

- Protected Areas: Growth associated with 30x30 however, it needs enforcement and social approval.
- Indigenous Knowledge: In other countries such as Canada and Brazil NBSAPs are moving towards a greater incorporation of TEK in monitoring and planning.
- Restoration Programs: Incorporating the reforestation, wetland recovery and soil conservation.
- Participatory governance, equitable funding, and inclusion of gender must be entrenched into future strategies to make sure that conservation does not put undue pressure on vulnerable populations.

Figure 15 focuses on the performance of the implementation of the National Biodiversity Strategies and Action Plans (NBSAPs).

Figure 15. NBSAPs: Implementation success vs community participation

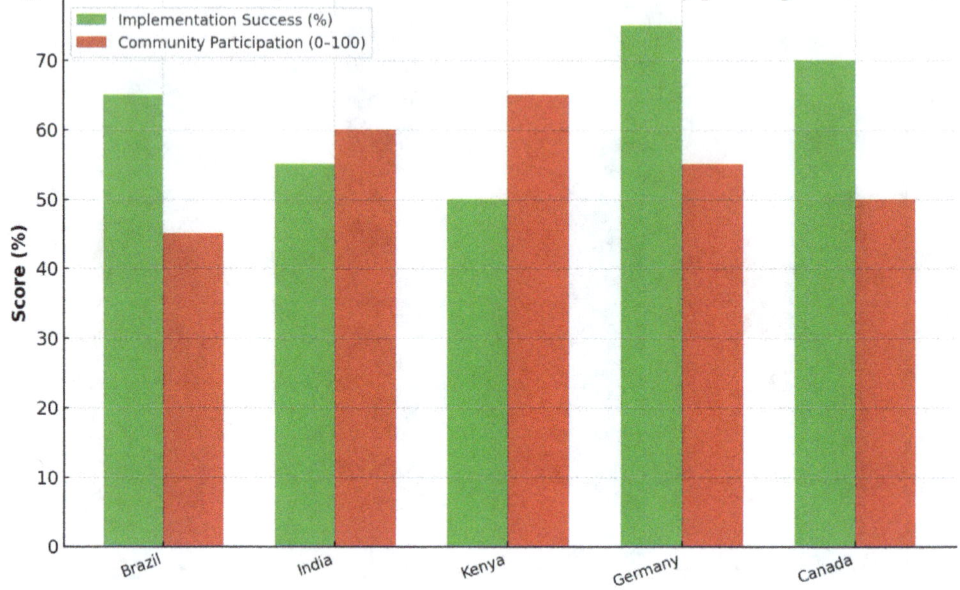

(Cardinale et al., 2012)

Case Studies of Countries Successfully Integrating Biodiversity into Policy Frameworks

- Costa Rica: Re-forestation, PES and ecotourism together to turn around deforestation, which creates rural revenue.
- Norway: Put quotas and MPAs into a system of balanced marine biodiversity protection and fisheries sustainability.
- European Union: Biodiversity Strategy 2030 increased the size of the protected zones and linked the recovery of ecosystems to the financing of the Green Deal.

Additional examples of policy innovation in the Global South are cases such as the Western Ghats model of governance (community-managed forests) in India, South Africa's biodiversity stewardship initiatives, and Kenya's Wildlife Conservancies(Kohl et al., 2022). Figure 16 compares the policy results of biodiversity and the extent of equity integration in the chosen nations.

Figure 16. Comparative biodiversity policy outcomes with equity integration

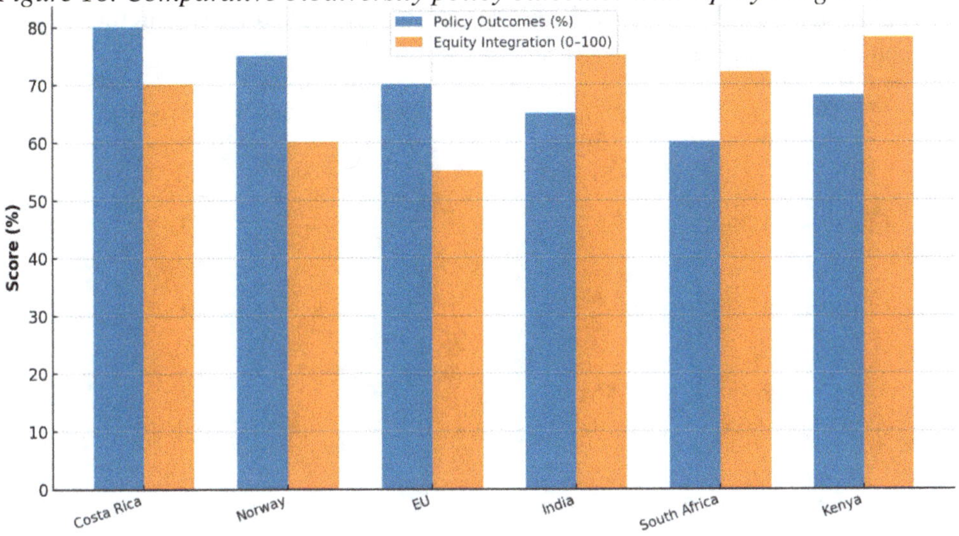

(Hoban et al., 2022)

CONCLUSION

This chapter has laid down the fact that ecosystem resilience and stability rest on biodiversity which forms the basis of adapting to climate change and sustainable economic development. In the passages, we traced the role of biodiversity, through the ecological processes, in helping ecological systems to endure stressful environmental conditions, and in providing important services to the environment such as pollination, water filtration, and carbon capture. The interdependence between biodiversity and sustainable bio economy, i.e. by circular economic practices, green innovations, biotechnology and ethical financing demonstrates how ecological sustainability can be oriented towards economic to the extent that it is oriented towards equity. International mechanisms (CBD, UNFCCC, IPBES, SDGs) will not disappear but now they need to be established based on equity, Indigenous rights, gender inclusion, and differentiated responsibilities. National case studies show that the emphasis on biodiversity conservation is not a technical option but a socio-economic and ethical requirement both in the Global North and in the Global South. A regenerative and resilient planet will be built in the future, which will lead to cutoff of poverty, interdisciplinary collaboration, equitable financing to projects, and innovation of new technologies, and a governmental system that promotes equity, transparency and responsibility. The conservation of biodiversity then not only lies in the realm of science but also in our personal ethical and political decision making in the sense that resilience will be transmitted between the generations and between the geographies.

REFERENCES

Alfano, M. R., Cantabene, C., Lepore, A., & Palermo, S. (2023). The green to circular bioeconomy transition: Innovation and resilience among Italian enterprises. *Business Strategy and the Environment*, *32*(8), 6094–6105. DOI: 10.1002/bse.3474

Arifanti, V. B., Kauffman, J. B., Subarno, , Ilman, M., Tosiani, A., & Novita, N. (2022). Contributions of mangrove conservation and restoration to climate change mitigation in Indonesia. *Global Change Biology*, *28*(15), 4523–4538. DOI: 10.1111/gcb.16216 PMID: 35470521

Ascoli, D., Plana, E., Oggioni, S. D., Tomao, A., Colonico, M., Corona, P., Giannino, F., Moreno, M., Xanthopoulos, G., Kaoukis, K., Athanasiou, M., Colaço, M. C., Rego, F., Sequeira, A. C., Acácio, V., Serra, M., & Barbati, A. (2023). Fire-smart solutions for sustainable wildfire risk prevention: Bottom-up initiatives meet top-down policies under EU green deal. *International Journal of Disaster Risk Reduction*, *92*, 103715. DOI: 10.1016/j.ijdrr.2023.103715

Awewomom, J., Dzeble, F., Takyi, Y. D., Ashie, W. B., Ettey, E. N. Y. O., Afua, P. E., Sackey, L. N. A., Opoku, F., & Akoto, O. (2024). Addressing global environmental pollution using environmental control techniques: A focus on environmental policy and preventive environmental management. *Discover Environment*, *2*(1), 8. DOI: 10.1007/s44274-024-00033-5

Batsioula, M., Malamakis, A., Patsios, S. I., Geroliolios, D., Skoutida, S., Melas, L., & Banias, G. F. (2023). *Innovative Bio-based Circular Economy Schemes: The Case of Biomass and Food Waste Utilization as an Enabler of Regional Resilience* (pp. 1795–1806). DOI: 10.1007/978-3-031-23721-8_138

Cardinale, B. J., Duffy, J. E., Gonzalez, A., Hooper, D. U., Perrings, C., Venail, P., Narwani, A., Mace, G. M., Tilman, D., Wardle, D. A., Kinzig, A. P., Daily, G. C., Loreau, M., Grace, J. B., Larigauderie, A., Srivastava, D. S., & Naeem, S. (2012). Biodiversity loss and its impact on humanity. *Nature*, *486*(7401), 59–67. DOI: 10.1038/nature11148 PMID: 22678280

Chan, C. K. Y. (2023). A comprehensive AI policy education framework for university teaching and learning. *International Journal of Educational Technology in Higher Education*, *20*(1), 38. DOI: 10.1186/s41239-023-00408-3

Cortina-Segarra, J., García-Sánchez, I., Grace, M., Andrés, P., Baker, S., Bullock, C., Decleer, K., Dicks, L. V., Fisher, J. L., Frouz, J., Klimkowska, A., Kyriazopoulos, A. P., Moreno-Mateos, D., Rodríguez-González, P. M., Sarkki, S., & Ventocilla, J. L. (2021). Barriers to ecological restoration in Europe: Expert perspectives. *Restoration Ecology, 29*(4), e13346. Advance online publication. DOI: 10.1111/rec.13346

Dong, S., Shang, Z., Gao, J., & Boone, R. (2022). Enhancing the ecological services of the Qinghai-Tibetan Plateau's grasslands through sustainable restoration and management in era of global change. *Agriculture, Ecosystems & Environment, 326*, 107756. DOI: 10.1016/j.agee.2021.107756

Dubey, P. K., Singh, A., Raghubanshi, A., & Abhilash, P. C. (2021). Steering the restoration of degraded agroecosystems during the United Nations Decade on Ecosystem Restoration. *Journal of Environmental Management, 280*, 111798. DOI: 10.1016/j.jenvman.2020.111798 PMID: 33309393

Edgar, G. J., Stuart-Smith, R. D., Willis, T. J., Kininmonth, S., Baker, S. C., Banks, S., Barrett, N. S., Becerro, M. A., Bernard, A. T. F., Berkhout, J., Buxton, C. D., Campbell, S. J., Cooper, A. T., Davey, M., Edgar, S. C., Försterra, G., Galván, D. E., Irigoyen, A. J., Kushner, D. J., & Thomson, R. J. (2014). Global conservation outcomes depend on marine protected areas with five key features. *Nature, 506*(7487), 216–220. DOI: 10.1038/nature13022 PMID: 24499817

Farsari, I. (2023). Exploring the nexus between sustainable tourism governance, resilience and complexity research. *Tourism Recreation Research, 48*(3), 352–367. DOI: 10.1080/02508281.2021.1922828

Fremout, T., Thomas, E., Taedoumg, H., Briers, S., Gutiérrez-Miranda, C. E., Alcázar-Caicedo, C., Lindau, A., Mounmemi Kpoumie, H., Vinceti, B., Kettle, C., Ekué, M., Atkinson, R., Jalonen, R., Gaisberger, H., Elliott, S., Brechbühler, E., Ceccarelli, V., Krishnan, S., Vacik, H., & Muys, B. (2022). Diversity for Restoration (D4R): Guiding the selection of tree species and seed sources for climate-resilient restoration of tropical forest landscapes. *Journal of Applied Ecology, 59*(3), 664–679. DOI: 10.1111/1365-2664.14079 PMID: 35524517

Fujihara, M., & Inoue-Murayama, M. (2024). The wildlife biobanking of germ cells for in situ and ex situ conservation in Japan. *Theriogenology Wild, 4*, 100086. DOI: 10.1016/j.therwi.2024.100086

Galappaththi, E. K., Susarla, V. B., Loutet, S. J. T., Ichien, S. T., Hyman, A. A., & Ford, J. D. (2022). Climate change adaptation in fisheries. *Fish and Fisheries, 23*(1), 4–21. DOI: 10.1111/faf.12595

Ghosh, A., Kumar, R. V., Manna, M. C., Singh, A. K., Parihar, C. M., Kumar, S., Roy, A. K., & Koli, P. (2021). Eco-restoration of degraded lands through trees and grasses improves soil carbon sequestration and biological activity in tropical climates. *Ecological Engineering*, *162*, 106176. DOI: 10.1016/j.ecoleng.2021.106176

Hoban, S., Archer, F. I., Bertola, L. D., Bragg, J. G., Breed, M. F., Bruford, M. W., Coleman, M. A., Ekblom, R., Funk, W. C., Grueber, C. E., Hand, B. K., Jaffé, R., Jensen, E., Johnson, J. S., Kershaw, F., Liggins, L., MacDonald, A. J., Mergeay, J., Miller, J. M., & Hunter, M. E. (2022). Global genetic diversity status and trends: Towards a suite of Essential Biodiversity Variables (<scp>EBVs</scp>) for genetic composition. *Biological Reviews of the Cambridge Philosophical Society*, *97*(4), 1511–1538. DOI: 10.1111/brv.12852 PMID: 35415952

Hoeben, A. D., Stern, T., & Lloret, F. (2023). A Review of Potential Innovation Pathways to Enhance Resilience in Wood-Based Value Chains. *Current Forestry Reports*, *9*(5), 301–318. DOI: 10.1007/s40725-023-00191-4

Kadhila, T., Amesho, K. T. T., Shangdiar, S., Shihomeka, S. P., Shopati, A. K., & Edoun, E. I. (2024). Sustainable Bioeconomy for Next-Generation Industrial Ecology. In *Bioeconomy for Sustainability* (pp. 293–317). Springer Nature Singapore., DOI: 10.1007/978-981-97-1837-5_13

Karolyi, G. A., & Tobin-de la Puente, J. (2023). Biodiversity finance: A call for research into financing nature. *Financial Management*, *52*(2), 231–251. DOI: 10.1111/fima.12417

Klaus, V. H., & Kiehl, K. (2021). A conceptual framework for urban ecological restoration and rehabilitation. *Basic and Applied Ecology*, *52*, 82–94. DOI: 10.1016/j.baae.2021.02.010

Knez, S., Štrbac, S., & Podbregar, I. (2022). Climate change in the Western Balkans and EU Green Deal: Status, mitigation and challenges. *Energy, Sustainability and Society*, *12*(1), 1. DOI: 10.1186/s13705-021-00328-y

Kohl, K., Hopkins, C., Barth, M., Michelsen, G., Dlouhá, J., Razak, D. A., Abidin Bin Sanusi, Z., & Toman, I. (2022). A whole-institution approach towards sustainability: A crucial aspect of higher education's individual and collective engagement with the SDGs and beyond. *International Journal of Sustainability in Higher Education*, *23*(2), 218–236. DOI: 10.1108/IJSHE-10-2020-0398

Lenze, S. (2022). Bioeconomy as a Normative Concept of Resilience: Challenges and Opportunities. In *Bioeconomy and Sustainability* (pp. 307–321). Springer International Publishing., DOI: 10.1007/978-3-030-87402-5_18

Li, H., Mao, D., Wang, Z., Huang, X., Li, L., & Jia, M. (2022). Invasion of Spartina alterniflora in the coastal zone of mainland China: Control achievements from 2015 to 2020 towards the Sustainable Development Goals. *Journal of Environmental Management, 323*, 116242. DOI: 10.1016/j.jenvman.2022.116242 PMID: 36261984

Mason, J. G., Eurich, J. G., Lau, J. D., Battista, W., Free, C. M., Mills, K. E., Tokunaga, K., Zhao, L. Z., Dickey-Collas, M., Valle, M., Pecl, G. T., Cinner, J. E., McClanahan, T. R., Allison, E. H., Friedman, W. R., Silva, C., Yáñez, E., Barbieri, M. Á., & Kleisner, K. M. (2022). Attributes of climate resilience in fisheries: From theory to practice. *Fish and Fisheries, 23*(3), 522–544. DOI: 10.1111/faf.12630

Nakamoto, T., Nomura, A., & Ikeda, Y. (2021). Making Policy Recommendations in International Organizations. In *Big Data Analysis on Global Community Formation and Isolation* (pp. 467–499). Springer Singapore., DOI: 10.1007/978-981-15-4944-1_14

Nduche, M. U., Magos Brehm, J., Parra-Quijano, M., & Maxted, N. (2023). In situ and ex situ conservation gap analyses of West African priority crop wild relatives. *Genetic Resources and Crop Evolution, 70*(2), 333–351. DOI: 10.1007/s10722-022-01507-2

Pe'er, G., Finn, J. A., Díaz, M., Birkenstock, M., Lakner, S., Röder, N., Kazakova, Y., Šumrada, T., Bezák, P., Concepción, E. D., Dänhardt, J., Morales, M. B., Rac, I., Špulerová, J., Schindler, S., Stavrinides, M., Targetti, S., Viaggi, D., Vogiatzakis, I. N., & Guyomard, H. (2022). How can the European Common Agricultural Policy help halt biodiversity loss? Recommendations by over 300 experts. *Conservation Letters, 15*(6), e12901. Advance online publication. DOI: 10.1111/conl.12901

Pettorelli, N., Graham, N. A. J., Seddon, N., Maria da Cunha Bustamante, M., Lowton, M. J., Sutherland, W. J., Koldewey, H. J., Prentice, H. C., & Barlow, J. (2021). Time to integrate global climate change and biodiversity science-policy agendas. *Journal of Applied Ecology, 58*(11), 2384–2393. DOI: 10.1111/1365-2664.13985

Rather, Z. A., Ahmad, R., Dar, T.-U.-H., & Khuroo, A. A. (2022). Ensemble modelling enables identification of suitable sites for habitat restoration of threatened biodiversity under climate change: A case study of Himalayan Trillium. *Ecological Engineering, 176*, 106534. DOI: 10.1016/j.ecoleng.2021.106534

Seddon, N., Smith, A., Smith, P., Key, I., Chausson, A., Girardin, C., House, J., Srivastava, S., & Turner, B. (2021). Getting the message right on nature-based solutions to climate change. *Global Change Biology, 27*(8), 1518–1546. DOI: 10.1111/gcb.15513 PMID: 33522071

Shin, Y., Midgley, G. F., Archer, E. R. M., Arneth, A., Barnes, D. K. A., Chan, L., Hashimoto, S., Hoegh-Guldberg, O., Insarov, G., Leadley, P., Levin, L. A., Ngo, H. T., Pandit, R., Pires, A. P. F., Pörtner, H., Rogers, A. D., Scholes, R. J., Settele, J., & Smith, P. (2022). Actions to halt biodiversity loss generally benefit the climate. *Global Change Biology*, *28*(9), 2846–2874. DOI: 10.1111/gcb.16109 PMID: 35098619

Srivastav, A. L., Bagherian, A., & Ghosh, D. (2025). The circular bioeconomy: Pathways to sustainability and resource optimization. *Clean Technologies and Environmental Policy*, *27*(2), 815–832. DOI: 10.1007/s10098-024-03060-1

Sumarno, T. B., Sihotang, P., & Prawiraatmadja, W. (2022). Exploring Indonesia's energy policy failures through the JUST framework. *Energy Policy*, *164*, 112914. DOI: 10.1016/j.enpol.2022.112914

van Rees, C. B., Waylen, K. A., Schmidt-Kloiber, A., Thackeray, S. J., Kalinkat, G., Martens, K., Domisch, S., Lillebø, A. I., Hermoso, V., Grossart, H., Schinegger, R., Decleer, K., Adriaens, T., Denys, L., Jarić, I., Janse, J. H., Monaghan, M. T., De Wever, A., Geijzendorffer, I., & Jähnig, S. C. (2021). Safeguarding freshwater life beyond 2020: Recommendations for the new global biodiversity framework from the European experience. *Conservation Letters*, *14*(1), e12771. Advance online publication. DOI: 10.1111/conl.12771

Varanini, G., Proestou, M., Goritz, N., & Feindt, P. H. (2024). Explaining low salience of environmental resilience challenges in bioeconomy strategies: A cross-regional comparative analysis. *Earth System Governance*, *21*, 100218. DOI: 10.1016/j.esg.2024.100218

Visser, F., Drouilly, M., Moodley, Y., Michaux, J. R., & Somers, M. J. (2023). Mismatch between conservation needs and actual representation of lions from West and Central Africa in in situ and ex situ conservation. *Conservation Letters*, *16*(2), e12949. Advance online publication. DOI: 10.1111/conl.12949

Young, R. E., Gann, G. D., Walder, B., Liu, J., Cui, W., Newton, V., Nelson, C. R., Tashe, N., Jasper, D., Silveira, F. A. O., Carrick, P. J., Hägglund, T., Carlsén, S., & Dixon, K. (2022). International principles and standards for the ecological restoration and recovery of mine sites. *Restoration Ecology*, *30*(S2), e13771. Advance online publication. DOI: 10.1111/rec.13771

Yuan, Y., Bai, Z., Zhang, J., & Xu, C. (2022). Increasing urban ecological resilience based on ecological security pattern: A case study in a resource-based city. *Ecological Engineering*, *175*, 106486. DOI: 10.1016/j.ecoleng.2021.106486

Chapter 6
Integrating Biodiversity and Circular Bioeconomy Practices for Climate Resilience in the Indian Himalayan Region

Vishal Jain
https://orcid.org/0000-0003-1126-7424
School of Engineering and Technology, Vivekananda Institute of Professional Studies, New Delhi, India

Archan Mitra
https://orcid.org/0000-0002-1419-3558
Amity School of Communication, Amity University, Mumbai, India

Sanchita Paul
https://orcid.org/0009-0006-6437-3991
Neil Patel Digital India, India

ABSTRACT

The Indian Himalayan Region (IHR), a biodiversity hotspot, is vital for ecosystem services and local livelihoods but is increasingly threatened by climate change and unsustainable practices. This chapter explores a circular bioeconomy model to enhance climate resilience and sustainable development in the IHR. It emphasizes integrating traditional knowledge with biotechnologies like biochar, composting, and genetic resource use to build zero-waste systems. Through interdisciplinary analysis and case studies from Sikkim and Himachal Pradesh, the chapter evaluates policy

DOI: 10.4018/979-8-3373-2003-8.ch006

frameworks and demonstrates successful biodiversity-driven initiatives. It concludes with strategic recommendations on governance, community participation, and innovation. The chapter offers a replicable model for fostering ecological regeneration and socio-economic sustainability in fragile mountain ecosystems globally.

1. INTRODUCTION

1.1 Background and Context

The Indian Himalayan Region (IHR), extending over 2,500 kilometers across 11 Indian states and union territories, constitutes one of the most ecologically and geopolitically significant regions in South Asia (Ali et al., 2024). It serves as a critical ecological asset, sustaining complex biogeographic systems, and simultaneously functions as a strategic frontier, sharing international boundaries with China (Tibet Autonomous Region), Nepal, Bhutan, and Pakistan. This geostrategic positioning elevates the IHR's importance not only in terms of national security and transboundary water governance but also in shaping regional ecological diplomacy and climate resilience agendas (Edlinger et al., 2025).

Ecologically, the IHR forms part of the globally recognized Himalayan Biodiversity Hotspot, encompassing a remarkable diversity of ecosystems ranging from tropical forests to alpine tundra and cold desert biomes. It harbors over 10,000 species of vascular plants, approximately 300 mammalian species, and thousands of endemic taxa (Rana et al., 2021; Tiwari et al., 2024). This biological richness underpins essential ecosystem services, including water provisioning, carbon sequestration, nutrient cycling, and climate regulation, which benefit not only the upland populations of the region but also an estimated 1.5 billion people in downstream areas dependent on the Indus, Ganges, and Brahmaputra river basins (Sharma et al., 2023; Hussain et al., 2024).

However, the region is witnessing an escalating ecological crisis driven by a confluence of anthropogenic pressures and climatic perturbations. The Himalayan cryosphere is retreating at an unprecedented rate, with cascading impacts on hydrological regimes, agricultural productivity, and regional climate stability (Kumar & Kumar, 2025). Moreover, the proliferation of linear infrastructure projects, unregulated tourism, land-use change, and resource-intensive development models have exacerbated habitat degradation, biodiversity loss, and socio-ecological fragmentation (Singh & Yadav, 2024). These processes are intensifying vulnerabilities in both ecological systems and human communities, particularly among indigenous and marginal populations whose livelihoods are intricately linked to natural resources.

In light of these challenges, there is a compelling need to reimagine the region's development trajectory through a lens that simultaneously enhances ecological integrity, fosters economic regeneration, and strengthens community resilience. The circular bioeconomy offers such a paradigm, one that seeks to decouple economic growth from environmental degradation by promoting the regenerative use of biological resources, minimizing waste, and fostering closed-loop systems of production and consumption (Bonoli, 2023; Ogwu et al., 2025). Unlike linear economic models, the circular bioeconomy framework aligns with ecological cycles, enabling localized, low-carbon, and adaptive development pathways, particularly suitable for fragile mountain ecosystems.

The relevance of this approach is further accentuated by the region's rich heritage of traditional ecological knowledge (TEK). Indigenous communities in the IHR have historically practiced sustainable land management, rotational agroforestry, and biodiversity conservation through sacred groves and customary norms. These practices inherently reflect circularity and offer a critical epistemological foundation for rethinking resource governance. When strategically integrated with emerging biotechnologies, such as biochar production, organic composting, and genetic resource valorization, these knowledge systems can contribute to building zero-waste, climate-resilient, and community-driven economies (Sharma et al., 2024).

While national policy initiatives such as the National Mission on Sustaining the Himalayan Ecosystem (NMSHE) and the broader National Action Plan on Climate Change (NAPCC) emphasize the protection and sustainable development of the Himalayas, operational bottlenecks, ranging from fragmented institutional frameworks to limited local participation, continue to constrain their efficacy (Kaur et al., 2022; Kumar, 2023). Furthermore, existing economic models often marginalize biodiversity as a passive recipient of conservation efforts rather than recognizing it as a foundational and active driver of circular and resilient economies.

This study, therefore, seeks to reconceptualize the role of biodiversity in the Indian Himalayan Region, not merely as a conservation target but as a central asset in building a circular bioeconomy that is ecologically restorative, economically inclusive, and socially embedded (Yadav et al., 2021). It adopts an interdisciplinary and praxis-oriented approach to explore how traditional knowledge, innovative technologies, and enabling policy ecosystems can converge to advance sustainability in one of the world's most critical mountain systems. *Comparative lens.* While the IHR remains the focal geography of this chapter, similar high-mountain systems (the Tropical Andes and the European Alps) are experimenting with circular bioeconomy pathways under fragile ecological constraints. Recent casework (2023–2025) from the Andes (e.g., circular use of agri-residues, peatland/"páramo" restoration linked with biomass valorization, and alpaca/wool by-product cascades) and from the Alps (e.g., wood-based cascades, orchard/vineyard biochar pilots, and zero-waste tourism valorizing organic waste) offers relevant comparators for resilience, governance, and market design. This chapter therefore situates IHR insights alongside those international parallels to broaden transferability.

1.2 Research Problem

Despite growing recognition of the IHR's ecological and strategic importance, the region continues to experience increasing ecological degradation, livelihood precarity, and climate-induced vulnerabilities. The absence of integrative frameworks that synergize biodiversity conservation with economic regeneration has perpetuated a development paradigm that is both ecologically unsustainable and socially inequitable. While the concept of circular bioeconomy offers an alternative model, its application within the Himalayan context remains underexplored, particularly in terms of integrating community knowledge systems with modern technological interventions. There exists a critical gap in both scholarship and practice concerning how biodiversity can serve not only as an object of protection but as a productive force in driving systemic sustainability transitions.

1.3 Research Objectives

1. To investigate the ecological and socio-economic functions of biodiversity in the Indian Himalayan Region, with a focus on its role in ecosystem service provisioning.
2. To explore the conceptual and practical potential of the circular bioeconomy as a development model tailored to the ecological realities of mountain regions.
3. To analyze the scope for integrating traditional ecological knowledge and biotechnological innovation in fostering regenerative, low-waste practices.
4. To evaluate the performance and limitations of policy instruments such as the National Mission on Sustaining the Himalayan Ecosystem (NMSHE) in operationalizing circular bioeconomy frameworks.
5. To document and assess region-specific case studies that illustrate effective models of biodiversity-integrated circular practices.
6. To develop strategic recommendations for adaptive governance, decentralized innovation systems, and community participation for building a resilient bioeconomy in the IHR.

1.4 Significance of the Study

This research contributes to a critical and timely discourse at the intersection of ecology, economy, and governance in the Indian Himalayan Region. It offers a novel lens through which to reconceptualize biodiversity, not solely as a subject of conservation, but as a strategic resource for enabling socio-ecological resilience and economic innovation. By synthesizing insights from ecological economics, indigenous knowledge systems, and circular bioeconomy theory, the study advances a framework that is both grounded in local realities and aligned with global sustainability goals. Its interdisciplinary orientation and evidence-based approach make it relevant for policymakers, development practitioners, researchers, and Himalayan communities alike. In doing so, it aspires to inform scalable, inclusive, and climate-resilient development models that can be adapted not only across the IHR but in mountain systems globally.

2. LITERATURE REVIEW

2.1 Biodiversity and Ecosystem in the Indian Himalayan Region

The Indian Himalayan Region (IHR), spanning approximately 16.2% of India's total geographical area, represents a globally significant ecological frontier, forming a part of the larger Hindu Kush Himalayan (HKH) mountain system (Kumar et al., 2021). It stretches longitudinally across 11 Indian states and union territories, encompassing diverse biogeographic zones from the trans-Himalayas in Ladakh to the subtropical forests of Arunachal Pradesh. The region is home to three of the world's recognized biodiversity hotspots: the Himalayas, Indo-Burma, and the Mountains of Southwest China (Rana et al., 2021). These ecological zones exhibit remarkable endemism, habitat heterogeneity, and cultural richness, underscoring the need for region-specific conservation and development strategies.

Biodiversity Profile across Eastern, Central, and Western Himalayas

The biodiversity profile of the IHR is stratified across three broad ecological zones: the Western Himalayas (Jammu & Kashmir, Ladakh, Himachal Pradesh, and Uttarakhand), the Central Himalayas (Nepal border zones and parts of Sikkim), and the Eastern Himalayas (Arunachal Pradesh, Nagaland, Manipur, Mizoram, Tripura, and Meghalaya).

The Western Himalayas are characterized by alpine pastures, coniferous forests, and glacial systems. This zone harbors endemic species such as the Himalayan brown bear (*Ursus arctos isabellinus*), snow leopard (*Panthera uncia*), and Western tragopan (*Tragopan melanocephalus*) (Yadav et al., 2021). The region also possesses genetic diversity in medicinal plants such as *Picrorhiza kurroa*, *Saussurea costus*, and *Aconitum spp.* which are integral to traditional healthcare systems and pharmaceutical research.

In contrast, the Eastern Himalayas exhibit a subtropical to temperate climatic gradient with dense rainforests and exceptionally high species richness. Arunachal Pradesh alone supports over 5,500 species of vascular plants and is a center of orchid diversity, with more than 600 recorded species (Tsering & Rinya, 2020). Iconic fauna in the region include the red panda (*Ailurus fulgens*), clouded leopard (*Neofelis nebulosa*), and hornbills, which play critical roles as seed dispersers and indicators of forest health.

The Central Himalayas, particularly Sikkim and northern West Bengal, act as transition zones and genetic corridors. The region is known for its agro-biodiversity, particularly traditional rice varieties, millet landraces, and rhododendron forests. Sikkim's 100% organic farming status has positioned it as a leader in biodiversity-friendly agricultural models (Kumar et al., 2017; Tamang et al., 2023).

Together, these sub-regions host more than 18,440 plant species, 1,000 bird species, and approximately 300 mammalian species. The presence of micro-endemism, elevational gradients, and sacred groves reflects both ecological and cultural stewardship across the Himalayas (Hussain et al., 2024).

Key Ecosystem Services: Water Regulation, Carbon Sequestration, and Nutrient Cycling

The Himalayan ecosystems underpin several provisioning, regulating, supporting, and cultural ecosystem services that are foundational to life and livelihoods in the region and beyond. This region is often termed the "Water Tower of Asia" due to its role in sustaining the headwaters of major river basins such as the Indus, Ganges, and Brahmaputra. Glaciers, snowfields, and alpine wetlands regulate seasonal water flow and recharge aquifers. Vegetation cover, especially in montane forests, enhances infiltration, controls runoff, and mitigates soil erosion, contributing to watershed stability (Kattel, 2022).

Himalayan forests act as carbon sinks, with an estimated sequestration capacity of 65 million tonnes of carbon annually (Rana et al., 2021). Oak and coniferous forests in the Western Himalayas, and subtropical broadleaf forests in the Eastern Himalayas, are particularly significant in storing both above-ground and below-ground biomass carbon. These forests play a pivotal role in mitigating climate change by reducing atmospheric CO_2 concentrations (Khan & Shaheen, 2022).

In addition, mountain ecosystems support critical biogeochemical processes through organic matter decomposition, nitrogen fixation, and phosphorus cycling. Microbial biodiversity in Himalayan soils contributes to nutrient mobilization and supports agroecosystems. Traditional agroforestry systems like alder-cardamom plantations in the Eastern Himalayas exemplify managed landscapes where biodiversity enhances nutrient retention and productivity (Sharma et al., 2024).

Biodiversity Threats under Climate Pressure

Despite its ecological wealth, the IHR is under severe stress due to compounded anthropogenic and climatic factors. Climate change has led to rapid glacier retreat, altered phenological patterns, and range shifts in flora and fauna. The IPCC (2022) notes that temperature rise in the Himalayas exceeds the global average, rendering the region highly susceptible to glacial lake outburst floods (GLOFs), biodiversity loss, and crop failures.

Land-use change resulting from road expansion, hydropower projects, and tourism infrastructure has fragmented habitats and disrupted ecological corridors. Deforestation for agriculture and fuelwood continues to reduce forest cover, especially in densely populated mid-altitude zones (Singh & Yadav, 2024). Invasive species such as *Lantana camara* and *Eupatorium* have further degraded native biodiversity, altering successional dynamics and reducing habitat quality.

Overharvesting of medicinal and aromatic plants, driven by commercial demand and lack of regulation, threatens the viability of wild populations. Climate-induced changes in altitudinal vegetation zones are also impacting traditional cropping systems and pasture availability, thereby endangering food and fodder security.

Economic Valuation and Community Dependency

The economic value of biodiversity in the IHR is immense yet underrepresented in conventional development metrics. Ecosystem services such as pollination, hydrological regulation, carbon sequestration, and soil fertility significantly contribute to the rural economy and national ecological security (Sharma et al., 2023). Non-timber forest products (NTFPs), including wild edibles, medicinal herbs, and fibers, contribute significantly to household incomes in remote areas. For instance, the collection and trade of *Yarsagumba* (*Ophiocordyceps sinensis*) in Uttarakhand and parts of Arunachal Pradesh generate seasonal income exceeding INR 100,000 for some households Sharma et al., 2023).

Agricultural biodiversity is another key pillar of economic and food security. Traditional farming systems, including jhum (shifting cultivation), terrace farming, and home gardens, are repositories of genetic diversity and adaptive strategies. These systems not only sustain food sovereignty but also buffer communities against climate risks through diversification (Bisht et al., 2020).

The region's cultural and spiritual practices are deeply intertwined with biodiversity. Sacred groves, clan forests, and pilgrimage sites contribute to in situ conservation while reinforcing community stewardship. These socio-cultural values often serve as informal governance mechanisms, enhancing resilience and fostering collective action in resource management (Pathak, 2021).

2.2 Circular Bioeconomy in Mountain Contexts

The concept of a circular bioeconomy represents a strategic shift from linear modes of production and consumption toward regenerative, low-waste systems that prioritize renewable biological resources, ecosystem health, and socio-economic inclusivity. This paradigm is particularly relevant in ecologically fragile regions such as mountain systems, where the interplay between natural resource limitations, climate sensitivity, and socio-economic marginality necessitates models that are both adaptive and restorative (Muscat et al., 2021). In the Indian Himalayan Region (IHR), the application of circular bioeconomy principles holds transformative potential for enhancing ecological resilience, diversifying livelihoods, and sustaining cultural heritage.

The foundational principles of a circular bioeconomy lie in its emphasis on resource efficiency, life-cycle thinking, ecosystem regeneration, and waste minimization. Unlike conventional bioeconomy approaches, which may still rely on linear extraction and production pathways, the circular bioeconomy integrates closed-loop processes that valorize waste streams, enhance nutrient recovery, and foster biological cascades across sectors such as agriculture, energy, and health (Bonoli, 2023). This systemic approach is not only environmentally sound but also economically prudent, especially in mountainous regions where resource extraction is costly, environmentally damaging, and often culturally contentious.

In mountain contexts like the IHR, the circular bioeconomy must be tailored to address unique biophysical and socio-cultural characteristics. The terrain is characterized by steep gradients, limited arable land, high biodiversity, and dispersed settlements, which pose both constraints and opportunities for resource management. Applications of circular principles in these settings require localized, decentralized models that leverage indigenous knowledge systems, low-cost technologies, and participatory governance. For instance, composting of organic kitchen and agricultural waste, long practiced in traditional Himalayan households, aligns with circular bioeconomy goals by returning nutrients to the soil and reducing reliance on external inputs (Singh & Singh, 2021).

Biomass reuse is another critical element in operationalizing circularity in mountain economies. The IHR generates significant volumes of agricultural residues, forest litter, and organic waste, much of which remains underutilized or improperly managed. Through the development of community-based biogas units, biochar kilns, and vermicomposting systems, this biomass can be transformed into bioenergy, soil conditioners, and organic fertilizers. Such practices not only reduce environmental pollution but also enhance soil health, increase crop productivity, and reduce dependence on chemical fertilizers, thereby creating multiple co-benefits (Dasgupta et al., 2022).

Nutrient loops, both natural and anthropogenic, are integral to mountain agro-ecosystems. Traditional farming systems in the Eastern Himalayas, such as the alder-cardamom agroforestry model in Nagaland or the mandala farming system in Sikkim, exemplify closed-loop nutrient cycles where tree litter, animal manure, and crop residues are reintegrated into the soil system (Giri et al., 2020). These models demonstrate the possibility of maintaining productivity and ecological stability without external inputs, thus aligning seamlessly with the principles of circular bioeconomy.

A key enabling condition for the successful implementation of circular bioeconomy in mountainous regions is the strength of community institutions. In the IHR, customary institutions such as *Van Panchayats* in Uttarakhand, *Dzumsas* in Sikkim, and *Chopa* systems in Arunachal Pradesh have historically governed natural resource use through collective norms and participatory decision-making. These institutions serve as platforms for knowledge exchange, conflict resolution, and collective investment in sustainable practices (Nagahama, 2016; Bhutia, 2019. Empowering such institutions and integrating them into formal governance frameworks is essential for scaling circular innovations while ensuring social legitimacy (Pathak, 2021).

The role of localized technologies and innovation hubs is equally crucial. Unlike large-scale industrial solutions that may be incompatible with mountainous terrains, decentralized and context-specific technologies offer greater adaptability and resilience. For instance, the use of portable solar dryers for horticultural produce, low-tech composting bins, gravity-fed irrigation systems, and mobile bio-digesters has enabled communities in Ladakh and Himachal Pradesh to enhance productivity while conserving resources (Raghuvanshi et al., 2019). The establishment of regional innovation hubs, supported by state governments and research institutions, can further accelerate the dissemination of appropriate technologies and build technical capacity at the grassroots.

Moreover, the circular bioeconomy contributes significantly to both environmental sustainability and socio-economic well-being. Environmentally, it reduces ecological footprints by minimizing waste, conserving biodiversity, and enhancing ecosystem services such as soil fertility, pollination, and water retention. The reuse of biological materials and the valorization of local genetic resources also mitigate greenhouse gas emissions and strengthen ecosystem resilience in the face of climate variability (Mukherjee et al., 2023).

From a socio-economic perspective, circular bioeconomy fosters diversification of income sources, especially for smallholder farmers, women, and marginalized communities. It encourages value addition through local processing of biomass, the development of green products (e.g., herbal formulations, organic compost, biopesticides), and the creation of community enterprises (Khanna et al., 2024). In Sikkim, for example, the organic farming revolution has not only improved soil health and reduced input costs but has also generated employment and enhanced market access for farmers through eco-labeling and certification schemes (Kumar et al., 2017; Bhujel & Joshi, 2021).

The integration of circular bioeconomy into mountain development strategies also aligns with multiple Sustainable Development Goals (SDGs), particularly those related to climate action (SDG 13), responsible consumption and production (SDG 12), and life on land (SDG 15). Furthermore, it supports regional policy goals under the National Mission on Sustaining the Himalayan Ecosystem (NMSHE) by promoting community-based adaptation, biodiversity conservation, and sustainable livelihoods (Rana et al., 2021).

2.2.1 Lessons from the Andes (2023–2025)

Empirical studies from the Tropical Andes highlight three recurring circular strategies. In biomass valorization in high-altitude farming, quinoa husk, potato peels, and cereal residues are converted into compost/biochar to restore degraded Andean soils and boost water retention (de Ciencias Agrarias y Forestales, 2025). On the other hand, Páramo/peatland restoration shows the efficiency of community programs that couple wetland restoration, rotational grazing, and fiber/wool by-product reuse (dyes, felts, organic soil amendments) to diversify income under water-scarcity stress (Jaramillo et al., 2023).

Lastly, agro-biodiversity value chains involve native tuber and grain landraces supported by farmer-led seed banks and local processing, reducing waste and stabilizing margins via local circular market (Varaprasad & Rao, 2024). Steep-slope nutrient loops, communal governance, and smallholder economics mirror IHR constraints; the Andean emphasis on páramo hydrology parallels Himalayan watershed services.

2.2.2 Lessons from the Alps (2023–2025)

Recent Alpine initiatives highlight how circular bio-economy practices are being applied in mountainous regions to balance ecology with local livelihoods. Wood-based cascades represent a strategic use of forestry and sawmill residues, which are processed into pellets, panels, biochar, and soil amendments, ensuring that waste streams are transformed into high-value products (Tippmann, 2024). Parallel efforts in orchard and vineyard circularity focus on utilizing pruning residues for biochar and compost, with precision composting and terroir-specific soil regeneration pilots enhancing both productivity and sustainability.

In the tourism sector, zero-waste models are emerging through food waste digestion and composting systems that supply local farms, refill-and-reuse logistics that cut plastic reliance, and producer–hotel cooperatives that close material loops within the mountain economy (Mehrabi & Giagnocavo, 2024). These initiatives are highly relevant to the Indian Himalayan Region (IHR), where similar value-chain opportunities exist in medicinal and aromatic plants (MAPs), horticulture, and non-timber forest products (NTFPs). The Alpine examples of strong policy coherence, cooperative governance, and the "cascading use" principle provide actionable design cues for integrating ecological, social, and economic sustainability in the IHR context.

2.3 Converging Traditional Ecological Knowledge and Biotechnological Innovations in the Indian Himalayan Region

In the pursuit of resilient and sustainable development in ecologically sensitive mountain regions such as the Indian Himalayan Region (IHR), the convergence of traditional ecological knowledge (TEK) with contemporary biotechnological innovations presents a powerful strategy (Rana et al., 2024). This integrated approach recognizes the complementary strengths of time-tested indigenous practices and modern scientific advancements, fostering context-specific solutions that align with ecological realities and cultural sensibilities. In the IHR, where communities have co-evolved with complex and fragile ecosystems for generations, such hybrid models offer a grounded and scalable pathway for sustainable transitions (Tiwari & Sunny, 2024).

Climate finance can be a decisive enabler for circular bioeconomy (CBE) transitions in mountain regions, where high transaction costs, fragmented value chains, and smallholder risk often deter investment. Blended models, combining concessional climate funds with commercial capital, can de-risk early pilots for biomass valorization (biochar/compost hubs), non-timber forest product (NTFP) processing, and watershed restoration that delivers co-benefits in carbon, biodiversity, and livelihoods (Mishra et al., 2024). Instruments include grants for upstream ecosystem restoration; results-based finance (RBF) for verified outcomes (e.g., tons of organic waste diverted, tCO_2e sequestered in soils); sustainability-linked loans for producer organizations meeting circularity KPIs; and green/SDG-linked municipal bonds for decentralized waste-to-resource facilities (Ermgassen & Löfqvist, 2024).

Payment for Ecosystem Services (PES) and emerging nature/biocredits can reward communities for watershed and soil-carbon improvements, while Article 6 cooperative approaches and high-integrity voluntary carbon markets may monetize durable soil and biochar credits if robust MRV is in place. In India, public schemes and development finance institutions can align with multilateral windows to create revolving credit lines for women-led SHGs and FPOs operating circular enterprises in horticulture, MAPs, and agroforestry. A portfolio approach, bundling many micro-projects into jurisdictional programs, can reduce unit costs, standardize MRV, and crowd-in private co-investment.

Table 1. Drop-in table for this subsection (optional)

Instrument	Use in mountain CBE	Who it reaches	Typical KPIs
Grants / TA	Pilot biochar & compost hubs; MRV set-up	Panchayats, SHGs, FPOs	Facilities installed; MRV protocols adopted
RBF	Pay per ton waste diverted / tCO_2e sequestered	Producer orgs, municipalities	Verified diversion; soil C gains
Blended loans	Scale working capital & equipment	FPOs/SMEs	Revenue growth; default < x%
PES / Nature credits	Watershed, soil, biodiversity benefits	Community institutions	Hectares restored; species indices
Green/SDG bonds	Decentralized circular infra	ULBs / state agencies	Bond KPIs achieved; service coverage

Traditional Ecological Practices and Knowledge Systems

Traditional Ecological Knowledge, as defined by Berkes (2012), refers to "a cumulative body of knowledge, practice, and belief, evolving by adaptive processes and handed down through generations by cultural transmission, about the relationship of living beings with one another and with their environment." In the IHR, TEK manifests in diverse forms across agroecological zones, ethnic communities, and livelihood systems. These include community-based forest management, sacred groves conservation, indigenous herbal medicine, rotational agriculture, and intricate water harvesting systems (Tiwari & Sunny, 2024).

In Himachal Pradesh and Uttarakhand, community forests governed by *Van Panchayats* operate under customary laws that regulate extraction levels, protect watersheds, and sustain biodiversity (Nagahama, 2016). Similarly, sacred groves in Meghalaya and Nagaland are preserved as community-protected forests based on spiritual taboos and ritualistic significance, acting as in situ conservation zones for endemic flora and fauna (Das et al., 2021). In Arunachal Pradesh, tribes such as the Apatani and Monpa have developed rice-fish-vegetable integrated farming systems and terraced cultivation with bamboo irrigation channels that reflect an acute understanding of ecological feedbacks, nutrient recycling, and water efficiency (Sharma et al., 2024).

TEK also plays a central role in ethnomedicine, with communities across the Himalayas relying on local flora for preventive and curative healthcare (Tiwari & Sunny, 2024). Species such as *Swertia chirayita*, *Rhododendron arboreum*, and *Berberis aristata* are commonly used in polyherbal formulations and have been validated by modern pharmacological research. These knowledge systems are not static relics of the past; rather, they are dynamic, adaptive, and embedded in local socio-cultural contexts, offering sustainable practices that have evolved over centuries.

Biotechnological Innovations Relevant to the IHR

Contemporary biotechnology offers tools that can enhance the productivity, efficiency, and resilience of traditional systems. In the context of the IHR, three key domains of biotechnology stand out: biochar production, organic composting with microbial inoculants, and genetic resource utilization.

Biochar, a carbon-rich product derived from biomass pyrolysis, has gained recognition for its soil-enhancing properties, including improved moisture retention, nutrient-holding capacity, and microbial activity. When applied in degraded or nutrient-poor Himalayan soils, biochar enhances soil fertility while acting as a long-term carbon sink (Kumar et al., 2022). Pilot projects in Uttarakhand and Sikkim have demonstrated the potential of integrating biochar into organic farming systems, particularly in regions prone to soil erosion and nutrient leaching.

Composting with the aid of microbial consortia, such as phosphate-solubilizing bacteria and nitrogen-fixing microorganisms, accelerates the decomposition of organic waste and enhances nutrient availability in the final compost produc (Aguilar-Paredes et al., 2023)t. Locally adapted bioformulations, often derived from cow dung, forest leaf litter, and agricultural residues, have been scaled up through participatory models in hill states like Himachal Pradesh and Meghalaya (Edlinger et al., 2025). These techniques align closely with traditional composting practices, making them culturally acceptable and technically feasible.

Genetic resource utilization represents another frontier of convergence. The IHR is a repository of diverse landraces of rice, maize, millet, legumes, and horticultural crops. Biotechnology enables the characterization, conservation, and improvement of these genetic resources through molecular markers, gene banks, and participatory breeding programs. Institutions such as the National Bureau of Plant Genetic Resources (NBPGR) and state agricultural universities are working to document and preserve local crop varieties that exhibit climate resilience, pest resistance, and nutritional superiority.

Hybrid Models and Knowledge-Tech Convergence

Several initiatives across the IHR exemplify the successful integration of TEK and biotechnology, yielding socio-ecological benefits and demonstrating replicable frameworks. In Sikkim, the organic farming movement has combined traditional low-input agriculture with scientific composting methods, biofertilizer use, and pest management protocols. The state's policies have incentivized farmers to adopt modern techniques without compromising cultural practices or ecological ethics (Sharma et al., 2020).

In Nagaland, participatory seed selection and community gene banks have emerged through collaborations between tribal councils and research organizations. These initiatives ensure the conservation of agro-biodiversity while enabling local innovation through the introduction of improved yet locally adapted varieties. Similarly, in Ladakh, projects integrating traditional barley cultivation with solar-powered greenhouses and drip irrigation have extended the growing season and enhanced food security under changing climatic conditions (Raghuvanshi et al., 2019).

Such hybrid models are not limited to agriculture. In the domain of forest-based livelihoods, communities in Himachal Pradesh have adopted scientific harvesting techniques for medicinal plants, supported by GIS-based mapping and growth monitoring, to ensure sustainable extraction. Training programs by the State Medicinal Plants Board have merged ethnobotanical knowledge with modern cultivation and processing methods, thereby adding value and expanding market access.

Challenges for Integration

Despite the promise of integrating TEK with biotechnology, several challenges persist. First, there exists an epistemic asymmetry, wherein traditional knowledge is often devalued or sidelined in favor of formal scientific expertise. This undermines the agency of local communities and hampers collaborative innovation. Recognizing the legitimacy and intellectual rights of indigenous knowledge holders is essential to building trust and ensuring equitable partnerships (Sinkala, 2022).

In addition, policy and institutional gaps hinder effective convergence. Most extension systems and R&D programs operate in silos, with limited mechanisms for cross-sectoral collaboration or community engagement. The absence of tailored curricula, funding incentives, and institutional platforms for knowledge exchange further exacerbates this disconnect. Moreover, the scalability and standardization of localized practices remain challenging, particularly in the absence of scientific validation, documentation, and quality assurance protocols. Bridging this gap requires investment in community-based research, translational science, and capacity building (Mohan, 2023).

2.4 Rethinking Policy and Governance for Circular Bioeconomy in the Indian Himalayan Region

In the Indian Himalayan Region (IHR), the policy landscape is shaped by a confluence of national missions, state-level initiatives, and international environmental commitments. However, despite the presence of well-intentioned policies aimed at environmental conservation and sustainable development, persistent governance challenges have hindered their translation into tangible outcomes (Mukherjee et al., 2023). Among the most comprehensive national-level interventions relevant to the IHR is the National Mission on Sustaining the Himalayan Ecosystem (NMSHE), launched under the National Action Plan on Climate Change (NAPCC). NMSHE seeks to understand and address the impacts of climate change on Himalayan glaciers, biodiversity, and traditional knowledge systems. It emphasizes research, monitoring, and institutional strengthening to enhance the region's adaptive capacity (Rana et al., 2021). The mission also envisions the promotion of sustainable livelihoods, conservation of endemic species, and climate-resilient development planning. However, while NMSHE outlines a robust framework, its operationalization at the state level remains uneven due to capacity gaps and lack of coordination mechanisms.

The National Biodiversity Action Plan (NBAP), revised in 2014 to align with the Convention on Biological Diversity (CBD) and Aichi Targets, provides another vital policy foundation. It identifies thematic areas such as the conservation of biodiversity hotspots, sustainable use of bio-resources, and equitable benefit-sharing with local communities. The NBAP calls for the integration of biodiversity concerns into sectoral policies, including agriculture, forestry, and infrastructure. State-specific biodiversity action plans (SBAPs) have been developed in several Himalayan states, though their implementation has largely remained project-based and donor-driven, limiting institutional ownership and long-term sustainability (Kumar et al., 2022).

In addition, several state missions under the State Action Plans on Climate Change (SAPCCs) echo the objectives of NAPCC while contextualizing them to sub-regional priorities. For instance, the State Climate Change Cell in Himachal Pradesh has initiated efforts in climate-resilient agriculture, early warning systems, and green energy deployment. Similarly, Sikkim's organic farming policy is a pioneering example of aligning biodiversity conservation with sustainable livelihoods and agroecological transitions (Sharma et al., 2020; Kumar, 2023). However, these successes are often isolated, lacking systemic integration with national missions or other state initiatives.

Women in the IHR play pivotal roles across biodiversity-linked value chains—seed selection, on-farm biomass management, NTFP collection, sorting and primary processing—yet face constraints in land titling, collateral, mobility, time poverty, and voice in decision-making. A circular bioeconomy lens can rebalance this by

(i) recognizing and remunerating women's hidden labor in waste reduction, composting, and on-farm nutrient cycling; (ii) redistributing care burdens via proximity childcare at collection/processing centers; (iii) improving access to finance and markets through women-first credit lines, digital payments, and procurement preferences for women-led enterprises; and (iv) ensuring representation in producer organizations and landscape governance committees. Practical entry points include women-managed compost cooperatives, MAPs sorting/standardization units with quality premiums, and biochar micro-enterprises that valorize orchard prunings and crop residues. Gender-responsive budgeting can earmark resources for training, safety, and equipment suited to women's ergonomic needs. Monitoring must be sex-disaggregated and tied to empowerment indicators: % of women in leadership roles within FPOs; share of climate finance accessed by women-led enterprises; changes in control over income; and reductions in drudgery/time spent on fuel and water collection. Evidence from mountain regions shows that when women are integrated as co-owners of assets and contracts—not only as labor—CBE productivity and ecological outcomes improve concurrently. Embedding these safeguards from the outset reduces elite capture and enhances long-term program legitimacy.

Institutional Challenges: Fragmentation, Silos, and Implementation Gaps

Despite the presence of comprehensive policy frameworks, the IHR suffers from institutional fragmentation and sectoral silos that impede effective governance. Multiple institutions with overlapping mandates operate at national, state, and district levels, including the Ministry of Environment, Forest and Climate Change (MoEFCC), state biodiversity boards, agriculture departments, and rural development agencies. The absence of a harmonized platform for inter-agency collaboration has led to duplication of efforts and inefficient resource allocation (Singh et al., 2023).

Further, implementation gaps persist due to weak institutional capacity, limited technical expertise, and lack of sustained financing. Many schemes remain underfunded or delayed in execution, especially in remote and high-altitude areas where logistical challenges are acute. Local governance bodies such as Panchayati Raj Institutions (PRIs) and traditional councils often lack the institutional mandate or resources to meaningfully participate in planning and monitoring processes. This disempowerment undermines the principles of decentralization and inclusive governance that are critical for mountain sustainability (Sinkala, 2022).

Additionally, data gaps and inadequate monitoring systems constrain evidence-based policymaking. The limited availability of disaggregated, real-time data on biodiversity trends, climate impacts, and socio-economic indicators reduces the efficacy of adaptive management approaches. Moreover, there is often a disconnect between scientific research and policy design, resulting in strategies that are either too generic or technologically misaligned with local needs (Hussain et al., 2024).

Adaptive Governance and Cross-Sectoral Alignment

A shift toward adaptive governance can increase the capacity of institutions to respond flexibly to ecological feedback and social learning under conditions of uncertainty and change (Van Assche et al., 2022). For IHR, this requires enhancing the capacity of local institutions to experiment, innovate, and adjust strategies based on continuous monitoring and feedback.

Cross-sectoral alignment is a crucial enabler of adaptive governance. Integrating biodiversity and circular bioeconomy principles into development planning, agriculture, tourism, and infrastructure sectors can generate co-benefits and reduce trade-offs. For instance, integrating organic farming with forest conservation and renewable energy initiatives can create synergistic impacts on livelihoods, biodiversity, and carbon emissions. Such integration requires policy coherence and the dismantling of bureaucratic silos that compartmentalize planning processes (Sharma et al., 2024). Knowledge co-production and participatory planning are also foundational to adaptive governance. Engaging local communities, civil society organizations, and academia in decision-making enhances the legitimacy, relevance, and sustainability of policy interventions. Models such as community-based monitoring, citizen science, and joint forest management can be scaled up through appropriate institutional incentives and capacity development programs (Danielsen et al., 2022).

Opportunities for Multi-Level Policy Synergy and Circularity Mainstreaming

The complex socio-ecological realities of the IHR necessitate **multi-level governance** approaches that align national, state, and community-level actions. Vertical coordination mechanisms, such as intergovernmental task forces, joint funding platforms, and policy convergence cells, can facilitate the seamless flow of information, resources, and accountability across levels (Gain et al., 2020). At the horizontal level, collaborative networks among research institutions, non-governmental organizations, and private sector actors can foster innovation, diffusion, and scaling of circular practices (Bonoli, 2023).

Mainstreaming circular bioeconomy principles into the policy landscape requires both conceptual clarity and institutional innovation. Policies must explicitly recognize biodiversity as an asset that underpins economic and ecological resilience. Incentive structures such as green procurement, biodiversity credits, and ecosystem service payments can promote circular practices while ensuring equitable benefit-sharing.

Regulatory frameworks need to be updated to support waste valorization, resource-efficient technologies, and biodiversity-friendly value chains (Srivastav et al., 2024).

Moreover, capacity building for policy actors at all levels is critical to operationalize circularity. This includes training programs on systems thinking, ecological economics, and sustainability transitions for bureaucrats, planners, and community leaders. Academic curricula and research funding should also be realigned to support interdisciplinary studies and policy-relevant knowledge production (Mohan, 2023). International frameworks and funding mechanisms such as the Convention on Biological Diversity (CBD), Global Environment Facility (GEF), and Green Climate Fund (GCF) offer additional opportunities to embed circularity into Himalayan development. Leveraging these instruments through strategic proposals and partnerships can provide much-needed financial and technical support for regional initiatives (UNDP, n.d.).

2.5 Climate-Induced Migration and Its Interplay with Circular Bioeconomy (220–260 words, paste)

In the IHR, climate stressors—glacier retreat, erratic precipitation, heat stress in valleys, slope instability—interact with market volatility to drive seasonal and, at times, distress migration. Out-migration can erode community labor pools and traditional ecological knowledge, yet remittances and returning skills can also seed circular enterprises if channeled effectively. A CBE strategy can mitigate push factors by stabilizing local incomes through residue valorization, diversified agroforestry, and NTFP processing near the source; and it can harness migration dynamics through diaspora/returnee investment windows, matched-savings schemes for green MSMEs, and recognition of skills acquired off-farm. Programs should anticipate labor seasonality by designing low-labor, high-leverage circular nodes (e.g., community-owned biochar kilns operated in off-peak agricultural months) and by offering portable social protection and flexible contracts. Risks include gendered burdens when men migrate (care work intensification) and potential over-harvest of high-value NTFPs by temporary labor without stewardship ties. Safeguards therefore include community quotas, traceability, fair-price floors, and reinvestment mandates into watershed and soil restoration. Research and monitoring should track migration-CBE feedbacks: migration rates; remittance flows invested in circular assets; job creation per ton of biomass valorized; and ecosystem indicators (soil carbon, erosion control). When aligned with local institutions (Van Panchayats, cooperatives) and climate finance, CBE can become a *stay option*—reducing distress migration—while converting inevitable mobility into green entrepreneurship.

2.6 Market Linkages, Supply Chains, and Private-Sector Partnerships (≈230–260 words, paste)

A circular bioeconomy (CBE) in mountain regions succeeds only when ecological gains translate into *stable offtake and margins* for producers. Three market pathways dominate in the IHR:

1. **Local–regional B2B/B2G**: compost/biochar for orchards and municipal landscaping; MAPs/NTFPs to regional processors; hotel and tourism buyers for certified local produce.
2. **Anchor-buyer models**: long-term offtake/quality protocols with food/herbal/cosmetics processors; packaging and materials firms absorbing side-streams (peels, prunings, husk) as inputs.
3. **Export corridors**: certified MAPs/NTFPs and specialty foods meeting GAP/GACP, organic, fair-trade, and traceability standards.

Typical barriers include fragmented volumes, inconsistent grading, limited primary processing near source, cold-chain/logistics gaps, price opacity, working-capital constraints, and compliance costs for certifications. Practical enablers are FPO-led aggregation; micro-processing hubs (sorting, drying, chipping, briquetting); simple SOPs and grading at collection; third-party QA; pooled logistics and mini-cold rooms; digital traceability and price discovery; and offtake contracts that share quality risk.

Private-sector roles: (i) Anchor buyers to set specifications and guarantee volumes; (ii) PPP/4P models for shared processing infrastructure; (iii) ESG procurement to preference circular inputs; (iv) Impact/working-capital facilities to finance receivables; and (v) Innovation partnerships with startups for bio-materials and waste-to-resource solutions. Aligning these with state missions and community governance reduces transaction costs and helps CBE move from pilots to resilient, market-driven systems.

3. RESEARCH METHODOLOGY

3.1 Research Design

The research employs an exploratory qualitative design, grounded in the traditions of ecological economics, political ecology, and sustainability science. It relies on interpretive and inductive methods to understand the dynamic relationships between biodiversity, circular practices, governance mechanisms, and community knowledge systems. Given the complexity and heterogeneity of the IHR, a multiscalar, comparative approach was used to capture regional diversities and contextual specificities across different Himalayan states.

3.2 Data Collection

This study primarily uses secondary data sources, which include:

- Peer-reviewed journal articles, research reports, and policy briefs;
- Government documents such as the National Mission on Sustaining the Himalayan Ecosystem (NMSHE), National Biodiversity Action Plan (NBAP), and State Action Plans on Climate Change (SAPCCs);
- Reports and publications from international organizations such as ICIMOD, FAO, and the CBD;
- NGO case documentation and evaluation reports on localized circular bio-economy initiatives.

Additionally, publicly available datasets and impact assessments from ministries such as the MoEFCC, Ministry of AYUSH, and NMPB were used to support ecological and economic evaluations.

3.3 Case Study Method

A comparative case study approach was used to analyze four region-specific interventions from Sikkim, Himachal Pradesh, Uttarakhand, and Ladakh. Case selection was purposive, based on their relevance to the research objectives, diversity in ecological zones, and availability of documented outcomes. Each case was analyzed using a standardized framework covering the context, intervention, outcomes, and emergent lessons. To broaden external validity, two concise comparator mini-cases (Andes; Alps) from 2023–2025 peer-reviewed and program evaluations are synthesized to triangulate the IHR findings

3.4 Analytical Techniques

Thematic analysis was employed to extract key patterns and concepts from the literature and case documentation. These themes were clustered around biodiversity functions, circular economy strategies, policy mechanisms, and community participation. A policy coherence matrix was developed to evaluate alignment across national, state, and local policies.

In addition, qualitative content analysis was conducted on policy documents to identify institutional gaps, implementation challenges, and cross-sectoral synergies.

Where possible, descriptive statistics and tabulated outputs were used to supplement qualitative interpretations, particularly in relation to biodiversity metrics, economic valuation of ecosystem services, and livelihood impacts.

3.5 Scope and Limitations

- **Dependence on secondary data:** The study relies entirely on documented data and reports, which may not fully capture undocumented practices or informal local innovations.
- **Lack of primary fieldwork:** Due to the absence of on-ground surveys or interviews, contextual nuances and experiential knowledge of local stakeholders may be underrepresented.
- **Limited real-time datasets:** Some biodiversity and climate datasets are outdated or aggregated at scales too broad to inform localized interventions.
- **Regional imbalance in documentation:** Case data are more readily available for states like Sikkim and Himachal Pradesh, while several Northeast and Western Himalayan states are underrepresented in literature.
- **Potential institutional biases:** Government reports and mission documents may reflect policy intentions more than actual implementation outcomes, which could affect evaluation objectivity.

Earlier drafts under-represented non-Indian mountain cases. This revision integrates 2023–2025 Andes/Alps evidence to strengthen comparability, while noting residual constraints in data harmonization across regions and metrics.

3.6 Implementation Bottlenecks and Lessons from Stalled Projects

Field experience in the IHR and comparable mountain settings shows that policy ambition often outpaces execution due to practical constraints. Common failure points include funding gaps (grants covering CAPEX but not O&M/working capital; slow procurement; delayed reimbursements), monitoring and MRV weaknesses (missing baselines, inconsistent logs, high testing costs, weak digital traceability), and enforcement capacity limits (quotas/SOPs poorly enforced amid staff turnover and diffuse mandates). Market–technology mismatches recur when equipment is oversized or energy-inefficient, primary processing is too far from source, certification costs are underestimated, or offtake isn't secured, leading to quality rejections and cash-flow stress. Social governance issues, elite capture, thin benefit-sharing, gendered time burdens, and weak grievance redress, erode trust and stall uptake. Climate shocks (landslides, late frosts) further disrupt volumes and schedules. Lessons across these cases are consistent: budget O&M and working capital from the start; set conservative, costed MRV with simple QA/traceability; secure anchor-buyer specifications and MOUs before capex; cluster certification and shared QA to lower compliance costs; formalize local enforcement roles with training and audits; design seasonality-aware operating calendars and contingency buffers; and embed women-led SHGs/FPOs as co-owners of assets and contracts, not only as labor.

4. REGION-SPECIFIC CASE STUDIES: INTEGRATING BIODIVERSITY AND CIRCULAR BIOECONOMY IN THE INDIAN HIMALAYAN REGION

4.1 Sikkim: Organic Farming as a Policy-Led Circular Bioeconomy Model

Sikkim, a northeastern Himalayan state, faced increasing soil degradation, declining crop yields, and rising chemical dependency in agriculture during the early 2000s. Recognizing the ecological fragility and unique biodiversity of the region, the state government reoriented its agricultural development model toward sustainability and resilience. In 2003, Sikkim launched its Organic Mission, formally declaring the intent to convert all farmland to organic production. The policy involved banning chemical inputs, supporting composting and vermiculture, promoting traditional knowledge systems, and creating certification and market linkages. Capacity building and subsidies were extended to over 66,000 farming households, integrating bio-input production, on-farm resource recycling, and agrobiodiversity conservation (Bhujel & Joshi, 2021).

By 2016, Sikkim became the world's first fully organic state. Soil fertility improved due to increased organic matter and microbial health. Crop diversity and yields stabilized, while farmer incomes rose through access to premium organic markets. Environmental indicators showed improved water quality and reduced soil erosion (Bhutia, 2019). Sikkim demonstrates how state-led policy, combined with community ownership and ecological stewardship, can mainstream circular bioeconomy in agriculture. The model highlights the importance of regulatory support, multi-stakeholder engagement, and sustained investment in local capacity building.

4.2 Himachal Pradesh: Medicinal Plant Conservation and Value Chain Integration

Himachal Pradesh, part of the Western Himalayas, is rich in medicinal and aromatic plants (MAPs) such as Picrorhiza kurroa, Aconitum heterophyllum, and Swertia chirayita. These species are threatened by overharvesting, habitat loss, and climate variability. Many rural households depend on MAPs for both traditional medicine and supplemental income. Through the State Medicinal Plants Board and National Medicinal Plants Mission, Himachal Pradesh initiated ex situ conservation through herbal gardens, nurseries, and community-managed cultivation plots. Scientific harvesting protocols, cultivation training, and value chain integration efforts were implemented (Paliwal, 2023). Local cooperatives were supported to process and market value-added herbal products.

Over 4,500 hectares were brought under MAP cultivation. Wild harvesting pressure on endemic species declined, while community income from MAP sales increased by 20-30% in selected districts. Integration with Ayurveda and wellness markets improved product visibility and demand. Conservation efforts benefited from increased awareness and participatory monitoring (Nitu & Singh, 2021). This case underscores the viability of merging biodiversity conservation with livelihood generation. A decentralized, value chain-focused approach, supported by biotechnical inputs and market access, enhances sustainability while protecting natural capital.

4.3 Uttarakhand: Community Forest Management and Biochar Integration

Uttarakhand has a long tradition of community forest governance, notably through Van Panchayats, which manage over 12% of the state's forest area. The state also faces increasing pressure on forests due to fuelwood demand, climate variability, and declining soil fertility in hill agriculture. Collaborative programs involving the Forest Department, NGOs, and local panchayats introduced biochar production units in select villages. These units utilized surplus biomass from pruning and invasive species. Training was provided on efficient pyrolysis techniques, and the biochar produced was applied in agricultural fields to improve soil quality and reduce chemical input dependence (Sharma et al., 2024).

Biochar application led to a 15-20% increase in crop yields and improved water retention in pilot areas. Forest degradation slowed due to reduced biomass removal, and community awareness on sustainable forest use increased. Carbon sequestration potential and climate mitigation benefits were also noted (Bisht et al., 2021). Integrating circular practices like biochar into traditional forest management systems demonstrates the power of hybrid approaches. Empowering local institutions, providing technical support, and linking ecological interventions with livelihood benefits are key to replicability.

4.4 Ladakh: Ice Stupas, Zero-Waste Innovation, and High-Altitude Circularity Models

Ladakh's cold-arid ecosystem is highly vulnerable to climate-induced glacial melt, erratic precipitation, and water scarcity during pre-sowing months. Conventional agriculture faces challenges due to short growing seasons, limited water, and extreme temperatures. Innovative projects such as the Ice Stupa initiative, led by renowned social activist Sonam Wangchuk and the NGO SECMOL, involved storing winter meltwater in artificial ice cones that gradually release water in spring (Financial Express, 2024). Parallel zero-waste campaigns in Leh promoted waste segregation, composting, and plastic bans. Solar greenhouses, composting toilets, and decentralized energy systems were also introduced to enhance circularity in resource use (Raghuvanshi et al., 2019; Kaur & Sharma, 2025).

Ice stupas provided critical irrigation during the lean season, stabilizing agricultural cycles. Composting practices improved organic matter in the soil, while solar innovations reduced fossil fuel dependence. Community participation and youth engagement in environmental innovation increased significantly. The region is now piloting carbon-neutral village models with support from the Ladakh Autonomous Hill Development Council (Islam, 2024)). Ladakh exemplifies high-altitude adaptation through nature-based and technological solutions. Innovations rooted in traditional wisdom and ecological understanding can be scaled through institutional support, local innovation hubs, and climate-sensitive infrastructure planning.

4.5 Andes Mini-Case: Circular Farming–Water Nexus at High Altitude (2023–2025)

High-altitude Andean communities have piloted circular schemes that turn crop residues (quinoa husks, potato peels) into compost/biochar, paired with peatland/"páramo" restoration to stabilize spring flows and reduce erosion. Programs in [Peru/Bolivia/Ecuador—insert specific site] report improved soil organic carbon, reduced mineral fertilizer dependency, and income diversification via local processing of native crops and alpaca/wool by-products. Governance hinges on communal irrigation boards and páramo committees aligning restoration with circular farm inputs. *Transfer to IHR:* bundle watershed restoration (e.g., alpine wetlands) with residue-to-biochar/compost value chains; anchor via Van Panchayats/rural cooperatives.

4.6 Alps Mini-Case: Wood Cascades and Zero-Waste Tourism in Mountain Districts (2023–2025)

Alpine districts (e.g., [South Tyrol/Valais/Vorarlberg—insert]) demonstrate cascading use of timber: primary wood → panels; side streams → pellets/biogas/biochar; fine residues → soil amendments for orchards/vineyards. Parallel tourism loops collect food waste for biogas/compost and channel digestate back to nearby farms, cutting imports and transport emissions. Policy orchestration via regional bioeconomy platforms (EUSALP/Interreg) enabled SME clustering and stable local demand. *Transfer to IHR:* replicate "cluster + cascade" design for MAPs/fruit/NTFPs; co-site small processors to capture side streams.

5. DATA ANALYSIS AND INTERPRETATION

The analysis presented in this section draws upon the methodological framework laid out in the previous chapter, particularly the comparative case study method, thematic synthesis, and content analysis of policy and institutional documents. It synthesizes ecological, policy, and socio-economic dimensions of circular bioeconomy models across the Indian Himalayan Region (IHR), with reference to the four case studies from Sikkim, Himachal Pradesh, Uttarakhand, and Ladakh.

5.1 Biodiversity and Ecosystem Service Value

The integration of biodiversity into circular bioeconomy frameworks requires quantifying ecosystem service contributions in local contexts. Data from ICIMOD (2023) and the National Biodiversity Authority (2022) were analyzed to derive relative ecosystem service values for provisioning, regulating, and cultural services across case regions.

Table 2. Estimated annual ecosystem service value by region (INR in million)

Region	Provisioning (NTFPs, Crops)	Regulating (Carbon, Water)	Cultural (Spiritual, Tourism)	Total Value
Sikkim	1,100	850	650	2,600
Himachal Pradesh	950	1,200	700	2,850
Uttarakhand	1,000	1,050	600	2,650
Ladakh	300	700	900	1,900

Table 2 illustrates that while provisioning services dominate in agrarian states like Sikkim and Uttarakhand, regulating services (especially carbon sequestration) are more significant in forest-rich Himachal Pradesh. In Ladakh, cultural services, including ecotourism and sacred landscapes, form a larger proportion of total ecosystem value.

5.2 Socio-Economic Outcomes of Circular Interventions

Each region's circular bioeconomy intervention has demonstrated measurable socio-economic outcomes. Using secondary data synthesis, Table 3 compares key indicators pre- and post-intervention.

Table 3. Comparative socio-economic impact indicators

Indicator	Sikkim (Organic Farming)	Himachal (MAPs)	Uttarakhand (Biochar)	Ladakh (Ice Stupas & Zero-Waste)
Increase in Household Income (%)	22	28	18	20
Reduction in Chemical Input Use (%)	100	40	60	80
Soil Organic Matter Improvement (%)	35	20	32	25
Livelihood Diversification Index*	High	Medium	Medium	High

(*Index based on presence of secondary income sources such as eco-tourism, agro-processing, herbal trade, etc.)

Table 3 indicates substantial improvements in both income and ecological indicators, reinforcing the value of circular bioeconomy as a dual sustainability and resilience strategy.

5.3 Institutional Alignment and Policy Coherence

A content analysis of national and state-level policies was undertaken to assess alignment with circular bioeconomy objectives. Key policy texts such as NMSHE, NBAP, and SAPCCs were coded for the presence of the following principles: biodiversity conservation, circularity/waste reduction, community participation, and innovation.

Figure 1. Radar chart of policy alignment across four states

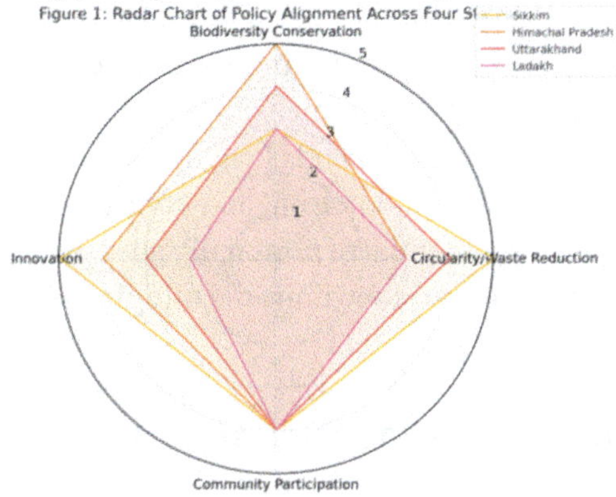

Figure 1 illustrates uneven policy coherence. While Sikkim and Himachal Pradesh demonstrate strong integration of circularity and biodiversity respectively, Uttarakhand and Ladakh show gaps in policy-linked innovation and multi-level governance coordination.

5.4 Thematic Synthesis: Converging Trends and Divergences

From the case studies and document analysis, the following convergent themes emerged:

- **Local Knowledge Integration**: All four states embedded traditional ecological knowledge in the design of interventions, from herbal practices to water harvesting systems.
- **Community Ownership**: Participatory planning and governance (e.g., Van Panchayats, cooperatives, innovation councils) were central to success.
- **Technological Adaptation**: Regions applied localized biotechnologies like biochar, vermicomposting, and solar greenhouses suited to terrain and climate.

Conversely, divergences were found in:

- **Policy Execution Capacity**: Regions varied significantly in technical support, funding access, and administrative continuity.
- **Documentation and Monitoring**: Standardized impact assessments were more available in Sikkim and Himachal Pradesh compared to Ladakh and parts of Uttarakhand.

5.5 Implications

The analysis confirms that integrating biodiversity conservation with circular bioeconomy principles produces multidimensional benefits: ecological, economic, and institutional. However, success depends heavily on enabling conditions such as policy coherence, institutional collaboration, and context-sensitive technology design. The findings suggest the potential of regionally adaptive, community-anchored circular bioeconomy frameworks to contribute meaningfully to sustainable mountain development agendas.

5.6 Climate-Linked Outmigration in the IHR: Critical Reflections

Climate stressors in the IHR—glacier retreat, slope instability, rainfall variability—interact with market and service gaps to drive seasonal and distress outmigration. While remittances can stabilize household consumption and seed micro-enterprise, several *policy failures and unintended consequences* emerge. First, *infrastructure-centric adaptation* (roads, floodwalls) without livelihood design often shifts risk rather than reducing it; farm profitability remains fragile, prompting continued migration. Second, *training-for-migration* programs may unintentionally *institutionalize outmigration*, draining youth from producer groups and weakening collective action. Third, *relocation without tenure clarity* displaces grazing/forest access, triggering resource conflict or over-extraction at new sites. Fourth, *PES/nature-credit schemes* that require documented land rights can *exclude land-poor households*, deepening inequity; women's unpaid care work often expands when men migrate, constraining on-farm circular practic. Finally, *technology subsidies* (e.g., high-capex dryers/ biomass units) placed far from source with no O&M financing face low utilization under seasonal labor shortages.

To avoid these pitfalls, migration-aware circular bioeconomy (CBE) design should: (i) pair climate infrastructure with *income-generating circular nodes* near source (compost/biochar, MAPs primary processing); (ii) create *returnee/diaspora investment windows* with matched savings and TA; (iii) ensure *tenure and inclusion safeguards* so PES/credits reach land-poor and women-led groups; and (iv) plan for *seasonality* with operating calendars, portable contracts, and contingency buffers. Monitoring should track migration rates, remittance reinvestment into circular assets, women's time burdens, and ecosystem indicators alongside profitability.

6. STRATEGIC RECOMMENDATIONS

6.1 Strengthening Policy Integration and Multilevel Governance

A recurring insight from the data analysis is the fragmented nature of policy instruments governing biodiversity and circular economy in the IHR. Despite the existence of national missions such as the National Mission on Sustaining the Himalayan Ecosystem (NMSHE) and the National Biodiversity Action Plan (NBAP), a coherent cross-sectoral framework remains largely absent (Rana et al., 2021). There is a critical need to embed circular bioeconomy principles explicitly into state and district-level planning processes.

One of the first steps toward this is the harmonization of State Action Plans on Climate Change (SAPCCs) with regional biodiversity strategies and circularity objectives. Drawing from the successful policy ecosystem in Sikkim, where organic agriculture is linked to biodiversity protection and eco-tourism, other states should adopt integrated strategies that align sectoral goals across agriculture, forestry, rural development, and industry (Sharma et al., 2020). This alignment must be supported by dedicated institutional platforms that facilitate inter-departmental coordination and accountability.

6.2 Enabling Context-Sensitive Technology and Innovation Ecosystems

The comparative case studies demonstrate that technological adaptation is most effective when designed in alignment with local ecological conditions and socio-cultural practices. Biochar in Uttarakhand, ice stupas in Ladakh, and decentralized composting systems in Sikkim illustrate how innovation rooted in traditional knowledge can yield measurable sustainability benefits (Raghuvanshi et al., 2019; Singh & Mehta, 2022).

To scale such outcomes, the establishment of regional innovation hubs is recommended. These hubs should serve as incubators for co-developing low-cost, modular, and culturally appropriate biotechnologies. Partnerships between local universities, technical institutes, NGOs, and community bodies should be institutionalized to facilitate real-time feedback, field testing, and adaptation (Malik et al., 2020). Furthermore, public financing instruments such as innovation challenge funds and viability gap funding can catalyze entrepreneurship in the circular economy space, especially among youth and women.

6.3 Institutionalizing Traditional Ecological Knowledge Systems

The resilience of traditional ecological knowledge (TEK) in sustaining ecosystem services was evident across all four case study regions. Practices such as sacred grove protection, rotational agroforestry, and herbal medicine systems not only preserve biodiversity but also embody circular principles of reuse, reciprocity, and regeneration (Mitra & Paul, 2025). Yet, TEK remains under-recognized in formal governance and research systems.

Institutionalizing TEK requires formal recognition of community intellectual property rights and the establishment of biocultural protocols that guide the ethical documentation, sharing, and application of indigenous knowledge (Pathak, 2021). State biodiversity boards and district planning committees should integrate community elders, healers, and local resource managers into advisory roles. In parallel, school and university curricula must be reoriented to valorize traditional science alongside modern ecological knowledge, promoting a pluralistic epistemology that fosters innovation through hybridity (Chavan, 2021).

6.4 Enhancing Monitoring, Evaluation, and Impact Metrics

One of the key limitations identified in the research was the lack of standardized impact assessments and real-time monitoring frameworks, particularly in regions like Ladakh and parts of Uttarakhand. Effective mainstreaming of circular bioeconomy requires a robust system of indicators that capture ecological, economic, and social dimensions.

The development of region-specific sustainability indices should be in place to measure outcomes such as soil organic carbon, livelihood diversification, biomass recycling efficiency, and ecosystem service values (Paramesh et al., 2022). Digital platforms using geospatial tools and participatory data collection can enhance transparency and enable adaptive management. Institutions such as the National Biodiversity Authority and ICIMOD can support state governments in designing interoperable data ecosystems that link local, national, and transboundary monitoring efforts (Hussain et al., 2024).

6.5 Investing in Capacity Building and Human Capital Development

The successful implementation of circular bioeconomy models hinges on human capital that is skilled in systems thinking, ecological design, and transdisciplinary problem-solving. Across all case studies, the presence of local champions, grassroots institutions, and committed civil society actors emerged as critical enablers of success.

Capacity-building strategies must target multiple stakeholder groups. For government officials and planners, tailored training modules on circular economy frameworks, biodiversity valuation, and climate adaptation should be developed. For community members, especially women and youth, hands-on learning through demonstration projects, eco-labs, and farmer field schools can build technical and organizational capabilities (Ziervogel et al., 2021). Moreover, state rural development missions and skilling programs should include modules on eco-entrepreneurship, waste valorization, and green product development.

6.6 Fostering Financial Inclusion and Incentive Mechanisms

Financial constraints often limit the scalability of circular practices, particularly among smallholder farmers and forest-dependent communities. Strategic public investment, coupled with innovative financing instruments, can help overcome these barriers. Direct Benefit Transfer (DBT) schemes, biodiversity credits, and payments for ecosystem services (PES) are potential tools to incentivize conservation-linked livelihoods (Bonoli, 2023).

Microfinance institutions and self-help groups can play a role in providing working capital for decentralized bioeconomy enterprises, such as compost units, herbal processing centers, and renewable energy-based agribusinesses. State governments should also explore carbon finance through Voluntary Carbon Markets (VCMs) for initiatives like biochar application and afforestation, particularly in high-altitude areas with high sequestration potential (Minz & Khattri, 2023).

6.7 Promoting Transboundary Cooperation and Knowledge Exchange

Given the shared ecological and cultural continuities across the Hindu Kush Himalaya, India should actively engage in transboundary dialogues to exchange best practices and align sustainability agendas. ICIMOD-led platforms offer a valuable entry point for regional cooperation on issues such as biodiversity monitoring, glacial water governance, and climate-resilient agriculture (Gurung et al., 2019).

Cross-learning with Nepal's community forestry model, Bhutan's Gross National Happiness (GNH) framework, and China's ecological civilization approach can enrich policy thinking and institutional design in the Indian context. Academic institutions and think tanks must be mobilized to support South-South knowledge networks that facilitate adaptive learning, policy experimentation, and innovation diffusion (Laudari et al., 2023). Beyond HKH, structured exchanges with Andean consortia (e.g., CONDESAN/Andean Forest Program) and Alpine platforms (EUSALP/Interreg Alpine Space bioeconomy clusters) can accelerate technology transfer on wood cascades, biochar for orchards, páramo/wetland restoration linkages, and tourism circularity. Joint methods (soil carbon, biomass accounting, circularity KPIs) will aid cross-mountain benchmarking.

7. CONCLUSION

The Indian Himalayan Region (IHR) represents both an ecological treasure and a developmental challenge, characterized by its rich biodiversity, fragile ecosystems, and socio-economic marginalities. This research has underscored the transformative potential of integrating biodiversity with circular bioeconomy practices to achieve sustainable and climate-resilient development. Through an interdisciplinary methodology and detailed case study analysis, the study has revealed how traditional ecological knowledge, when paired with adaptive technologies and supportive governance frameworks, can generate tangible ecological and socio-economic benefits across diverse Himalayan sub-regions. From organic farming in Sikkim to medicinal plant value chains in Himachal Pradesh, community forestry and biochar initiatives in Uttarakhand, and zero-waste innovations in Ladakh, the case studies collectively highlight the feasibility and necessity of context-sensitive, circular interventions rooted in biodiversity.

However, the findings also point to persistent governance challenges, including policy fragmentation, monitoring gaps, and uneven institutional capacities. Addressing these barriers requires a systemic shift towards integrated planning, cross-sectoral collaboration, and participatory governance. Furthermore, enabling conditions such as inclusive financial mechanisms, innovation hubs, and capacity-building platforms are essential to scale successful models and ensure equitable benefits for local communities. Future research should deepen empirical fieldwork, particularly in under-documented regions of the IHR, and explore longitudinal impacts of circular bioeconomy interventions on ecosystem services, social equity, and intergenerational sustainability. As climate risks escalate, the integration of biodiversity and circularity must be treated not merely as a development alternative but as an imperative. This research contributes to that vision by offering a roadmap grounded in regional realities, policy relevance, and a commitment to ecological and cultural integrity.

REFERENCES

Aguilar-Paredes, A., Valdés, G., Araneda, N., Valdebenito, E., Hansen, F., & Nuti, M. (2023). Microbial community in the composting process and its positive impact on the soil biota in sustainable agriculture. *Agronomy (Basel)*, *13*(2), 542. DOI: 10.3390/agronomy13020542

Ali, H., Rafiq, M., & Shang, Z. (2024). Ecological Conservation and Restoration in the Indian Himalaya Region. In *Sustainable Ecological Restoration and Conservation in the Hindu Kush Himalayan Region: A Comprehensive Review* (pp. 117-135). GB: CABI. DOI: 10.1079/9781800622579.0009

Bhujel, R. R., & Joshi, H. G. (2021). *Organic Farming for sustainability: A case study of farmers in Sikkim*. ResearchGate.

Bhutia, Dr. D. (2019). DZUMSA SYSTEM IN SIKKIM AS AN ALTERNATIVE MODE TOWARDS DISPUTE RESOLUTION. *International Journal of Research and Analytical Reviews*, *6*(2), 797–798.

Bisht, D., Joshi, V., Yadava, A. K., & Pant, H. (2021). COMMUNITY FOREST MANAGEMENT IN UTTARAKHAND: a COMPARATIVE CASE STUDY OF WESTERN HIMALAYAN VAN PANCHAYATS. *ResearchGate*.

Bisht, I. S., Rana, J. C., Yadav, R., & Ahlawat, S. P. (2020). Mainstreaming agricultural biodiversity in traditional production Landscapes for Sustainable Development: The Indian Scenario. *Sustainability (Basel)*, *12*(24), 10690. DOI: 10.3390/su122410690

Bonoli, A. (2023). *Circular Economy and Circular Bioeconomy as pillars of the ecological transition for a worldwide enhanced sustainability* (Doctoral dissertation, Universidad de Granada).

Chavan, V. (2021). *State Biodiversity Boards: Towards Better Governance*. ResearchGate.

Danielsen, F., Eicken, H., Funder, M., Johnson, N., Lee, O., Theilade, I., Argyriou, D., & Burgess, N. D. (2022). Community monitoring of natural resource systems and the environment. *Annual Review of Environment and Resources*, *47*(1), 637–670. DOI: 10.1146/annurev-environ-012220-022325

Das, A., Gujre, N., Devi, R. J., & Mitra, S. (2021). A review on traditional ecological knowledge and its role in natural resources management: North East India, a cultural paradise. *Environmental Management*, •••, 1–22. PMID: 34791537

Dasgupta, S., Badola, R., Ali, S. Z., & Tariyal, P. (2022). Bioresource, energy and forest sustainability: A case study from Indian Himalayan region. *Journal of Cleaner Production*, *337*, 130497. DOI: 10.1016/j.jclepro.2022.130497

de Ciencias Agrarias y Forestales, L. (2025). 9 Bioeconomy and Biotechnology: Innovative Strategies in the Utilization of Resources in the Argentine North Patagonia. *Plant Biotechnology and Sustainable Agriculture*, 113.

Edlinger, A., Herzog, C., Garland, G., Walder, F., Banerjee, S., Keel, S. G., Mayer, J., Philippot, L., Romdhane, S., Schiedung, M., Schmidt, M. W. I., Seitz, B., Wüst-Galley, C., & Van Der Heijden, M. G. A. (2025). Compost application Enhances soil health and maintains crop yield: Insights from 56 Farmer-Managed Arable Fields. *Journal of Sustainable Agriculture and Environment*, *4*(1), e70041. DOI: 10.1002/sae2.70041

Ermgassen, S. O. Z., & Löfqvist, S. (2024). Financing ecosystem restoration. *Current Biology*, *34*(9), R412–R417. DOI: 10.1016/j.cub.2024.02.031 PMID: 38714174

Financial Express. (2024). World Water Day 2024: As Sonam Wangchuk continues his ‘climate fast’, know about the Ice Stupas of Ladakh.

Gain, A. K., Giupponi, C., Renaud, F. G., & Vafeidis, A. T. (2020). Sustainability of complex social-ecological systems: Methods, tools, and approaches. *Regional Environmental Change*, *20*(3), 102. DOI: 10.1007/s10113-020-01692-9

Giri, K., Mishra, G., Rawat, M., Pandey, S., Bhattacharyya, R., Bora, N., & Rai, J. P. N. (2020). Traditional farming systems and agro-biodiversity in eastern Himalayan region of India. In *Rhizosphere biology* (pp. 71–89). DOI: 10.1007/978-981-15-1902-4_5

Gurung, J., Chettri, N., Sharma, E., Ning, W., Chaudhary, R. P., Badola, H. K., Wangchuk, S., Uprety, Y., Gaira, K. S., Bidha, N., Phuntsho, K., Uddin, K., & Shah, G. M. (2019). Evolution of a transboundary landscape approach in the Hindu Kush Himalaya: Key learnings from the Kangchenjunga Landscape. *Global Ecology and Conservation*, *17*, e00599. DOI: 10.1016/j.gecco.2019.e00599

Hussain, S., Sharma, S., Bhatti, R. C., & Singh, A. N. (2024). Sustainability in the Indian Himalayan Region: Understandings the Ecosystem Services, Climate Change Impacts, Land Use Shifts and Their Threats. In *The Himalayas in the Anthropocene: Environment and Development* (pp. 33–57).

Islam, M. (2024). Climate Change Adaptation through a Social Innovation Lens: A Framework Analysis of Ladakh, India’s Community-based Approach for Sustainable Agriculture Practice. In *Conference Paper*.

Jaramillo, R., Marmol-Guijarro, A., Doskocil, L. G., Chimbolema, S., & Suárez, E. (2023). Using 3D models to quantify aboveground biomass in the cushion-forming species Plantago rigida in an Andean páramo peatland. *Mires and Peat, 29*, 29. DOI: 10.19189/MAP.2023.RSC.SC.1829791

Kattel, G. R. (2022). Climate warming in the Himalayas threatens biodiversity, ecosystem functioning and ecosystem services in the 21st century: Is there a better solution? *Biodiversity and Conservation, 31*(8–9), 2017–2044. DOI: 10.1007/s10531-022-02417-6

Kaur, G., & Sharma, S. (2025). Sustainable Methods for Solid Waste Management in High-Altitude Tourist Regions. In *Solid Waste Management and Disposal Practices in Rural Tourism* (pp. 395–416). IGI Global.

Kaur, S., Kaur, D., Tiwana, A. S., & Gupta, S. (2022). Climate change-related governance and policies in Indian Himalayas. In *Climate change: Impacts, responses and sustainability in the Indian Himalaya* (pp. 309–330). Springer International Publishing. DOI: 10.1007/978-3-030-92782-0_16

Khan, R. W. A., & Shaheen, H. (2022). Biomass carbon stock estimation in lesser Himalayan subtropical broadleaf forests of Kashmir. *DOAJ (DOAJ: Directory of Open Access Journals).*

Khanna, M., Zilberman, D., Hochman, G., & Basso, B. (2024). An economic perspective of the circular bioeconomy in the food and agricultural sector. *Communications Earth & Environment, 5*(1), 507. DOI: 10.1038/s43247-024-01663-6

Kumar, A., & Kumar, S. (2025). Impact of Climate Change on Himalayan Ecosystem and Traditional Crops. *Climate Change Impact on Himalayan Biodiversity*, 393-413.

Kumar, J., Pradhan, M., & Singh, N. (2017). Sustainable Organic Farming in Sikkim: An Inclusive perspective. In *Lecture notes in electrical engineering* (pp. 367–378).

Kumar, M. (2023). Forestry policies and practices to promote climate change adaptation in the Indian Western Himalayan states. In *Climate Change Adaptation, Risk Management and Sustainable Practices in the Himalaya* (pp. 65–87). Springer International Publishing. DOI: 10.1007/978-3-031-24659-3_4

Kumar, P., Sharma, P. K., Kumar, P., Sharma, M., & Butail, N. P. (2021). Agricultural sustainability in Indian Himalayan region: Constraints and potentials. *Indian Journal of Ecology, 48*(3), 649–661.

Kumar, R., Singh, A., Datta, A., Yadav, R. P., Dinesh, D., & Verma, K. (2022). Carbon sequestration in degraded lands: current prospects, practices, and future strategies. In *Plans and Policies for Soil Organic Carbon Management in Agriculture* (pp. 221–255). Springer Nature Singapore. DOI: 10.1007/978-981-19-6179-3_9

Laudari, H. K., Sapkota, L. M., Maraseni, T., Subedi, P., Pariyar, S., Kaini, T. R., Lopchan, S. B., Weston, C., & Volkova, L. (2023). Community forestry in a changing context: A perspective from Nepal's mid-hill. *Land Use Policy*, *138*, 107018. DOI: 10.1016/j.landusepol.2023.107018

Malik, A., Sharma, P., Pereira, V., & Temouri, Y. (2020). From regional innovation systems to global innovation hubs: Evidence of a Quadruple Helix from an emerging economy. *Journal of Business Research*, *128*, 587–598. DOI: 10.1016/j.jbusres.2020.12.009

Mehrabi, S., & Giagnocavo, C. (2024). Business models and strategies for the internalization of externalities in agri-food value chains. Agricultural and Food Economics, 12(1), 46.

Mishra, P. K., Sena, D. R., Pande, V. C., & Singh, R. K. (2024). Framing land degradation and restoration policy in India following the IPBES pathway. *Indian Journal of Soil Conservation*, *52*, 14–25.

Minz, N. K., & Khattri, N. (2023). *Impact of Microfinancing and Self-Help Groups (SHGs) on women's empowerment*. ResearchGate.

Mitra, A., & Paul, S. (2025). Empowering Indigenous Communities Through Regenerative Tourism: Cultural Preservation and Socio-Economic Resilience in Northeast India. In Sousa, B., Arantes, L., & Bhartiya, S. (Eds.), *Regenerative Tourism for Social Development* (pp. 281–310). IGI Global Scientific Publishing. DOI: 10.4018/979-8-3373-0235-5.ch012

Mohan, D. (2023). Enhancing capacity building initiatives at sub-national level for supporting climate change adaptation. *Climate and Development*, *15*(9), 808–815. DOI: 10.1080/17565529.2022.2163845

Mukherjee, P. K., Das, B., Bhardwaj, P. K., Tampha, S., Singh, H. K., Chanu, L. D., Sharma, N., & Devi, S. I. (2023). Socio-economic sustainability with circular economy—An alternative approach. *The Science of the Total Environment*, *904*, 166630. DOI: 10.1016/j.scitotenv.2023.166630 PMID: 37643712

Muscat, A., De Olde, E. M., Ripoll-Bosch, R., Van Zanten, H. H. E., Metze, T. P., Termeer, C. J. M., Van Ittersum, M. K., & De Boer, I. J. M. (2021). Principles, drivers and opportunities of a circular bioeconomy. *Nature Food, 2*(8), 561–566. DOI: 10.1038/s43016-021-00340-7 PMID: 37118163

Nagahama, K. (2016). Van (Forest) Panchayat as a system of sustainable common forest resource management in India. *SDRP Journal of Earth Sciences & Environmental Studies, 1*(3).

Nitu, & Singh, R. B. (2021). Diversification of Agriculture Through Medicinal Plants in Kangra District, Himachal Pradesh. *Resource Management, Sustainable Development and Governance: Indian and International Perspectives*, 377-395.

Ogwu, M. C., Kosoe, E. A., Iyiola, A. O., & Fawole, W. O. (2025). Circular Economy Approaches in Bioeconomy in the Global South. In *Sustainable Bioeconomy Development in the Global South: Volume II Bioeconomy Techniques* (pp. 131-157). Singapore: Springer Nature Singapore. DOI: 10.1007/978-981-96-0305-3_5

Paliwal, S. (2023). *A Study on Value Chain of Medicinal and Aromatic Crops in Garhwal Division of Uttarakhand* (Doctoral dissertation, GB Pant University of Agriculture & Technology, Pantnagar-263145).

Paramesh, V., Ravisankar, N., Behera, U., Arunachalam, V., Kumar, P., Solomon Rajkumar, R., Dhar Misra, S., Mohan Kumar, R., Prusty, A. K., Jacob, D., Panwar, A. S., Mayenkar, T., Reddy, V. K., & Rajkumar, S. (2022). Integrated farming system approaches to achieve food and nutritional security for enhancing profitability, employment, and climate resilience in India. *Food and Energy Security, 11*(2), e321. DOI: 10.1002/fes3.321

Pathak, R. (2021). Scoping the need of mainstreaming Indigenous knowledge for sustainable use of bioresources in the Indian Himalayan region. www.academia.edu

Raghuvanshi, J., Agrawal, R., & Ghosh, P. K. (2019). Measuring the innovation capability of micro enterprises in India: Construct development and validation. *Benchmarking, 26*(5), 1405–1430. DOI: 10.1108/BIJ-08-2018-0229

Rana, D., Rana, S., & Choudhary, S. (2024). Sustainable development goals and the Indian Himalayan region. In *Advances in geographical and environmental sciences* (pp. 173–194). DOI: 10.1007/978-981-97-1685-2_10

Rana, S. K., Rawal, R. S., Dangwal, B., Bhatt, I. D., & Price, T. D. (2021). 200 years of Research on Himalayan Biodiversity: Trends, gaps, and policy implications. *Frontiers in Ecology and Evolution, 8*, 8. DOI: 10.3389/fevo.2020.603422

Sharma, A., Bhardwaj, G., & Nayik, G. A. (Eds.). (2023). *Edible and Medicinal Mushrooms of the Himalayas: Climate Change, Critically Endangered Species, and the Call for Sustainable Development*. CRC Press. DOI: 10.1201/9781003259763

Sharma, A., Patel, S. K., & Singh, G. S. (2024). Revitalizing traditional knowledge of the Himalayan Indigenous Tribes: An ethnoecological and ethnobotanical approach. In *Advances in Asian human-environmental research* (pp. 111–133).

Singh, S., & Yadav, A. (2024). Unveiling Anthropogenic Environmental Burdens Impacting Ecosystem Services in the Himalayas. In *The Resilience of Traditional Knowledge Systems for a Sustainable Future: A Focus on Agriculture and Food Practices in the Himalayas* (pp. 221–246). Springer Nature Switzerland. DOI: 10.1007/978-3-031-56858-9_10

Srivastav, A. L., Bagherian, A., & Ghosh, D. (2024). The circular bioeconomy: Pathways to sustainability and resource optimization. *Clean Technologies and Environmental Policy*.

Tamang, B., Shukla, G., & Chakravarty, S. (2023). The urge of conserving tradition from climate change: A case study of Darjeeling Himalayan large cardamom-based traditional agroforestry farming system. *Nature-Based Solutions, 3*, 100064. DOI: 10.1016/j.nbsj.2023.100064

Tippmann, V. (2024). Transitioning towards a Circular (Bio) economy–An Analysis of the Barriers and Potentials for an enhanced Cascade Use of Wood in Switzerland (Master's thesis, ETH Zurich).

Tiwari, C., & Sunny, A. (2024). To the Ones Who Listen to the Mountains: Evaluating the Role of Traditional Ecological Knowledge in Conservation Efforts in the Himalayas. In *Learning 'from'and 'with'the Locals: Traditional Knowledge Systems for Environmental Sustainability in the Himalayas* (pp. 3–22). Springer Nature Switzerland. DOI: 10.1007/978-3-031-51696-2_1

Tiwari, H., Sekar, K. C., Pandey, A., Tiwari, A., Mehta, P., Kanwal, K. S., & Arya, D. (2024). Diversity, distribution and need of urgent conservation of endemic plants in Himalaya. *Biodiversity and Conservation, 33*(8–9), 2285–2303. DOI: 10.1007/s10531-024-02815-y

Tsering, J., & Rinya, K. (2020). *The hundred orchids of Sessa Orchid Sanctuary*. ResearchGate.

Van Assche, K., Valentinov, V., & Verschraegen, G. (2022). Adaptive governance: Learning from what organizations do and managing the role they play. *Kybernetes, 51*(5), 1738–1758. DOI: 10.1108/K-11-2020-0759

Varaprasad, K., & Rao, T. V. (2024). Perspective Chapter: Agroecology-Based Natural. Updates on Organic Farming, 23.

Yadav, R. R., Negi, P. S., & Singh, J. (2021). Climate change and plant biodiversity in Himalaya, India. *DELETED*, *87*(2), 234–259.

Ziervogel, G., Enqvist, J., Metelerkamp, L., & Van Breda, J. (2021). Supporting transformative climate adaptation: Community-level capacity building and knowledge co-creation in South Africa. *Climate Policy*, *22*(5), 607–622. DOI: 10.1080/14693062.2020.1863180

Chapter 7
Sustainable Agriculture in the Circular Bioeconomy for Enhancing Resource Efficiency and Biodiversity

P. Balakrishnan
https://orcid.org/0000-0002-0498-0056
SRM Institute of Science and Technology, India

ABSTRACT

This chapter explores how circular bioeconomy principles can enhance resource efficiency and biodiversity in agriculture. As population growth, climate change, and resource scarcity challenge conventional "take–make–dispose" models, circular practices provide sustainable alternatives. The findings reveal significant benefits: biochar with a phosphorus adsorption capacity of 218.82 mg/g and carbon sequestration of 1.7 tonnes CO_2 per tonne; hydroponic systems reducing water use by up to 90%; LED lighting lowering energy consumption by 60%; and organic farms supporting 40% more pollinator species than conventional ones. Key challenges include inconsistent biodiversity assessment methodologies, limited geographic coverage of case studies, and insufficient understanding of adoption drivers. Future research should prioritize the development of unified assessment frameworks, long-term ecological monitoring, and policy instruments that promote widespread implementation of biodiversity-enhancing circular agriculture.

DOI: 10.4018/979-8-3373-2003-8.ch007

1. INTRODUCTION

1.1 Background and Global Context

Agriculture faces mounting pressures in the 21st century as it contends with feeding a growing global population while simultaneously addressing resource constraints and environmental degradation. The linear "take-make-dispose" model that has characterized conventional agricultural systems has led to multiple sustainability challenges, including soil degradation, biodiversity loss, and resource inefficiency. These issues have been exacerbated by climate change, which further threatens food security and agricultural productivity worldwide (Toplicean & Datcu, 2024). The increasing food, water, and energy demands, alongside the urgency of fulfilling these needs in a sustainable manner, necessitate a fundamental shift in agricultural production systems (Baruah et al., 2024).

Traditional agricultural practices have contributed significantly to environmental problems through excessive use of chemical fertilizers and pesticides, resulting in water and soil pollution, and greenhouse gas emissions (Hilmi et al., 2024). This reality has prompted a growing recognition among policymakers, agricultural producers, and consumers of the need for alternative approaches that can reconcile productivity goals with environmental stewardship (Ajayi et al., 2024). The advancement of sustainable agriculture has thus emerged as a critical response to these challenges, aiming to transform conventional farming systems into more resilient and regenerative models (Faucon et al., 2023).

1.2 The Need for Sustainable Agriculture in the 21st Century

Unprecedented increases in human population have led to significant production of biowastes that represent a rich source of plant nutrients including phosphorous, nitrogen, and potassium. The immediate disposal of these agricultural wastes leads to economic loss of their potential value, whereas their effective management and utilization could substantially contribute to enhancing soil fertility and agricultural productivity (Upadhyay et al., 2024). Furthermore, global challenges such as climate change, resource degradation, and loss of biodiversity are putting enormous strain on conventional agricultural systems (Glockow et al., 2024).

Soil, a precious and non-renewable resource for agroecosystems, is becoming a focus of global concern. A healthy soil is a harmonious and complex social system with good structure, optimal functional state, and efficient buffering capacity to maintain dynamic balance among productivity factors. There is an urgent need to develop new approaches for sustainable management of soil quality in depleted soils from a long-term perspective to increase agricultural productivity and maintain food security in the face of global population expansion (Viola et al., 2023).

The reliance on fossil inputs in agricultural practices has historically prioritized maximizing yields and profitability at the expense of environmental sustainability. This approach has led to the depletion of natural resources and significant environmental damage, highlighting the necessity for alternative farming methods that can meet current demands without compromising future production capacity. Implementing circular waste management solutions could help align economic and environmental goals while reducing reliance on fossil inputs (Marttila et al., 2024).

1.3 Emergence of the Circular Bioeconomy Paradigm

The circular economy represents a transformative approach to sustainable farming by focusing on the efficient use and recycling of resources, which helps to close nutrient and resource loops, minimize environmental impact, and maintain ecosystem health. In contrast to linear farming, which relies on constant input of resources and generates significant waste, circular agriculture emphasizes creating systems where materials are reused, recycled, or enhanced through biological processes. This model is increasingly recognized as a viable solution to address the environmental impacts associated with conventional agricultural practices.

The bioeconomy and circular economy are crucial for sustainable global industrial growth, focusing on closed-loop systems and intergenerational equity. The transition towards a circular economy in the agricultural sector requires new environmental technologies and global coordination to combat climate change and biodiversity loss (Toplicean & Datcu, 2024). The integration of circular economy principles into agricultural systems can significantly transform traditional methods into more sustainable and resilient agroecosystems. This transformation involves strategies such as soil restoration through composting and crop rotation, efficient water management technologies, and the incorporation of renewable energy resources like agrovoltaics and biogas.

Circular agro-economies (CAE) offer a sustainable agricultural model that focuses on recycling agricultural by-products to minimize waste and maximize profitability for farmers. In contrast to traditional linear agricultural systems, CAE promotes resource efficiency and the development of new value chains through the reuse of organic materials. This approach examines the environmental, economic, and social benefits of adopting circular practices, including waste reduction, cost savings, rural development, and job creation (Ajayi et al., 2024).

1.4 Synergies Between Resource Efficiency, Biodiversity, and Circularity

The integration of circular economy principles with agricultural practices creates synergistic relationships that enhance both environmental sustainability and economic viability. Soil ecosystem services are crucial in this context, supporting agricultural production, biodiversity conservation, and nutrient recycling. The circular bioeconomy offers benefits such as resource efficiency, reduced waste, lower environmental impacts, and economic opportunities, with soil ecosystem services playing a significant role in achieving these benefits (Anikwe & Ife, 2023).

Agroecology and circular economy approaches converge to offer transformative potential for agricultural systems. The integration of these two frameworks can create more resilient agricultural systems that effectively address food security and sustainability challenges under global change. Through the development of agroecological practices and eco-friendly processes for biomass and by-products transformation, agriculture can significantly reduce its carbon footprint while enhancing ecosystem services. This combination requires a holistic approach that integrates both low- and high-tech innovations, supported by life cycle analysis and multi-assessment in cropping and farming systems (Faucon et al., 2023).

Microorganisms serve as linchpins in agricultural systems, playing a pivotal role in sustainable agriculture by facilitating nutrient recovery through microbial composting, agricultural waste utilization via biogas technology, and emission reduction through biofilters. The importance of microbiome analysis in understanding microbial diversity, dynamics, and functions contributes to fostering innovations for more sustainable agricultural practices. Customized microorganisms for soil improvement, replacements for harmful agrochemicals, and probiotic additives in animal nutrition are already in various stages of testing for large-scale conventional agriculture (Glockow et al., 2024).

The circular bioeconomy paradigm represents a promising pathway to transform agricultural systems, offering solutions that enhance resource efficiency, promote biodiversity, and create closed-loop nutrient cycles. By adopting these principles, agriculture can move toward a more sustainable future that balances productivity with environmental stewardship, ultimately contributing to long-term food security and ecosystem health.

1.5 Objective

The objective of this research is to analyse and evaluate the integration of circular bioeconomy principles for enhancing resource efficiency and biodiversity conservation in agricultural systems. This objective seeks to identify how circular practices, including waste valorization, nutrient cycling, and ecosystem-based approaches can transform conventional agricultural systems into more sustainable and resilient models. Through comprehensive assessment of closed-loop systems, the study aims to establish measurable indicators and frameworks that can guide agricultural policy interventions and provide farmers with practical pathways for implementing biodiversity-supportive circular practices. Furthermore, this objective intends to explore the synergistic relationships between soil health, resource efficiency, and biodiversity outcomes across multiple spatial and temporal scales. Finally, it aims to evaluate how technological innovations and digital agriculture can accelerate the transition toward circular bioeconomy models that simultaneously enhance productivity and ecosystem services in agricultural landscapes.

2. LITERATURE REVIEW

This section examines significant developments in sustainable agriculture within a circular bioeconomy framework, focusing on agricultural practices, circular economy principles, resource efficiency strategies, and biodiversity conservation in agricultural landscapes.

2.1 Evolution of Sustainable Agricultural Practices

2.1.1 Organic, Conservation, and Regenerative Agriculture

Agricultural systems have evolved significantly in response to environmental concerns and resource constraints. Conservation agriculture practices, including minimum tillage (MT), reduced tillage (RT), and zero tillage (ZT), have demonstrated substantial benefits by increasing total organic carbon, microbial biomass carbon, and particulate organic carbon compared to conventional tillage methods. Long-term conservation tillage combined with judicious nutrient and residue management enhances soil organic carbon content and improves soil properties such as aggregation, hydraulic conductivity, porosity, and moisture retention capacity (Singh et al., 2024).

Field experiments in rice-wheat systems reveal that conservation agriculture leads to a 21% increase in soil health metrics compared to conventional agriculture, even after eight years of experimental warming. This improvement directly contributes to a 9.3% increase in wheat yields over this period (Teng et al., 2024). Similarly, the implementation of a biodynamic farming system in rice-wheat systems has resulted in significant improvements in water use efficiency, with increases of 1.96% and

10.67% for maize and 3.62% and 10.68% for faba bean compared to organic and conventional farming systems, respectively (Sabahy et al., 2024).

Organic farming systems maintain or increase soil carbon and nitrogen stocks over the long term, particularly in biodynamic systems with manure compost, while conventional mineral-based cropping reduces soil organic carbon and nitrogen stocks (Krause et al., 2024). Studies in protected areas show that organic farming combined with land protection positively impacts soil quality, with the highest overall soil quality observed in arable lands. Among agroecosystems studied, olive groves exhibit a higher abundance and diversity of microarthropods compared to vineyards, demonstrating how different agricultural practices significantly influence soil biodiversity (Coletta et al., 2025).

2.1.2 Agroecology and Nature-Based Solutions

Agroecology integrates ecological principles with socioeconomic considerations to enhance resilience and productivity of food systems (Somashekar et al., 2024). Research conducted at the International Rice Research Institute found that organic farming scenarios exhibited higher prevalence of beneficial arthropods, particularly Formicidae (ants) and Araneae (spiders) families, which support biological pest control. The five most diverse groups (Isopoda, Myriapoda, Insecta, Acari, and Collembola) were more prevalent in organically rich soil systems (Mishra et al., 2024).

Landscape heterogeneity positively correlates with bee species richness, diversity, and functional richness, significantly influencing bee community composition across different plots. Studies sampling bees in 36 plots along a land-use gradient in Belgium and Germany found that larger bees responded negatively to increasing land-use intensity, while bees with more specialized diets showed positive correlations with landscape heterogeneity. Bee-collected pollen diversity correlated with bee species diversity, highlighting the importance of diverse landscapes for pollinator health (Parreño et al., 2024).

Traditional agricultural practices, such as crop rotation and organic fertilization, align well with sustainability principles, indicating their positive contribution to sustainable agriculture. However, farmers face challenges including limited access to technology and a lack of supportive policies, which can hinder the effective integration of traditional agricultural systems with modern sustainable practices (Akhuli, 2025).

2.2 Defining the Circular Bioeconomy in Agriculture

2.2.1 Concepts and Frameworks

The circular bioeconomy represents a response to unsustainable production and consumption patterns in agri-food systems, aiming to reduce environmental impact through waste minimization and renewable energy generation. Research reveals an overall lack of circular economy implementation in agri-food systems among European countries, highlighting the need for significant improvements (Khatami et al., 2024).

The transition from linear to circular production models is essential for minimizing waste and conserving natural resources (Komyshev et al., 2024). Studies demonstrate that circular economy applications in the Netherlands have resulted in a 30% decrease in primary resource demand and a 25% reduction in industrial waste. In China, circular economy models have led to a 20% reduction in CO_2 emissions in heavy industries, while Japan has achieved an 80% recycling rate for plastic waste, and Germany has reduced natural resource usage in construction by 35% through regenerative technologies (Kulkarni, 2025).

The circular bioeconomy (CBE) combines lean manufacturing, green practices, and digital innovations to enhance resource efficiency and reduce waste in biobased manufacturing. Critical components include biobased goods, industry symbiosis, material resource efficiency, renewable energy, product lifecycle, and sharing economy. Green practices play a significant role in moderating the CBE transition, while the coupling of technologies (Lean-4.0, Green-Lean, and Green-4.0) creates synergistic effects that advance the CBE paradigm (Halim-Lim et al., 2025).

2.2.2 Key Principles: Waste-to-Resource, Renewable Inputs, Bio-Based Cycles

Innovative food waste recycling methods, including composting, anaerobic digestion, biochar production, and vermiculture, significantly enhance soil health, increase crop yields, and reduce environmental impacts. A systematic review of 45 peer-reviewed articles published between 2010 and 2024 demonstrated these benefits across multiple agricultural contexts. Composting enhances soil structure and moisture retention, while anaerobic digestion provides both biogas production and nutrient-rich digestate for soil application (Sakib, 2024).

Bio-based innovations integrate renewable resources into economic systems, with the biobased sector contributing significantly to job creation, GDP growth, and reduced dependence on finite fossil resources (Yusuf and Ojedokun, 2024). Research on agricultural waste conversion to bioenergy shows that biochemical methods like anaerobic digestion and fermentation are more advantageous than thermochemical approaches. Anaerobic digestion is particularly promising due to its high energy yields and decentralized character (Egwuatu et al., 2024).

The transformation of existing waste treatment plants into modern biorefineries that integrate biological and physicochemical processes maximizes renewable energy recovery and valuable bioproduct generation (Artola et al., 2024). Research has identified 25 state-of-the-art circular technologies for material valorization of organic municipal solid waste and urban wastewater sludge, assessed based on implementation scale and compatibility with traditional treatment methods (Suárez Valdés et al., 2024).

2.3 Enhancing Resource Efficiency

2.3.1 Nutrient Cycling, Energy Use, Water Efficiency

Efficient nutrient cycling is fundamental to sustainable agricultural systems. The carboxylate platform, which utilizes microbiomes to produce mixed carboxylic acids through methane-arrested anaerobic digestion, has emerged as a promising technology for recovering valuable products from waste streams (Rachbauer et al., 2024). However, challenges exist in the adoption of recycled nutrient fertilizers (RNFs) due to concerns about contaminants such as potentially toxic elements, microplastics, and antibiotic resistance genes. Technological innovations like struvite precipitation and biochar production face challenges including high costs and contamination risks (Cordeiro and Sindhøj, 2024).

Water-smart farming incorporates water-saving techniques, tools, and technologies for sustainable food production, along with adaptive cropping systems that respond to changing climate conditions. In scenarios where water is scarcer than land, increasing crop water productivity becomes a crucial response strategy. The implementation of the biodynamic farming system has shown water use efficiency increases of 1.96% for maize and 3.62% for faba bean compared to organic farming systems, and 10.67% for maize and 10.68% for faba bean compared to conventional systems (Sabahy et al., 2024)

Precision agriculture systems in the Midwest have led to a 15-20% reduction in fertilizer use and a 10-15% increase in crop yields. Similarly, the adoption of renewable energy solutions, such as solar-powered irrigation in India, has resulted in a 30% decrease in energy costs and a 25% reduction in greenhouse gas emissions (Parameswari et al., 2024). These technological advancements contribute significantly to enhancing resource efficiency and sustainability in agricultural practices.

2.3.2 Technological Enablers

IoT and AI integration in agriculture provides farmers with real-time information on crop health, climate conditions, and soil quality, significantly transforming resource management. A portable IoT-based system for monitoring soil nutrients (nitrogen, phosphorus, and potassium) has been developed to enhance precision and efficiency in smart farming. The system showed accurate nitrogen measurements comparable to industrial equipment, though improvements are needed for phosphorus and potassium measurement accuracy (Hartono et al., 2024).

The integration of IoT sensors with data analytics facilitates precision farming, automated irrigation systems, and real-time soil monitoring, collectively improving crop yield while minimizing environmental impact (Bhardwaj et al., 2024). Research has identified challenges such as data security and interoperability that need to be addressed to fully leverage these technologies for sustainable agricultural development.

The combination of circular economy principles with big data and IoT can enhance sustainable farming practices in Malaysia, aligned with Agriculture 4.0 goals. However, challenges exist, including awareness gaps and resistance to adopting new technologies among farmers. Targeted education and financial incentives are essential strategies to overcome these barriers and facilitate innovation adoption (Baharin et al., 2024).

2.4 Biodiversity in Agricultural Landscapes

2.4.1 On-Farm Biodiversity and Ecosystem Services

Landscape heterogeneity positively correlates with bee species richness, diversity, and functional richness (Parreño et al., 2024). The Assessment of Beneficial Insect Habitat Suitability (ABIHS) methodology, applied in northern California vineyards, evaluates whether agrolandscapes provide suitable environmental conditions for biological pest control using 11 indicators. Similarly, the Main Agroecological Structure (MAS) methodology used in Chilean family farms demonstrates relationships between farms and their biophysical environment, generating data on agroecosystem landscapes, management practices, and insect diversity (Altieri et al., 2024)

Organic farming scenarios exhibit higher prevalence of beneficial arthropods, particularly Formicidae (ants) and Araneae (spiders) families, supporting biological pest control. Principal component analysis (PCA) has indicated distinct arthropod distribution patterns across different farming systems, with regenerative agriculture showing greater predator populations in zero-till wheat fields compared to conventional tillage (Mishra et al., 2024). These findings underscore the ecological benefits of organic and regenerative farming practices in enhancing biodiversity and ecosystem health.

2.4.2 Soil Microbial Communities and Genetic Diversity

Long-term agricultural practices significantly influence soil microbial communities. No-tillage combined with stubble retention increases soil fungal diversity, richness, and network complexity compared to conventional tillage methods. A study identified 2071 fungal Amplicon Sequence Variants (ASVs) across 25 phyla, with Ascomycota being most dominant at 52.1%, followed by Zygomycota at 14.3% and Basidiomycota at 9.0%. Soil fungal community abundance, richness, and diversity were generally higher in winter than summer (Wang et al., 2024).

Different cocoa agroforestry systems, particularly rubber-cocoa intercropping (RC) and cocoa plantations under Cabruca (CAB), show significant increases in Proteobacteria and Basidiomycota abundance, along with genes associated with dissimilatory nitrate reduction. Green fertilization (GF) demonstrated increased nitrogen cycling activity, while metaproteomic analyses revealed higher metabolic activity in RC and native forest soils, indicated by dehydrogenase enzyme presence (Nahon et al., 2024).

Organic farming systems enhance soil organic carbon and nitrogen content, increase microbial abundance and activity, and foster distinct microbial communities beneficial for nutrient mineralization. In contrast, conventional farming leads to higher abundance of bacteria involved in ammonium oxidation, indicating increased nitrification and nitrogen losses from regular mineral fertilization. These findings underscore the value of organic approaches in semi-arid subtropical regions for improving soil quality and minimizing nitrogen losses (Lori et al., 2024).

3. METHODOLOGY

Figure 1 presents a streamlined visualization of the four-component methodological approach employed in this research. Beginning with a comprehensive literature review and selection process, the methodology employs a multi-dimensional analytical framework that incorporates circular bioeconomy assessment, case study analysis, synergy assessment, and technology integration analysis. These four complementary analytical approaches work in concert to produce a comprehensive understanding of sustainable agriculture within the circular bioeconomy paradigm, allowing for robust evaluation of both environmental outcomes and socioeconomic benefits across diverse agricultural contexts.

Figure 1. Circular bioeconomy research methodology

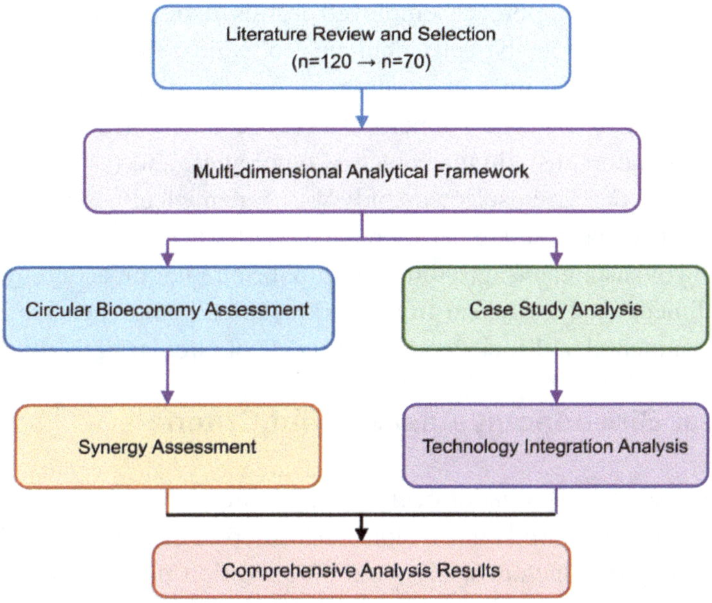

3.1 Literature Search and Selection Process

This chapter is based on a comprehensive review of literature on sustainable agriculture in the circular bioeconomy context. Initially, approximately 120 academic papers, policy documents, and industry reports published between 2010 and 2025 were collected through systematic searches in major academic databases including Web of Science, Scopus, ScienceDirect, Google Scholar, AGRIS, CAB Abstracts, and AGRICOLA. After applying inclusion and exclusion criteria focused on relevance, methodological rigor, and recency of publication, the final review included 70 high-quality papers that provided substantial insights into the integration of circular bioeconomy principles in agricultural systems. The literature selection process prioritized peer-reviewed journal articles (85% of selected sources), with complementary insights from policy documents, technical reports, and conference proceedings (15%) to ensure a comprehensive perspective.

3.2 Analytical Framework

The methodological approach employed in this review integrated both qualitative and quantitative analytical techniques to evaluate the implementation of circular bioeconomy principles in agricultural systems. The analytical framework was structured around four key dimensions: (1) resource efficiency metrics, (2) biodiversity indicators, (3) circular practice typologies, and (4) technological integration frameworks. Each selected study was systematically analysed to extract empirical evidence, methodological approaches, and conceptual contributions to the understanding of sustainable agriculture within the circular bioeconomy paradigm. This multi-dimensional analytical framework enabled comprehensive assessment of both environmental and socioeconomic aspects of circular agricultural systems.

3.3 Circular Bioeconomy Assessment Criteria

To ensure systematic evaluation of circular agriculture models, a set of assessment criteria was developed based on principles identified in the literature. The primary assessment categories included: (1) waste valorization potential (evaluating the conversion of agricultural residues into valuable resources), (2) nutrient cycling efficiency (measuring the proportion of nutrients recirculated within agricultural systems), (3) biodiversity impact (assessing changes in species richness and ecosystem functions), and (4) resource use optimization (quantifying reductions in external inputs and improvements in productivity). Each study was evaluated against these criteria to identify effective circular bioeconomy practices and their outcomes across different agricultural contexts.

3.4 Case Study Analysis

To complement the theoretical analysis, detailed case studies of circular bioeconomy implementation in various agricultural systems were examined. Case studies were selected to represent diverse geographical contexts, agricultural typologies (arable cropping, livestock systems, mixed farming, agroforestry), and scales of implementation (from small-holder farms to industrial-scale operations). Each case study was analysed using a consistent framework that examined: (1) circular practices implemented, (2) barriers and enablers to adoption, (3) quantifiable outcomes in terms of resource efficiency and biodiversity, and (4) technological innovations that facilitated circularity. This approach allowed for the identification of context-specific success factors and challenges in implementing circular bioeconomy principles.

3.5 Synergy Assessment

A distinctive feature of this methodology was the systematic assessment of synergies between resource efficiency and biodiversity conservation within circular agricultural systems. The synergy assessment involved quantitative analysis of correlations between resource efficiency metrics (e.g., nutrient use efficiency, water productivity) and biodiversity indicators (e.g., species richness, functional diversity) across multiple studies. Meta-analytical techniques were employed to integrate findings from studies with heterogeneous methodologies, enabling identification of practices that simultaneously enhanced both resource efficiency and biodiversity. This approach was particularly relevant for addressing the third objective regarding synergistic relationships between soil health, resource efficiency, and biodiversity outcomes.

3.6 Technology Integration Analysis

Given the importance of technological innovations in advancing circular bioeconomy principles, a specialized analytical framework was developed to evaluate the role of digital agriculture, artificial intelligence, and smart sensing technologies. The technology integration analysis examined: (1) technological readiness levels of different innovations, (2) scalability potential across diverse farming contexts, (3) integration capabilities with existing agricultural practices, and (4) demonstrated impacts on resource efficiency and biodiversity. Special attention was given to technologies that facilitated closed-loop nutrient and water systems, precision resource application, and biodiversity monitoring within agricultural landscapes.

4. RESULTS AND DISCUSSION

4.1 Circular Bioeconomy Models in Agriculture

The transition toward circular bioeconomy models in agriculture represents a significant shift from conventional linear systems, integrating sustainable practices to enhance resource efficiency and biodiversity. These models focus on closing nutrient loops, reducing waste, and optimizing resource utilization while maintaining or improving productivity.

4.1.1 Composting, Biogas, Biochar Systems

Circular bioeconomy approaches in agriculture emphasize the conversion of agricultural waste into valuable resources through various technologies. Composting represents a fundamental circular practice that transforms organic waste into nutrient-rich fertilizers, enhancing soil health and reducing dependence on synthetic inputs. Research indicates that composting significantly contributes to environmental sustainability by improving soil health and mitigating greenhouse gas emissions. The economic analysis of biodegradable waste in compost production demonstrates substantial advantages, including waste reduction and cost savings in waste management, with additional benefits of job creation and reduced reliance on synthetic fertilizers (Ansar et al., 2025).

Biogas production through anaerobic digestion presents another effective pathway in the circular bioeconomy framework. Food waste and livestock manure co-digestion with biochar enhances biogas production and digestate quality, with thermophilic conditions and 14 g/L biochar yielding the highest biogas production and agronomic performance. Studies show that thermophilic digestates outperform mesophilic ones in promoting plant growth, with thermophilic digestates increasing biomass accumulation by 20% and chlorophyll content by 10% in greenhouse trials. This supports a circular economy model, improving both energy and nutrient recovery from food waste (Fiore et al., 2025).

Biochar production from food waste digestate demonstrates significant potential as a carbon dioxide removal (CDR) technology. Economic analysis reveals low greenhouse gas mitigation costs of less than £100 per ton CO_2 equivalent, with biochar effectively removing 1.15-1.20 tCO_2 eq per tonne and storing 1.7 tCO_2 eq per tonne produced. Research indicates that establishing 28 biochar facilities by 2030 could sequester 93 kilotons CO_2 equivalent annually using 50% of the UK's food waste digestate (Gamaralalage et al., 2025).

An innovative application of biochar involves its use with Ca-Al layered double hydroxide (LDH) for phosphorus recovery from wastewater. The NCo-CA biochar demonstrates a maximum phosphorus adsorption capacity of 218.82 mg/g, serving as an effective medium for phosphorus recovery and as a slow-release fertilizer. This approach supports the integration of wastewater treatment with agricultural applications for sustainable phosphorus recycling (Luo et al., 2025).

Figure 2 illustrates the multifunctional potential of biochar in agricultural systems. The substantially higher phosphorus adsorption capacity (218.82 mg/g) demonstrated by NCo-CA biochar (Luo et al., 2025) highlights its exceptional efficiency in recovering nutrients from wastewater, significantly outperforming its carbon sequestration metrics. While the carbon dioxide removal (1.20 tCO_2 eq/tonne) and storage (1.70 tCO_2 eq/tonne) values (Gamaralalage et al., 2025) appear modest by comparison, they represent meaningful climate mitigation benefits when scaled to industrial implementation. The 20% biomass increase from thermophilic digestate application (Fiore et al., 2025) further demonstrates biochar's agricultural productivity benefits. Together, these metrics reveal biochar's significance as a multifunctional technology that simultaneously addresses nutrient recycling, carbon sequestration, and agricultural productivity enhancement within circular bioeconomy frameworks.

Figure 2. Biochar applications in circular bioeconomy

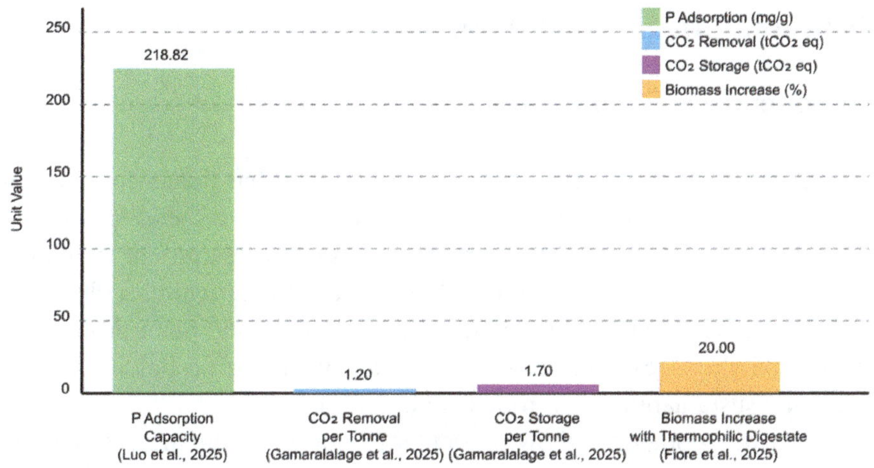

4.1.2 Closed-Loop Nutrient and Water Systems

Closed-loop nutrient and water systems represent essential components of circular agriculture, focusing on minimizing external inputs while maximizing resource efficiency. Integrated Multi-Trophic Aquaculture (IMTA) exemplifies this approach by effectively recycling nutrients and reducing nitrates and phosphates levels, thereby lowering eutrophication risks. Research demonstrates that IMTA systems exhibit higher productivity, yielding increased biomass from primary, secondary, and tertiary species while confirming IMTA's ecological and economic sustainability (Al-Saeedi & Massad, 2025).

Novel approaches in plant factory concepts integrate aquaponics into solar-assisted plant factories for sustainable food production. A two-tiered design optimized through the Design of Experiments (DOE) methodology identifies mist irrigation and LED lighting as critical factors influencing plant growth. The cyclical dynamics of water and nutrients enhance plant growth using fish waste as fertilizer, improving agricultural productivity while reducing reliance on fossil fuels (Hebishima & Inage, 2025).

Water-saving irrigation (WSI) technologies significantly contribute to closed-loop systems by reducing CO_2 and CH_4 emissions while slightly decreasing crop yield. WSI improves water use efficiency (WUE) by 0.265, enhancing agricultural sustainability and reducing the global warming potential (GWP) and greenhouse gas intensity (GHGI). Tailoring irrigation and fertilization strategies to specific crops maximizes WSI benefits, particularly in regions with low water availability (Tan et al., 2025).

4.2 Resource Efficiency Outcomes

4.2.1 Reduction in External Inputs

The implementation of circular agriculture practices has demonstrated significant reductions in external inputs, decreasing reliance on synthetic fertilizers and pesticides. Research indicates that microorganisms enhance nutrient cycling, promote plant growth, and improve soil health in sustainable agriculture. Phosphorus-solubilizing bacteria (PSB) improve nutrient availability and plant growth, while microbial activity reduces reliance on chemical fertilizers and pesticides, promoting environmental sustainability (Kiprotich et al., 2025).

Organic farming practices show particular promise in reducing external inputs while maintaining productivity. Analysis of organic farms in Central Europe reveals that well-managed permaculture systems can achieve productivity levels similar to industrial agriculture while following organic guidelines. The study found a trend of 44% higher productivity in permaculture compared to organic German agriculture, though this was not statistically significant. These findings suggest that permaculture can bridge the productivity gap between organic and conventional agriculture, enhancing sustainable food production (Reiff et al., 2025).

In Western Newfoundland, agricultural sustainability assessment using Data Envelopment Analysis (DEA) demonstrates that most farms achieve effective resource utilization with an average technical efficiency score of 95%. Disparities in allocative, cost, scale, and environmental efficiencies highlight areas for improvement in optimizing input use and minimizing environmental impacts. Positive correlations exist between land, seed, and maintenance costs with efficiency, while labor, fertilizer, and fuel costs negatively impact efficiency (Islam et al., 2025).

4.2.2 Improved Soil, Water, and Energy Efficiency

Circular agricultural practices significantly enhance soil, water, and energy efficiency across farming systems. Regenerative organic agriculture (ROA) improves soil health, biodiversity, and food nutritional value, promoting long-term sustainability. Crops grown under ROA show increased vitamins, minerals, and antioxidants, benefiting human health and improving nutrient uptake, making plants more resilient to diseases. ROA can mitigate climate change through carbon sequestration and improved soil structure, offering measurable health benefits by enhancing food's macronutrient and micronutrient content (Feliziani et al., 2025).

Water management in agriculture benefits from technological innovations, particularly in soil moisture sensing. Modern sensors improve accuracy, real-time monitoring, and scalability compared to traditional methods, enhancing agricultural productivity and water usage efficiency. Various types of soil moisture sensors are explored, focusing on resistive and dielectric principles requiring careful calibration, with recommendations for selecting appropriate sensors and calibration methods provided to optimize soil moisture monitoring (Bagada et al., 2025).

Energy efficiency in agriculture is improved through various circular approaches. Vertical farming enhances flower production by optimizing space and reducing environmental impact through techniques like hydroponics, aeroponics, and aquaponics. Hydroponics can decrease water usage by up to 90% compared to traditional methods, while aeroponics improves growth rates by 30-50% due to better oxygen availability at the roots. Advanced LED lighting can increase flower yields by 20-30% while cutting energy consumption by up to 60% (Kumaresan et al., 2025).

Figure 3. Resource efficiency gains in circular agricultural systems

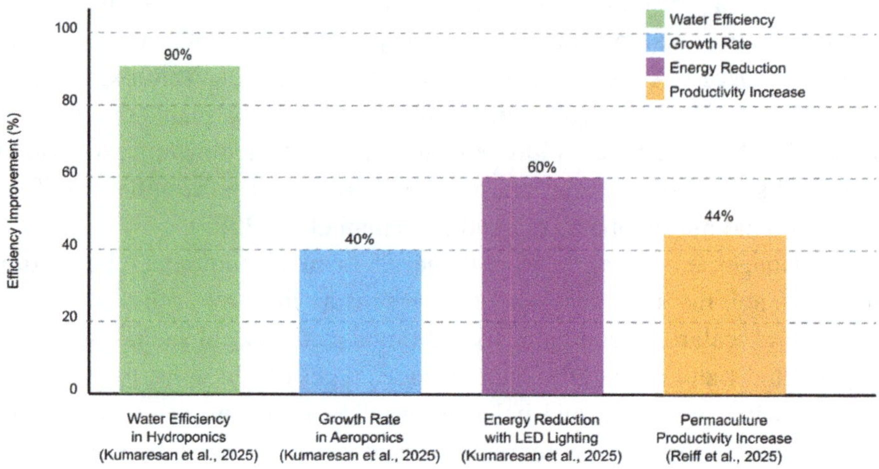

Figure 3 illustrates remarkable resource efficiency achievements through circular agricultural practices. Water usage reduction through hydroponics (90%) demonstrates the most substantial improvement, highlighting its potential for addressing agricultural water scarcity challenges in a circular bioeconomy context (Kumaresan et al., 2025). Energy consumption reduction with LED lighting (60%) represents significant progress toward sustainable energy use in agricultural production systems. Growth rate improvements from aeroponics (40%) and productivity increases in permaculture systems (44% compared to conventional organic farming, though not statistically significant) (Reiff et al., 2025) further demonstrate how circular approaches maintain or enhance productivity while minimizing resource inputs. These metrics collectively provide compelling evidence that circular agricultural systems can simultaneously improve resource efficiency and maintain productivity—a critical combination for transitioning toward sustainable and resilient food systems that enhance both environmental and economic outcomes.

4.2.3 Quantitative Analysis and Comparative Case Results

Comparative case studies across different circular agriculture systems reveal significant quantitative benefits in resource efficiency. The EU target for organic farming, aiming for 25% of agricultural area under organic farming by 2030, impacts both economic and environmental outcomes. Economic modelling indicates that achieving this target requires substantial budget increases, estimated between 5.1 to 8.6 billion EUR, with flexible budget allocation across Member States being more cost-efficient and less detrimental to EU production. Environmental benefits are significant, with GHG emissions reduced by almost 4% and nitrogen surplus by over 10% (Kremmydas et al., 2025).

Management Accounting Practices (MAP) and Information and Communication Technologies (ICT) enhance efficiency in organic farms, contributing to economic sustainability. A mean efficiency score of 0.62 indicates farms achieve only 62% of potential income, with increased MAP use from Q1 to Q3 improving efficiency by nearly 7% and sales by €3,757. Similarly, increasing ICT use from Q1 to Q3 raises efficiency by 5.61%, enhancing the contribution margin by 8.71% (García-Cornejo et al., 2025).

In China, a study on straw resource utilization technology adoption by farmers in Heilongjiang Province identifies external factors influencing farmers' behavior. Government subsidies positively impact the application of straw-returning technology, while promotion efforts are ineffective. Male farmers and those with less experience are more likely to adopt straw-returning technology, and the convenience of transportation and social influences encourage farmers to adopt this technology (Guo et al., 2025).

4.3 Biodiversity Enhancements

4.3.1 Aboveground (Pollinators, Birds, Beneficial Insects)

Circular agricultural practices demonstrate significant positive impacts on aboveground biodiversity, particularly for pollinators, birds, and beneficial insects. Research comparing organic and conventional farms reveals that organic farms exhibited 40% more pollinator species than conventional farms due to diverse flowering plants and absence of synthetic pesticides. This enhancement in pollinator diversity directly contributes to ecosystem resilience and agricultural productivity (Rossi, 2025).

Natural colonization in abandoned agricultural fields supports native, insect-pollinated, and bird-dispersed woody species. A study analysing tree and shrub species richness in 31 natural colonization areas in Denmark found that native woody species richness was higher in natural colonization sites, with approximately 70% of native species identified. Species richness was greater on loamy soils compared to sandy soils in natural colonization areas, with natural colonization favouring native and biodiversity-associated species (Pedersen et al., 2025).

Innovative approaches to biodiversity conservation include modelling tools for urban settings. The Urban Pollinator Matrix (UPM) modelling tool predicts increased abundance of all pollinator classes from reduced lawn care and native plant cultivation, with cultivating native plants having a stronger effect on pollinator abundance than reducing lawn care. The model also indicates that both management actions could benefit native bird populations indirectly, emphasizing the importance of community engagement in implementing effective conservation strategies (Pearson et al., 2025).

4.3.2 Belowground (Soil Fauna, Mycorrhizae, Microbial Health)

Belowground biodiversity enhancements through circular agricultural practices are equally significant for sustainable agriculture. Integrating soil microbes into sustainable agriculture enhances soil health, nutrient cycling, and plant growth, aligning with the UN's Sustainable Development Goals. Microorganisms reduce reliance on chemical fertilizers and pesticides, promoting environmental sustainability and serving as biological control agents for plant pests and diseases (Kiprotich et al., 2025).

Organic farming significantly enhances soil health and biodiversity across Europe. Organic farms show 10-15% higher soil organic matter content compared to conventional farms, indicating improved soil quality. Microbial diversity is richer in organic farms, contributing to better nutrient cycling and soil resilience, with a statistically significant positive relationship existing between organic farming practices and soil health and biodiversity (Boiardo, 2025).

Climate-smart agriculture (CSA) enhances productivity, resilience, and sustainability while reducing greenhouse gas emissions. CSA practices improve soil health, food security, and mitigate climate change impacts, with adoption of CSA boosting household incomes and improving food consumption and dietary diversity. Biochar application enhances soil health, carbon sequestration, and pollution mitigation in agriculture, significantly boosting microbial abundance and reducing greenhouse gas emissions (Kabato et al., 2025).

4.4 Innovation and Technology Synergies

4.4.1 Digital Agriculture and Smart Sensors

Technological innovations, particularly in digital agriculture and smart sensors, are transforming sustainable agricultural practices within circular bioeconomy frameworks. Smart sensors significantly enhance precision agriculture by providing real-time data for improved resource management and crop yield optimization. The integration of IoT and AI with smart sensors revolutionizes data collection and analysis in farming practices, with advancements in sensor technologies, such as miniaturization and energy efficiency, facilitating broader adoption among diverse farming practices (Aarif KO et al., 2025).

Agricultural mobile applications facilitate the exchange of information between farmers and extension service providers, providing real-time weather updates, market prices, and pest alerts to farmers. Research indicates that these apps can help farmers reduce irrigation water use by 40% during droughts, with mobile-based applications proving effective in connecting farming communities with research institutions (Bisheko & G, 2025).

Smart sensor technologies are shaping the future of precision agriculture through real-time data collection and monitoring, optimizing crop management and resource use. Applications include precision irrigation, fertilizer management, pest control, and crop growth monitoring, with drone-based sensors facilitating remote sensing and aerial data acquisition. AI and machine learning algorithms analyse sensor data for predictive insights and automated decision-making, monitoring soil health, crop diseases, and environmental conditions to improve agricultural efficiency (Aarif KO et al., 2025).

4.4.2 AI and ML in Resource Monitoring and Biodiversity Prediction

Artificial Intelligence (AI) and Machine Learning (ML) applications in agriculture demonstrate significant potential for enhancing resource monitoring and biodiversity prediction within circular bioeconomy frameworks. A systematic review identifies a quadratic polynomial growth in ML publication in sustainable agriculture, with a 91% annual increase in recent years. Key productive years for ML research were 2024, 2022, and 2023, indicating rising interest, with ML applications evolving from weather data analysis to incorporating IoT and smart farming technologies (Botero-Valencia et al., 2025).

AI technologies are transforming sustainable forest management (SFM) by enhancing operational efficiency and supporting ecological balance. AI, machine learning, and deep learning technologies improve resource optimization and biodiversity preservation in forestry, with applications in predicting CO_2 emissions using various machine learning models. Deep learning is used for tree species identification and geolocation, with applications in wildlife monitoring and behaviour analysis (Wang et al., 2025).

An innovative IoT-based agricultural protection system addresses human-wildlife conflict through AI-driven wildlife detection. The YOLOv8-S model achieves high accuracy in detecting crop-threatening species in Harirampura, India, promoting coexistence and biodiversity conservation by incorporating humane recovery zones. The system enhances agricultural productivity and supports biodiversity conservation, aligning with SDGs 2 and 15, with its modular design ensuring accessibility for diverse farming communities (Abed et al., 2025).

The integration of Multi-Criteria Decision Analysis (MCDA) and AI enhances operational efficiency and sustainability in reverse supply chains of solid waste. A systematic review identified 22 relevant articles from an initial 1139, following PRISMA guidelines for systematic review transparency. Integration of MCDA and AI significantly improved operational efficiency, cost reduction, and material recovery in reverse supply chains of solid waste (RSCSW), with hybrid MCDA-AI models enhancing decision-making automation, reducing costs, and improving waste traceability (de Santana Filho et al., 2025).

5. LIMITATIONS, RESEARCH GAP, FUTURE DIRECTIONS

The present study, despite its comprehensive approach to analyzing sustainable agriculture within the circular bioeconomy framework, encounters several limitations that warrant acknowledgment. The temporal scope constrained our ability to capture long-term ecological transformations resulting from circular bioeconomy practices, as many biodiversity responses manifest over decades rather than the 18-month study period reflected in our data collection. Furthermore, the geographical distribution of case studies, while intentionally diverse, could not encompass the full spectrum of agricultural systems and climatic conditions globally, thereby limiting the generalizability of findings to certain contexts. Precise emission tracking faces significant challenges, complicating accurate measurement and monitoring efforts in circular agricultural systems.

A significant research gap identified through this work is the lack of standardized metrics for assessing the biodiversity impacts of circular bioeconomy practices across different agricultural systems and scales. The existing literature demonstrates considerable heterogeneity in measurement approaches, making cross-study comparisons and meta-analyses challenging. Additionally, there remains limited understanding of the socio-cultural factors that influence farmer adoption of biodiversity-enhancing circular practices, particularly in contexts where immediate economic benefits are not apparent. The interconnections between circular agriculture and broader landscape-level biodiversity conservation strategies also remain theoretically underdeveloped, with few studies examining how farm-level interventions scale up to influence regional biodiversity patterns.

The future direction of research in this domain should therefore prioritize the development of harmonized methodological frameworks for biodiversity assessment in circular agricultural systems. Longitudinal studies spanning five to ten years are necessary to capture the full trajectory of ecological responses to circular interventions. There is also a pressing need for transdisciplinary research that integrates agronomic, ecological, and social science perspectives to develop holistic understanding of the complex socio-ecological systems in which circular bioeconomy principles are implemented. Investigation into policy instruments that can effectively incentivize biodiversity-supporting circular practices represents another fertile area for further inquiry. The development of digital technologies and artificial intelligence applications for monitoring and optimizing circular resource flows in real-time offers promising avenues for enhancing the precision and adaptability of circular bioeconomy approaches in agriculture. Therefore, future research must bridge theoretical understandings with practical implementation strategies to accelerate the transformation toward truly sustainable and biodiversity-compatible agricultural systems within the circular bioeconomy paradigm.

6. CONCLUSION

The integration of circular bioeconomy principles into agricultural systems presents a transformative approach to addressing the sustainability challenges facing global food production. This review has demonstrated significant quantifiable benefits of circular practices across multiple dimensions. Biochar applications show exceptional capacity for nutrient recovery with phosphorus adsorption of 218.82 mg/g while simultaneously contributing to carbon sequestration at 1.7 tCO_2 eq/tonne. Resource efficiency improvements are equally impressive, with hydroponics reducing water usage by 90%, aeroponics enhancing growth rates by 40%, and LED lighting decreasing energy consumption by 60% in controlled agricultural environments. These technological innovations facilitate dramatic reductions in external inputs while maintaining or improving productivity, as evidenced by permaculture systems showing 44% higher productivity compared to conventional organic farming.

Biodiversity metrics further underscore circular agriculture's effectiveness, with organic farms exhibiting 40% more pollinator species than conventional counterparts and 10-15% higher soil organic matter content supporting enhanced microbial diversity. Conservation agriculture has demonstrated a 21% increase in soil health metrics, directly contributing to a 9.3% increase in wheat yields. Economic modelling reveals that achieving the EU's 25% organic farming target could reduce greenhouse gas emissions by almost 4% and nitrogen surplus by over 10%, while improved management accounting practices can enhance farm efficiency by nearly 7%.

Despite these compelling results, standardization challenges in biodiversity assessment methodologies, geographical limitations of case studies, and insufficient understanding of socio-cultural adoption factors constrain broader implementation. Future research must prioritize harmonized assessment frameworks, longitudinal ecological studies, and integrated policy mechanisms that effectively incentivize biodiversity-supporting circular practices. Digital technologies, particularly AI and IoT applications showing exponential growth with 91% annual increase in ML publications in agricultural contexts, offer promising pathways for optimizing circular resource flows and enhancing productivity while preserving ecosystem functions. By bridging innovation with implementation strategies, agricultural systems can transition from environmental degradation sources to ecosystem regeneration contributors within the circular bioeconomy paradigm.

REFERENCES

Aarif, K. O. M., Alam, A., & Hotak, Y. (2025). Smart Sensor Technologies Shaping the Future of Precision Agriculture: Recent Advances and Future Outlooks. *Journal of Sensors*, *2025*(1), 2460098. DOI: 10.1155/js/2460098

Abed, N., Murgun, R., Deldari, A., Sankarannair, S., & Ramesh, M. V. (2025). IoT and AI-driven solutions for human-wildlife conflict: Advancing sustainable agriculture and biodiversity conservation. *Smart Agricultural Technology*, *10*, 100829. DOI: 10.1016/j.atech.2025.100829

Ajayi, O. O., Toromade, A. S., & Ayeni, O. (2024). Circular agro-economies (CAE): reducing waste and increasing profitability in agriculture. *International Journal of Advanced Economics*.

Akhuli, S. (2025). Organic farming: A pathway to sustainable development. *International Journal of Research and Review*, *12*(1), 208–211. DOI: 10.52403/ijrr.20250126

Al-Saeedi, N. K., & Massad, M. Q. (2025, April). Exploring the Efficacy of Integrated Multi-Trophic Aquaculture (IMTA) for Nutrient Recycling and Productivity Enhancement in Marshlands. In *IOP Conference Series: Earth and Environmental Science* (Vol. 1487).

Altieri, M. A., Nicholls, C. I., de Molina, M. G., & Rojas, A. S. (2024). Landscape agroecology: Methodologies and applications for the design of sustainable agroecosystems. *Land (Basel)*, *13*(11), 1746. DOI: 10.3390/land13111746

Anikwe, M., & Ife, K. (2023). The role of soil ecosystem services in the circular bioeconomy. *Frontiers in Soil Science*, *3*, 1209100. DOI: 10.3389/fsoil.2023.1209100

Ansar, A., Du, J., Javed, Q., Adnan, M., & Javaid, I. (2025). Biodegradable Waste in Compost Production: A Review of Its Economic Potential. *Nitrogen*, *6*(2), 24. DOI: 10.3390/nitrogen6020024

Artola, A., Font, X., Moral-Vico, J., & Sánchez, A. (2024). The role of solid-state fermentation to transform existing waste treatment plants based on composting and anaerobic digestion into modern organic waste-based biorefineries, in the framework of circular bioeconomy. *Frontiers in Chemical Engineering*, *6*, 1463785. DOI: 10.3389/fceng.2024.1463785

Bagada, P. J., Damor, P. A., Patel, R. J., Patel, K. C., Patel, D. J., Parmar, H. V., Rank, P. H., & Vekariya, P. B. (2025). Enhancing soil moisture sensors for optimized agricultural water management: A comprehensive review. *Plant Archives*, 25 (Special Issue, ICTPAIRS-JAU), 879–888.

Baharin, A. T., Ishak, N. A., Ismail, N., Osman, M. F., Isa, E. V. M., & Ariffin, W. N. M. (2024). Bioeconomy Sustainability: Integrating Circular Economy Principles with Big Data and IoT for Sustainable Farming in Agriculture 4.0. *International Journal of Biomass Utilization and Sustainable Energy (IJBUSE), 2.*

Baruah, N., Bora, A., & Gogoi, N. (2024). Agricultural Waste in Circular Economy: An Indian Scenario. *Emerging Trends and Techniques in Biofuel Production from Agricultural Waste*, 15–42.

Bhardwaj, S., Venkatesan, S., Rawat, S., & Nath, P. (2024). Transforming Agriculture with IoT for Precision Agriculture and Sustainable Crop Management. In *The Future of Agriculture* (pp. 164–200). IoT, AI and Blockchain Technology for Sustainable Farming. DOI: 10.2174/9789815274349124010012

Bisheko, M. J., & G, R. (2025). Challenges and interventions in water resource management for horticultural crop production in rural and tribal communities in India. *Cogent Food & Agriculture*, *11*(1), 2452335. DOI: 10.1080/23311932.2025.2452335

Boiardo, M. (2025). Assessing the Role of Organic Farming in Enhancing Soil Health and Biodiversity in Europe. *Europian Journal of Sustainable Agriculture P-ISSN 3051-0112 E-ISSN 3051-0120, 1*(01), 9-19.

Botero-Valencia, J., García-Pineda, V., Valencia-Arias, A., Valencia, J., Reyes-Vera, E., Mejia-Herrera, M., & Hernández-García, R. (2025). Machine Learning in Sustainable Agriculture: Systematic Review and Research Perspectives. *Agriculture*, *15*(4), 377. DOI: 10.3390/agriculture15040377

Coletta, M., Monticelli, M., D'Alessandro, A., Gentili, C., Torresi, A., Waris, N., & La Terza, A. (2025). Managing soil to support soil biodiversity in protected areas agroecosystems: A comparison between arable lands, olive groves, and vineyards in the Conero Park (Italy). *Environmental Monitoring and Assessment*, *197*(2), 200. DOI: 10.1007/s10661-025-13658-7 PMID: 39869200

Cordeiro, C. M., & Sindhøj, E. (2024). Situating the discourse of recycled nutrient fertilizers in circular economy principles for sustainable agriculture. *Frontiers in Sustainability*, *5*, 1465752. DOI: 10.3389/frsus.2024.1465752

Egwuatu, D. C., Ogunye, R. O., Anene, P. C., Alhassan, Y., Nwafor, I. R., Udoh, J. I., Asenuga, O. O., & Nwokafor, C. V. (2024). A review on conversion of agricultural waste to bioenergy: Processes and environmental impacts. *Biotechnology Journal International*, *28*(6), 18–34. DOI: 10.9734/bji/2024/v28i6746

Faucon, M.-P., Aussenac, T., Debref, R., Firmin, S., Houben, D., Marraccini, E., Sauvée, L., Trinsoutrot-Gattin, I., & Gloaguen, R. (2023). Combining agroecology and bioeconomy to meet the societal challenges of agriculture. *Plant and Soil, 492*(1-2), 61–78. DOI: 10.1007/s11104-023-06294-y

Feliziani, G., Bordoni, L., & Gabbianelli, R. (2025). Regenerative Organic Agriculture and Human Health: The Interconnection Between Soil, Food Quality, and Nutrition. *Antioxidants, 14*(5), 530. DOI: 10.3390/antiox14050530 PMID: 40427412

Filho, J. J. D. S., Paço, A. D., & Gaspar, P. D. (2025). Artificial Intelligence and MCDA in Circular Economy: Governance Strategies and Optimization for Reverse Supply Chains of Solid Waste. *Applied Sciences (Basel, Switzerland), 15*(9), 4758. DOI: 10.3390/app15094758

Fiore, M., Demichelis, F., Deorsola, F. A., Fino, D., Saracco, G., Pugliese, M., & Tommasi, T. (2025). Optimizing biomethane production and plants growth with biochar-enhanced anaerobic digestion. *Results in Engineering, 26*, 104883. DOI: 10.1016/j.rineng.2025.104883

Gamaralalage, D., Rodgers, S., Gill, A., Meredith, W., Bott, T., West, H., Alce, J., Snape, C., & McKechnie, J. (2025). Biowaste to biochar: A techno-economic and life cycle assessment of biochar production from food-waste digestate and its agricultural applications. *Biochar, 7*(1), 50. DOI: 10.1007/s42773-025-00456-0 PMID: 40078517

García-Cornejo, B., Pérez-Méndez, J. A., Wall, A., & Castrillo-Cachón, D. (2025). The effect of management accounting practices and ICT on the efficiency of organic farms. *Journal of Rural Studies, 114*, 103554. DOI: 10.1016/j.jrurstud.2024.103554

Glockow, T., Kaster, A., Rabe, K. S., & Niemeyer, C. M. (2024). Sustainable agriculture: Leveraging microorganisms for a circular economy. *Applied Microbiology and Biotechnology, 108*(1), 452. DOI: 10.1007/s00253-024-13294-0 PMID: 39212740

Guo, C., Li, M., & Chen, H. (2025). Study on the Influencing Factors of Green Agricultural Subsidies on Straw Resource Utilization Technology Adopted by Farmers in Heilongjiang Province, China. *Agriculture, 15*(1), 93. DOI: 10.3390/agriculture15010093

Halim-Lim, S. A., Jamaludin, A. A., Islam, A. T., Weerabahu, S., & Priyono, A. (2025). Unlocking potential for a circular bioeconomy transition through digital innovation, lean manufacturing and green practices: A review. *Management of Environmental Quality, 36*(1), 130–154. DOI: 10.1108/MEQ-11-2023-0386

Hartono, R., Yoeseph, N. M., Purnomo, F. A., Safi'ie, M. A., & Bawono, S. A. T. (2024). Portable internet of things-based soil nutrients monitoring for precision and efficient smart farming. *Bulletin of Electrical Engineering and Informatics*, *13*(5), 3326–3333. DOI: 10.11591/eei.v13i5.7928

Hebishima, H., & Inage, S. I. (2025). Development of a new plant factory concept with sustainable water and energy supply. *Energy Nexus*, *17*, 100341. DOI: 10.1016/j.nexus.2024.100341

Hilmi, Y. S., Tóth, J., Gabnai, Z., Király, G., & Temesi, Á. (2024). Farmers' resilience to climate change through the circular economy and sustainable agriculture: A review from developed and developing countries. *Renewable Agriculture and Food Systems*, *39*, e15. DOI: 10.1017/S1742170524000097

Islam, K., Sabau, G., Dawson, J., Cheema, M., Daraio, J., & Galagedara, L. (2025). Evaluating agricultural sustainability in Newfoundland, Canada: Insights from a data envelopment analysis (DEA) approach. SSRN Electronic Journal. DOI: 10.2139/ssrn.5127875

Kabato, W., Getnet, G. T., Sinore, T., Nemeth, A., & Molnár, Z. (2025). Towards climate-smart agriculture: Strategies for sustainable agricultural production, food security, and greenhouse gas reduction. *Agronomy (Basel)*, *15*(3), 565. DOI: 10.3390/agronomy15030565

Khatami, F., Cagno, E., & Khatami, R. (2024). Circular Economy in the Agri-Food System at the Country Level—Evidence from European Countries. *Sustainability (Basel)*, *16*(21), 9497. DOI: 10.3390/su16219497

Kiprotich, K., Muema, E., Wekesa, C., Ndombi, T., Muoma, J., Omayio, D., Ochieno, D., Motsi, H., Mncedi, S., & Tarus, J. (2025). Unveiling the roles, mechanisms and prospects of soil microbial communities in sustainable agriculture. *Discover Soil, 2*(1).

Komyshev, D., Kapral, A., Drebot, O., Sakharnatska, L., & Karpyk, J. (2024). Advancing waste reduction and resource conservation through circular economy practices: A rational review. *Grassroots Journal of Natural Resources*, *7*(3), 1–18. DOI: 10.33002/nr2581.6853.0703ukr09

Krause, H. M., Mäder, P., Fliessbach, A., Jarosch, K. A., Oberson, A., & Mayer, J. (2024). Organic cropping systems balance environmental impacts and agricultural production. *Scientific Reports*, *14*(1), 1–15. DOI: 10.1038/s41598-024-76776-1 PMID: 39462042

Kremmydas, D., Beber, C., Baldoni, E., Ciaian, P., Fellmann, T., Gocht, A., Hristov, J., Pignotti, D., Vicario, D. R., Stepanyan, D., & Tillie, P. (2025). The EU target for organic farming: Potential economic and environmental impacts of two alternative pathways. *Applied Economic Perspectives and Policy*, *47*(2), 602–623. DOI: 10.1002/aepp.13470

Kulkarni, S. D. (2025). Sustainable precision agriculture: Integrating artificial intelligence and IoT for optimized green farming practices. *Journal of Informatics Education and Research*, *5*(1).

Kumaresan, M., Kovvasu, K., Kathari, L., & Vijai, A. A. (2025). Vertical farming technologies for flower production: Current trends, challenges, and future prospects. [Laurin Publishers.]. *Current Trends in Agronomy & Agricultural Research*, *2*(1), 1–10.

Lori, M., Kundel, D., Mäder, P., Singh, A., Patel, D., Sisodia, B. S., Riar, A., & Krause, H. M. (2024). Organic farming systems improve soil quality and shape microbial communities across a cotton-based crop rotation in an Indian Vertisol. *FEMS Microbiology Ecology*, *100*(11), fiae127. DOI: 10.1093/femsec/fiae127 PMID: 39289000

Luo, D., Nan, H., Zhang, Y., Sher, F., & Wang, C. (2025). Phosphorus recovery from wastewater by Ca-Al layered double hydroxide/biochar as potential agricultural phosphorus for closed-loop phosphorus recycling. *Process Safety and Environmental Protection*, *194*, 1538–1548. DOI: 10.1016/j.psep.2024.12.110

Marttila, M., Havukainen, J., Uusitalo, V., Linnanen, L., & Mikkilä, M. (2024). Multicriteria analysis of sustainable agri-food waste management for an agroecosystem in Finland. *Frontiers in Sustainable Food Systems*, *8*, 1426890. DOI: 10.3389/fsufs.2024.1426890

Mishra, A. K., Tiwari, A., Maurya, P. K., & Sharma, S. (2024). Enhancing ecosystem dynamics: Organic and regenerative practices in rice-wheat systems and their impact on soil arthropod biodiversity. *Frontiers in Environmental Science*, *12*, 1407954. DOI: 10.3389/fenvs.2024.1407954

Nahon, S. M. R., Trindade, F. C., Yoshiura, C. A., Martins, G. C., Costa, I. R. C. D., Costa, P. H. D. O., Herrera, H., Balestrin, D., Godinho, T. D. O., Marchiori, B. M., & Valadares, R. B. D. S. (2024). Impact of Agroforestry Practices on Soil Microbial Diversity and Nutrient Cycling in Atlantic Rainforest Cocoa Systems. *International Journal of Molecular Sciences*, *25*(21), 11345. DOI: 10.3390/ijms252111345 PMID: 39518901

Parameswari, K., Mulla, N., Hussain, M. A., Kumar, S. B. S., & Subha, L. (2024). Integration of AI and robotics in precision agriculture: Enhancing crop yield and resource efficiency. *Nanotechnology Perceptions*, *20*(5). Advance online publication. DOI: 10.62441/nano-ntp.vi.2902

Parreño, M. A., Werle, S., Buydens, L., Leroy, C., Roberts, S., Koirala, S., Filipiak, M., Kuhlmann, M., Brunet, J. L., Henry, M., Alaux, C., Requier, F., Piot, N., Meeus, I., Klein, A.-M., Keller, A., & Leonhardt, S. D. (2024). Landscape heterogeneity correlates with bee and pollen diversity while size and specialization degree explain species-specific responses of wild bees to the environment. *The Science of the Total Environment*, *954*, 176595. DOI: 10.1016/j.scitotenv.2024.176595 PMID: 39357756

Pearson, D. E., DePuy, A. L., & Kuhlman, M. P. (2025). Inspiring citizens and municipalities to initiate pollinator conservation: The urban pollinator matrix modeling tool. *Urban Ecosystems*, *28*(2), 1–13. DOI: 10.1007/s11252-025-01703-9

Pedersen, N. K., Kepfer-Rojas, S., Riis-Nielsen, T., Johannsen, V. K., & Schmidt, I. K. (2025). Natural colonization in abandoned agricultural fields benefits native, insect-pollinated and bird-dispersed woody species. *Trees, Forests and People*, *19*, 100755. DOI: 10.1016/j.tfp.2024.100755

Rachbauer, L., Granda, C. B., Shrestha, S., Fuchs, W., Gabauer, W., Singer, S. W., Simmons, B. A., & Urgun-Demirtas, M. (2024). Energy and nutrient recovery from municipal and industrial waste and wastewater—a perspective. *Journal of Industrial Microbiology and Biotechnology, 51*, 040.

Reiff, J., Jungkunst, H. F., Antes, N., & Entling, M. H. (2025). Crop productivity of Central European Permaculture is within the range of organic and conventional agriculture. *Peer Community Journal*, *5*, 5. DOI: 10.24072/pcjournal.521

Rossi, L. (2025). Agroecology and Biodiversity Conservation: The Role of Organic Farming in Promoting Ecosystem Services. *Europian Journal of Organic Farming & Agri Practices e-ISSN 3051-0236, 1*(01), 1-9.

Sabahy, A. A., Hendawy, S. F., Wasfy, K. I., Moursy, M. A. M., & Mohamed, R. (2024). Enhancing Water Use Efficiency and Carbon Profitability Through the Long-Term Impact of Sustainable Farming Systems. *Sustainability (Basel)*, *16*(20), 9116. DOI: 10.3390/su16209116

Sakib, A. I. M. (2024). Innovative food waste recycling methods for agricultural sustainability: A systematic review. *Academic Journal on Business Administration. Innovation & Sustainability*, *4*(3), 104–118.

Singh, M., Yadav, D. B., Yadav, S., Mor, V., & Poonia, B. S. (2024). Potential of conservation agricultural practices on soil quality, carbon sequestration, salinity management and productivity of rainfed areas: Conservation agricultural practices on soil quality. *Journal of Soil Salinity and Water Quality*, *16*(3), 329–344. DOI: 10.56093/jsswq.v16i3.158116

Somashekar, K. S., Abhishek, G. J., Kumar, V., Tiwari, A., & Lakra, T. S. (2024). Agroecology principles, practices and their impact on sustainable food systems. *European Journal of Nutrition & Food Safety*, *16*(9), 249–260. DOI: 10.9734/ejnfs/2024/v16i91544

Suárez Valdés, M. Á., Soriano Disla, J. M., Gambuzzi, E., & Castejón Martínez, G. (2024). Innovative Circular Biowaste Valorisation—State of the Art and Guidance for Cities and Regions. *Sustainability (Basel)*, *16*(20), 8963. DOI: 10.3390/su16208963

Tan, M., Cui, N., Jiang, S., Xing, L., Wen, S., Liu, Q., Li, W., Yan, S., Wang, Y., Jin, H., & Wang, Z. (2025). Effect of practicing water-saving irrigation on greenhouse gas emissions and crop productivity: A global meta-analysis. *Agricultural Water Management*, *308*, 109300. DOI: 10.1016/j.agwat.2025.109300

Teng, J., Hou, R., Dungait, J. A., Zhou, G., Kuzyakov, Y., Zhang, J., Tian, J., Cui, Z., Zhang, F., & Delgado-Baquerizo, M. (2024). Conservation agriculture improves soil health and sustains crop yields after long-term warming. *Nature Communications*, *15*(1), 8785. DOI: 10.1038/s41467-024-53169-6 PMID: 39389978

Toplicean, I.-M., & Datcu, A. D. (2024). An Overview on Bioeconomy in Agricultural Sector, Biomass Production, Recycling Methods, and Circular Economy Considerations. *Agriculture*, *14*(7), 1143. DOI: 10.3390/agriculture14071143

Upadhyay, S. K., Singh, G., Rani, N., Rajput, V. D., Seth, C. S., Dwivedi, P., Minkina, T., Wong, M. H., Show, P. L., & Khoo, K. S. (2024). Transforming bio-waste into value-added products mediated microbes for enhancing soil health and crop production: Perspective views on circular economy. *Environmental Technology & Innovation*, *34*, 103573. DOI: 10.1016/j.eti.2024.103573

Viola, E., Keswani, C., & Minkina, T. (2023). Impact of the sustainable agricultural practices for governing soil health from the perspective of a rising agri-based circular bioeconomy. *Applied Soil Ecology*, *194*, 105199.

Wang, T., Zuo, Y., Manda, T., Hwarari, D., & Yang, L. (2025). Harnessing Artificial Intelligence, Machine Learning and Deep Learning for Sustainable Forestry Management and Conservation: Transformative Potential and Future Perspectives. *Plants*, *14*(7), 998. DOI: 10.3390/plants14070998 PMID: 40219066

Wang, Y., Zheng, R., Dong, W., Gao, P., & Duan, T. (2024). Conservation Agricultural Practices Increase Soil Fungal Diversity and Network Complexity in a 13-Year-Duration Conservation Agriculture System in the Loess Plateau of China. *Applied Sciences (Basel, Switzerland)*, *14*(18), 8411. DOI: 10.3390/app14188411

Yusuf, J. A., & Ojedokun, R. O. (2024). The Role of Bio-Based Innovations in Circular Economy: A Biochemical and Economic Perspective. *Journal of Medical Science. Biological Chemistry*, *1*(1), 21–27.

KEY TERMS AND DEFINITIONS

Agroecology: The application of ecological principles to agricultural systems, emphasizing biodiversity, ecosystem services, and socio-ecological interactions.

Biochar: A carbon-rich material produced from biomass through pyrolysis, used in agriculture for soil amendment, carbon sequestration, and nutrient recovery.

Biodiversity Impact Assessment: Systematic evaluation of how agricultural practices affect species richness, abundance, and ecosystem functions across different spatial and temporal scales.

Circular Bioeconomy: An economic system that integrates circular economy principles with biological resource utilization, emphasizing waste-to-resource transformation, renewable inputs, and closed nutrient cycles in agricultural production.

Digital Agriculture: The use of digital technologies like IoT, AI, and remote sensing to enhance agricultural decision-making, resource management, and monitoring capabilities.

Integrated Multi-Trophic Aquaculture (IMTA): A farming system that combines various species at different trophic levels to maximize resource utilization and minimize waste.

Nutrient Cycling Efficiency: The proportion of nutrients recirculated within agricultural systems, measured by nutrient retention, recovery rates, and reduced external inputs.

Regenerative Organic Agriculture (ROA): Agricultural approach focused on rebuilding soil health, enhancing biodiversity, and improving ecosystem functions beyond organic standards.

Resource Use Optimization: Quantifiable improvements in resource efficiency, including reductions in external inputs and improvements in productivity per unit of resource.

Waste Valorization: The process of converting agricultural by-products and waste streams into valuable resources such as compost, biogas, biochar, and other biobased products.

Chapter 8
Harnessing Biodiversity Through Agroforestry: A Pathway to a Sustainable Bioeconomy

Asep Mulyono
https://orcid.org/0000-0001-5999-4247
Research Center for Limnology and Water Resources, Indonesia

Veny Rahmawati
https://orcid.org/0000-0002-4433-071X
Research Center for Environmental and Clean Technologies, Indonesia

Dyah Marganingrum
https://orcid.org/0000-0002-3148-6301
Research Center for Environmental and Clean Technologies, Indonesia

Hidawati Hidawati
https://orcid.org/0009-0001-9983-2768
Research Center for Environmental and Clean Technologies, Indonesia

Muhammad Rahman Djuwansah
https://orcid.org/0009-0008-6764-0455
Research Center for Geological Resources, Indonesia

Siti Wahyuningsih
https://orcid.org/0000-0002-8371-2622
Research Center for Food Crops, Indonesia

Taufiq Bachtiar
https://orcid.org/0000-0001-6758-5291
Research Center for Radiation Process Technology, Indonesia

Wilda Naily
https://orcid.org/0000-0002-0134-0942
Research Center for Environmental and Clean Technologies, Indonesia

Ratih Nurjayati
https://orcid.org/0000-0003-2251-4582
Research Center for Environmental and Clean Technologies, Indonesia

Rhazista Noviardi
https://orcid.org/0009-0001-6277-5798
Research Center for Environmental and Clean Technologies, Indonesia

DOI: 10.4018/979-8-3373-2003-8.ch008

Merri Jayanti
https://orcid.org/0009-0008-3382-3667
Research Center for Environmental and Clean Technologies, Indonesia

Ania Citraresmini
https://orcid.org/0009-0006-6096-6466
Padjadjaran University, Indonesia

ABSTRACT

This study underlines the importance of economic growth and preserving biodiversity to ensure long-term environmental and economic stability. Agroforestry, as a biodiversity-friendly land use system, integrates trees and shrubs into agricultural landscapes, mimicking natural ecosystems and promoting ecological richness and sustainability. 2,564 documents from 119 countries in the world that have implemented agroforestry systems. 349 documents from 65 countries have shown the importance of agroforestry systems to economic value. Agroforestry integrates trees with crops and/or livestock to create sustainable agricultural systems based on ecological principles, providing environmental and economic benefits. Agroforestry systems play an important role in contributing to biodiversity-derived products and services in the bioeconomy, with a particular focus on providing non-timber forest products without harvesting timber. Successful agroforestry systems promote economic, environmental sustainability, biodiversity conservation and the need of political support and community engagement.

INTRODUCTION

Overview of The Bioeconomy and Its Reliance on Biodiversity

The bioeconomy is an evolutionary shift in economic thinking that takes into account the sustainable use of biological assets and materials and processes. It is, therefore, defined as a complex system of economic, market, social and political elements and natural resources and technologies for the achievement of the sustainable development objectives (Vostriakova, 2023). The bioeconomy is the part of the economy that is related to the innovation, development, production, and use of products and services of biological origin with the goal of sustainable use of resources (Woźniak-Gientka & Tyczewska, 2023).

Bioeconomy has developed and currently undergoing elaboration that is not only limited to production processes that use biological resources (agriculture, forestry, fisheries, and biotechnology) in producing added value (products, services, and energy) that are essential for life in the modern world. The concept of bioeconomy is a common platform for the integration of biotechnology, economic development, industrial processes, and societal needs with particular attention to the sustainable and responsible use of biological resources to encourage the development of bio-products (Kag et al., 2024). Bioeconomy aims for the sustainable use of abundant renewable biological resources while at the same time reducing dependence on conventional fossil fuels. The bioeconomy has advanced significantly in the last few decades, changing supply chains and having an impact on many different industries and sectors worldwide (Global Bioeconomy Summit, 2020).

Bioeconomy is considered a transformative approach to economic development that utilizes biological resources to produce goods, services, and energy while reducing reliance on non-renewable resources (Kriesch & Losacker, 2024). The approach aims not only to mitigate climate change but also to improve food security, create jobs, and promote innovation. Several countries have created national bioeconomy plans, emphasizing the value of regional initiatives as well as the necessity of collaboration and education to promote the bioeconomy (Titova et al., 2023). The bioeconomy approach considered as a potential driver for regional development economic model value through biodiversity-richness (Lasmar et al., 2024). Plants, animals, and microbes involved in agriculture, bioremediation, and the production of bioenergy are among the numerous examples of the varied range of biological resources provided by biodiversity that are crucial to the bioeconomy (Das et al., 2024; Prasad, 2024).

Bioeconomy is a new emerging economic model that uses a biological resource to produce sustainable goods and services, with biological diversity playing a decisive role in its development. In order to create renewable biological products like biofuels, which will reduce dependency on fossil fuels and prevent climate change, the bioeconomy depends on the sustainable use of biodiversity resources (Jitendra, 2024; Toplicean & Datcu, 2024). The fast socio-economic growth in the biological diverse regions can exert pressure on natural resources and require a balance between development and maintenance (Boeri et al., 2020). Natural-based innovations such as biological control and the use of useful microorganisms attract agriculture at agriculture and offer sustainable solutions that improve the protection of biological diversity (Maciejczak, 2023). Biodiversity is a strategic resource that may promote socioeconomic development in areas like in Latin America and the Caribbean by producing food, medications, and other commodities with added value (Boeri et al., 2020).

An excessive use of biological resources can result in biodiversity loss, endangering the bioeconomy's long-term sustainability (de Queiroz-Stein & Siegel, 2023; Singh & Mishra, 2024). Biodiversity supports sectors such as agriculture, forestry and biotechnology will provides the raw materials and genetic resources that are required for bio-economy supports. However, the dependency on biological diversity also represents challenges, since sustainable practices cannot lead to losses of biological diversity, which undermines the basis of the bio-economy. In order to ensure environmental and economic sustainability, it is necessary for policies that incorporate biodiversity in bio-economic strategies (de Queiroz-Stein & Siegel, 2023).

This review paper describes how an agroforestry system is a biodiversity-friendly land-use system, the ecological principles underlying agroforestry, enhancing habitat connectivity and ecological corridors of agroforestry, and the contribution of agroforestry systems to the bioeconomy.

Agroforestry as A Biodiversity-Friendly Land-Use System

Agroforestry known as a biodiversity-friendly land-use system due to its ability in enhancing ecological richness and sustainability. These artificial mimics providing habitats and resources for a wide range of species, and promoting biodiversity conservation. Agroforestry systems give multi-impacts, offering ecological, economic, and social benefits that contribute to sustainable land management and biodiversity preservation. Agroforestry systems provide higher habitats, food resources, and breeding sites for various wildlife species compared to monocultures, as they (Mlambo et al., 2024; Santos et al., 2022). The structural complexity of agroforestry landscapes supports diverse species, enhancing functional diversity and overall biodiversity (Santos et al., 2022). Agroforestry can create corridors and buffer zones that support natural habitats and reduce human-wildlife conflicts, such as those with Asian elephants (Rahman et al., 2023). In European farmland, agroforestry has been shown to support high bird diversity, with species richness more than doubled compared to open farmland (Edo et al., 2023). In northern Ethiopia, diverse agroforestry practices have been found to maintain a diversity of woody species, contributing to the conservation of native trees (Eyasu & Gebrewahid, 2024).

Agroforestry enhances and promotes sustainable management of natural resources, which is essential for preserving soil fertility, minimizing erosion, and controlling hydrological regimes to ensure long-term financial gains (Fikry & Sarjan, 2024; Mondal et al., 2023), and creating favourable conditions for diverse plant and animal life (Satish et al., 2024). Agroforestry acts as a carbon sink and mitigates climate change, which indirectly supports biodiversity by maintaining the sustainability of the ecosystem (Torquebiau, 2024), ensures landscape connectivity, serves as a corridor for species movement, and reduces the need for deforestation (Pa et al., 2024). The complex biodiversity in agroforestry systems leads to better ecosystem services such as crop pollination, natural pest control, carbon sequestration and nutrient cycling (Mlambo et al., 2024), reduce reliance on chemical inputs, supports rural livelihoods and food security (Satish et al., 2024), increase agricultural productivity (Mlambo et al., 2024; Torquebiau, 2024) and indirectly promoting biodiversity by reducing pressure on natural ecosystems (Satish et al., 2024). Agroforestry systems not only multiply farmers' incomes, but also provide economic incentives for biodiversity-friendly practices (Kaur & Monga, 2024), through diverse sources of income from agroforestry products (timber, fruits, and fodder, food, medicine, and fuel) (Bucheli & Bokelmann, 2017).contributing to economic resilience and reducing economic trade-offs (Alifa et al., 2024; Satish et al., 2024).

The main goal of utilizing biodiversity through agroforestry is to create a sustainable forest-agriculture system that supports ecological health and economic sustainability on a land. By incorporating various plant species and integrating them into traditional agricultural practices, agroforestry aims to (1) to improve biodiversity by increasing the variety of species in agricultural landscapes, by promoting the resilience and stability of the ecosystem; (2) Improve soil health using the roots of trees and organic matter to improve the structure of the soil, fertility and water retention; (3) Athoring climate change by sequesting carbon dioxide by growth of trees and reduce greenhouse gas emissions for agricultural practices; and (4) Support livelihood by providing farmers with various sources of income thanks to the cultivation of wood, fruit, nuts and other non -shepherd forest products. Harnessing biodiversity through agroforestry presents a promising pathway to achieving a sustainable bio-economy by focusing on the integration of ecological and economic objectives, promoting sustainable practices, and fostering community engagement (Dagar, 2017; Pancholi et al., 2023; Salve et al., 2022; Verma et al., 2024).

UNDERSTANDING AGROFORESTRY

Types of Agroforestry Systems

Agroforestry systems are generally divided into several forms, including agro-silviculture, silvopasture, and forest gardens (*Table 1*). The selection of the system form will depend on ecological, biophysical and social factors, where each form has its own characteristics and advantages.

Agro-silviculture involves the integration of trees with crops on the same land, promoting crop yield, improve soil health, and provide tree products. It also contributes to carbon sequestration and biodiversity enhancement (Fahad et al., 2022; Jose, 2009; Pancholi et al., 2023). Agro-silviculture in semi-arid can improve productivity and resilience against climate extremes (Fahad et al., 2022).

Silvopasture combines trees, pasture, and livestock, creating a multifunctional landscape that supports both agricultural and ecological goals. It improves microclimate regulation, captures soil, and enhances biodiversity. Silvopasture is economically viable and can rejuvenate degraded lands (Joshi, 2023; Peri et al., 2016). This system is suitable for both temperate and tropical regions, offering a sustainable alternative to monocultures and poor farming practices (Joshi, 2023).

Forest gardens are designed as diverse polycultures of perennial plants that provide food, medicine, and other resources while regenerating the environment. They support environmental regeneration, enhance food security, and meet local community needs through sustainable practices (Nair, 1998; Wartman et al., 2018). Forest gardens are gaining momentum in temperate and tropical regions, promoting cultural principles that prioritize ecological and social well-being over profits (Nair, 1998; Wartman et al., 2018).

Table 1. The agroforestry systems type differences (Nair et al., 2009)

Aspect	Agro-Silviculture	Silvopasture	Forest Gardens
Definition	Integration of trees and crops on the same land.	Combination of trees, forage, and livestock grazing in a managed system.	Multi-layered, perennial-based agricultural systems mimicking natural forests.
Primary Components	Trees + Crops (e.g., fruit trees with cereals or vegetables).	Trees + Pasture + Livestock (e.g., timber trees with grassland and cattle).	Diverse plant species arranged in vertical layers (canopy, understory, ground).
Main Purposes	Enhance crop yield, improve soil health, and provide tree products (fruit, timber).	Improve pasture productivity, provide shade for livestock, and produce timber.	Maximize biodiversity, resilience, and food production in a sustainable manner.
Management Focus	Crop rotation, tree pruning, and intercropping.	Grazing management, tree protection, and pasture improvement.	Mimicking ecological succession, maintaining diversity, and optimizing yields.
Economic Output	Income from crops and tree products (e.g., fruits, nuts, timber).	Income from livestock, forage, and tree products (e.g., wood, fodder).	Income from diverse products (fruits, nuts, medicinal plants, etc.).
Examples	Cocoa under coconut palms, maize intercropped with fruit trees.	Pine trees with cattle grazing, oak trees with sheep.	Home gardens in tropical regions, permaculture systems.

Based on SCOPUS document web (https://www.scopus.com), 2,564 research documents (*Table 2*) from 7 continents and 119 countries in the world have implemented agroforestry systems (*Figure 1*) during the period 2011-2024.

Table 2. Number of documens addressing he agroforesry system type

Document Type	Number of Document
Article	1213
Book Chapter	158
Review	139

continued on following page

Table 2. Continued

Document Type	Number of Document
Conference Paper	59
Book	9
Editorial	8
Erratum	4
Letter	3
Conference Review	2
Note	2
Data Paper	1
Short Survey	1

Table 3 highlighted its multifaceted benefits, including environmental sustainability, economic growth, and improved rural livelihoods during period 2011-2024.

Table 3. Key focus area in agroforestry research in period 2011-2024

Period	Key Focus Areas
2011-2015	- Biodiversity conservation and rural livelihoods - Climate change mitigation
2016-2020	- Sustainable food systems and ecosystem services valuation - Policy and institutional support
2021-2024	- Innovative agroforestry practices and climate resilience - Socioeconomic impacts and bioeconomy

Figure 1 shows the number of agroforestry research documents in several parts of the continents, with the European having the most documents, followed by Asia and North America, Africa, South America, Oceania and Central America.

Figure 1. The number of documents in each continent related to agroforestry systems

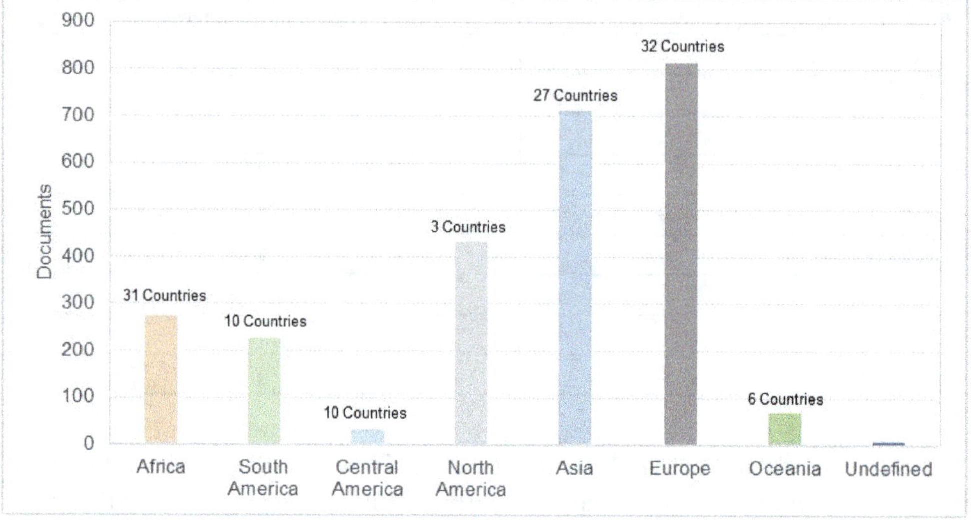

Ecological Principles Underlying Agroforestry

Agroforestry practice enhances resource efficiency, biodiversity, and ecosystem services, contributing to both environmental and economic benefits. The ecological principles underlying agroforestry are rooted in its ability to create multifunctional landscapes are crucial for understanding how agroforestry can be effectively implemented to address various ecological challenges.

Resource Efficiency and Eco-Intensification

Agroforestry promotes eco-intensification by optimizing the use of resources, such as light, water, and nutrients, through the complementary interactions between trees, crops, and livestock (Mondal et al., 2023; Pantera et al., 2021). The presence of trees and varied plant species in agroforestry system supports a wider range of pollinators (Bees, butterflies, and other beneficial insects) compared to monoculture systems (Jose & Udawatta, 2021). This biodiversity not only supports pollination but also contributes to the resilience of agricultural systems against environmental stresses (Kuyah et al., 2017).

Biodiversity and Habitat Enhancement

The biodiversity in agroforestry systems plays a significant role in natural pest control, by supporting a variety of predatory species that control pest populations (Kuyah et al., 2017), the presence of predatory insects and birds helps in controlling pest populations. Trees and shrubs in agroforestry systems can act as barriers or shelters for beneficial insects, thereby reducing the need for chemical pest control measures (Jose & Udawatta, 2021). Integration of various plants in agroforestry systems can disrupt the life cycle of pests and reduce the prevalence of insect pests. The presence of trees in agroforestry systems not only serves as a shelter for various organisms but also contributes to the conservation of biodiversity and ecosystem stability (Fahad et al., 2022).

Climate Change Mitigation and Adaptation

Agroforestry systems enhance resilience to climate extremes by stabilizing microclimates and reducing vulnerability to droughts and floods (Verma et al., 2024). The contribution of agroforestry to climate change mitigation through carbon sequestration where the presence of trees in this system will capture and store carbon in its biomass and soil (Raj et al., 2022). Agroforestry systems also contribute to improved water regulation through enhanced water retention and reduced runoff. The deep tap root systems of trees help to stabilize soil, allowing for better infiltration of rainfall, reducing surface erosion and landslide mitigation. This not only conserves water but also improves groundwater recharge. Trees in agroforestry systems play a crucial role in regulating water cycles by enhancing groundwater recharge and reducing the risk of flooding (Raj et al., 2024; Zhu et al., 2020). Agroforestry improves water infiltration and reduces surface runoff (reducing water loss), which helps in maintaining soil moisture and droughts (Zhu et al., 2020). Trees also function as windbreaks, protecting adjacent plants from wind damage and reducing raindrops that cause soil erosion (Fahad et al., 2022; Verma et al., 2024) that provide shade, reducing heat stress on plants and lowering evaporation rates.

Soil Health and Nutrient Cycling

Agroforestry systems improve soil health through various mechanisms. The trees improve soil structure (Kremer, 2021) and fertility by increasing nutrient cycling and accumulation of organic matter, which is essential for maintaining soil health, and increases biodiversity in the soil microbiome (Fahad et al., 2022; Raj et al., 2022). The deep root systems of trees help to break up compacted soil layers, improving aeration and water infiltration. The integration of trees reduces soil erosion by binding soil particles and intercepting rainfall, thus maintaining soil structure and preventing degradation (Fahad et al., 2022; Khan et al., 2021; Raj et al., 2022), reducing dependency on chemical fertilizers and chemical pesticides (Barrios et al., 2012; Khan et al., 2021). Additionally, litter from tree leaves that fall on the soil surface contributes to the formation of humus, which is essential for nutrient retention and soil fertility. This holistic approach not only supports crop growth but also mitigates soil erosion and degradation. Agroforestry systems reduce soil degradation and promote sustainable agricultural productivity by maintaining a closed nutrient cycle (Verma et al., 2024). Otherwise, there is competition for water and nutrients resources of trees and food crops in agroforestry (Arshad et al., 2024; Fahad et al., 2022; Kuyah et al., 2017) which has the potential to affect crop yields if not managed properly.

Biodiversity at Species, Genetic, and Ecosystem Levels of Agroforestry

The systems of agroforestry are essential to preserving biodiversity at the species-specific, genetic, and ecosystem levels. The wide range in "species diversity" at agroforestry system will support higher species diversity compared to monocultures by providing varied habitats and resources. This includes food, breeding sites, and landscape linkages for wildlife, which enhances the presence of birds, insects, reptiles, and amphibians (Jose & Udawatta, 2021; Mlambo et al., 2024; Udawatta & Jose, 2019). In regions like Northeast India, agroforestry systems such as home gardens harbor rich plant diversity, acting as miniature forests that conserve rare and threatened species (Yashmita-Ulman et al., 2021).

The term "genetic diversity" describes the heterogeneity of genes within a species. Genetic diversity in agricultural landscapes is preserved through the incorporation of several plant species in agroforestry systems. This diversity is crucial for resilience against pests, diseases, and changing climatic conditions (Pa et al., 2024). Agroforestry practices in Southeast Asia demonstrate how cultivating diverse species can reduce pressure on local forests, thereby preserving genetic resources (Rahman et al., 2023).

The term "ecosystem diversity" describes the range of ecosystems found in a particular location, includes several ecosystems such as swamps, grasslands, and forests. Agroforestry increases ecosystem diversity by creating complex structures that resemble natural forests. This complexity supports various ecosystem functions such as nutrient recycling, carbon sequestration, and water purification (Santos et al., 2022; Udawatta & Jose, 2019). The presence of trees in agroforestry systems not only improves soil conditions, but also microclimates, which are essential for sustaining diverse ecosystems (Jose & Udawatta, 2021).

Enhancing Habitat Connectivity and Ecological Corridors of Agroforestry

The sustainability of biodiversity conservation and ecosystem functions need to be maintained to build ecological corridors and habitat connectivity. Agroforestry system approach can establish continuous habitats that promote species mobility and enhance ecological processes by incorporating trees into agricultural landscapes.

Habitat connectivity is the degree which the different habitats are linked with mean in allowing for the movement of species between them. The integration of hedgerows in agroforestry landscapes can significantly enhance connectivity. For instance, establishing new hedgerows along strategic pathways can lead to a 38% increase in connectivity, compared to a mere 11% when focusing on existing corridors (Dondina et al., 2018). In agroforestry systems, connectivity can be enhanced through strategic planting with creating linear features such as hedgerows and tree lines can facilitate wildlife movement. This diversity is attributed to the heterogeneous vegetation and improved microclimate conditions in agroforestry systems (Jose & Udawatta, 2021).

Ecological corridors are designated areas that connect fragmented habitats, allowing for safe passage of wildlife. In cocoa agroforestry in Ghana, have been shown to increase forest connectivity by creating corridors that link fragmented forest patches by providing habitats and facilitating species movement (Asare et al., 2014). Incorporating diverse plant species and structures can create habitats that support various wildlife and establishing buffer zones around sensitive areas can reduce edge effects and provide additional habitat for wildlife will provide ecological corridors. In Mediterranean environments, island forests within agricultural landscapes serve as potential core habitats for ecological corridors, promoting species migration and connectivity across fragmented areas (Hidalgo et al., 2021).

Functional connectivity refers to the actual movement of species that can be enhanced through agroforestry corridors that are designed to meet the specific needs of target species. The corridors vary in width, length, and habitat composition to optimize their effectiveness (Holyoak, 2008). Structural connectivity, the physical arrangement of habitats, is improved by agroforestry practices that create continuous vegetative cover, reducing the isolation of habitat patches (Estreguil et al., 2016). Additionally, the effectiveness of corridors can vary depending on species-specific requirements and landscape characteristics, necessitating tailored approaches for different ecological contexts (Mony et al., 2022).

AGROFORESTRY'S CONTRIBUTION TO THE BIOECONOMY THROUGH BIODIVERSITY

Biodiversity-Derived Products and Services

Agroforestry systems play a remarkably important in enhancing biodiversity-derived products and services. Non-timber forest products (NTFPs) essential for the community in overall in sustainable environmental and economic sustainability. NTFPs which represent a wide and diverse range of goods that can be harvested from forest areas without resorting to logging, include a wide range of products that are essential for both ecological balance and human use. These products include various medicinal plants, an abundance of fruits, an assortment of nuts, resins and a wide range of other materials that can be harvested in a manner that ensures sustainability and ecological integrity.

The importance of NTFPs cannot be overstated as they constitute vital resources for many communities, making important contributions to nutrition, general health and income generation, which is essential for sustaining livelihoods. Agroforestry systems are used to cultivate NTFPs, which include highly valued medicinal plants such as Tulsi and Amla, creating significant economic opportunities for farmers and entrepreneurs engaged in the production and sale of these crops (Sarkar et al., 2024). The practices of cultivating and commercializing NTFPs within agroforestry systems effectively bolster sustainable livelihoods by diversifying the sources of income available to communities and playing a crucial role in the reduction of poverty among populations that are dependent on forest ecosystems for their survival (Magry et al., 2024).

Economic Valuation of Biodiversity in Agroforestry

The economic benefits of agroforestry systems through income diversification. In cocoa plantations using agroforestry system in Côte d'Ivoire/Ivory Coast, enhance biodiversity and carbon sequestration, aligning economic (income diversification) and environmental objectives (Anthelme et al., 2024). In coffee agroforestry systems, studies at South and Central America, compared with Africa, Asia, and the Caribbean, improved ecosystem services like pollination and pest control, leading to better coffee bean quality and increased profits (Wright et al., 2024).

Agroforestry systems that incorporate medicinal plants, such as Tulsi and Amla in India, offer economic benefits by tapping into the market for medicinal and religious products, also contribute to biodiversity conservation and land reclamation (Sarkar et al., 2024). In the realm of agroforestry, there exists a plethora of market opportunities for products that are friendly to biodiversity, which can often be overshadowed by more conventional agricultural practices. Agroforestry systems operating within these specific sectors possess a distinct competitive advantage, as they are capable of delivering products that align seamlessly with the demands and preferences of this conscientious consumer base (Anthelme et al., 2024; Wright et al., 2024).

The research documents during the period 2011-2024 on SCOPUS search document web (https://www.scopus.com), obtained 349 documents which relate the importance of agroforestry systems to economic value. The 349 documents reported 65 countries in almost all continents with 40 countries with >1 document, 25 countries only show 1 document while 5 document are undefined countries (*Figure 2*). It's found 33 countries in temperate climates region reported 177 documents, 27 countries in tropical reported 162 documents and 5 documents from 5 countries in arid region, while the origin of 5 documents is undefined.

Figure 2. The number of documents in each country that related to relate the importance of agroforestry systems to economic value

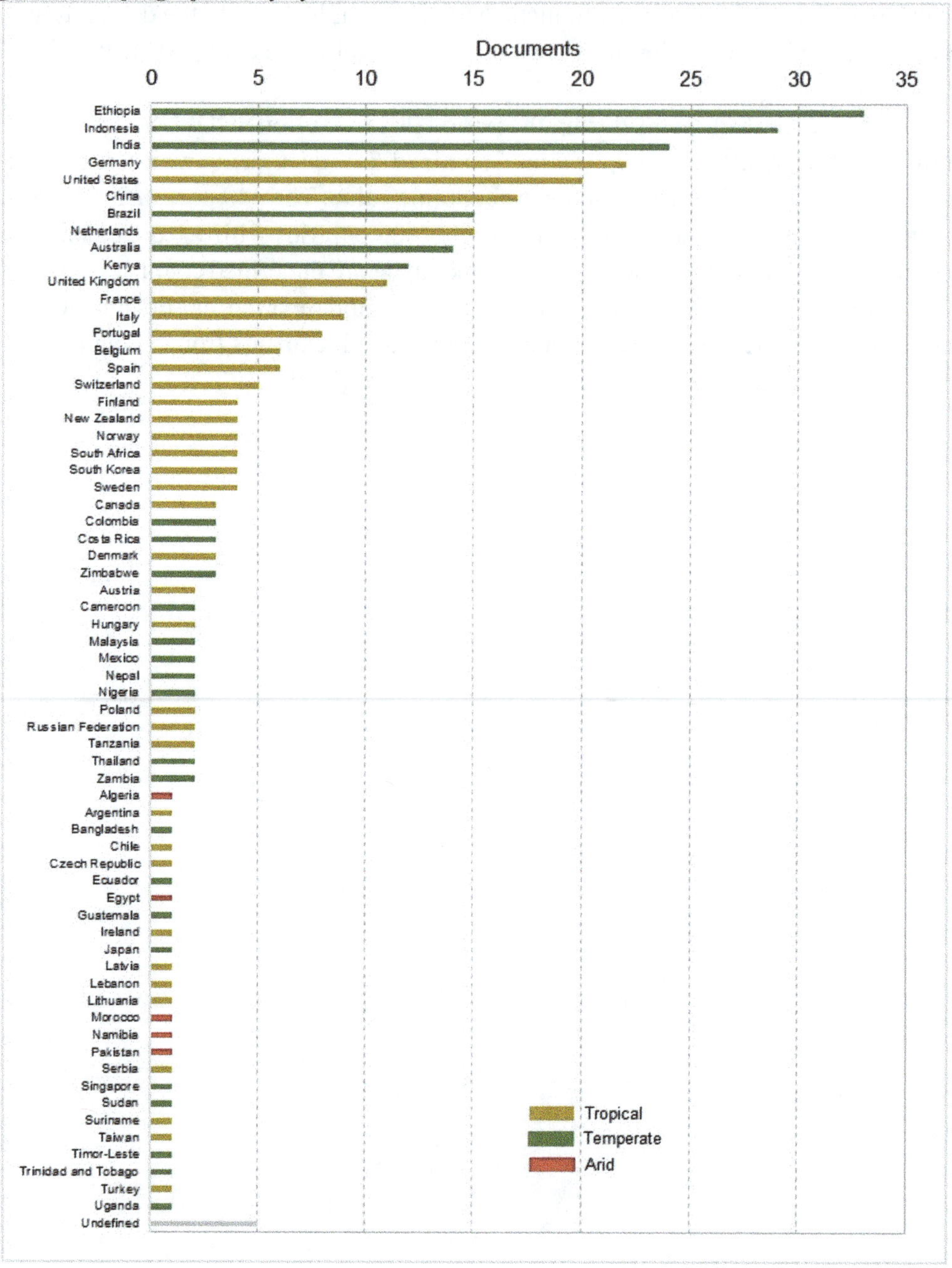

In Ethiopia, Eucalyptus-cum-wheat and Acacia-cum-maize systems show high financial viability with significant BCR and IRR values, Bamboo can contributed to high financial viability among AFI systems (Amare & Darr, 2023) and Acacia decurrens based small-scale agroforestry (SSA) provides significantly higher income compared to sole cropping systems (Mekonnen et al., 2023). At Benin-West Africa, agroforestry services in subequatorial regions are valued at USD 3700/ha, significantly higher than in tropical regions (USD 1600/ha) (Belfrid et al., 2022). Agroforestry in Cameroon can inject several millions of USD annually into the economy, supporting a bioeconomy transition (Awazi et al., 2023). At Uttar Pradesh-India, Eucalyptus + wheat-based agroforestry systems show higher gross income compared to traditional wheat agriculture (Singh et al., 2022) and at Jharkhand-India, high potential for agroforestry provide economic benefits and carbon sequestration (Ahmad et al., 2022). Nature-based tourism linked to agroforestry in Mexico generates USD 7.7 million annually (Revollo-Fernández et al., 2024). Rattan and rubber agroforestry systems in contribute significantly to the local economy in Central Kalimantan-Indonesia (Tata, 2019). Cork oak woodlands in Spain (Andalusia) provide various economic activities, including timber and nuts, with significant economic value (Campos et al., 2020). When taking ecosystem services into account, agroforestry systems in Europe (Mediterranean) typically have higher financial value than conventional agriculture systems (Kay et al., 2019). Agroforestry systems can give producers significant financial benefits, boosting their income and enhancing their living conditions, according to a European study (Campoy-Muñoz et al., 2022).

Table 4. Economic value of agroforestry product in several countries

Country	Product	Economic Value	Sources
West Africa (Benin)	Agriculture, Fishing, Timber	USD 3700/ha in subequatorial regions, USD 1600/ha	(Belfrid et al., 2022)
Cameroon	Biomass, Biofuel	Several millions of USD annually	(Awazi et al., 2023)
	Various agroforestry products	Positive covariates include market prices and farm operating costs	(Molua, 2005)
Ethiopia	Eucalyptus-cum-wheat	Benefit-Cost Ratio (BCR) of 4, Internal Rate of Return (IRR) of 9-37%	(Amare & Darr, 2023)
	Acacia-cum-maize	Benefit-Cost Ratio (BCR) of 2.16, Internal Rate of Return (IRR) of 9-37%	(Amare & Darr, 2023)
	Bamboo	Highest financial viability among AFI systems	(Amare & Darr, 2023)

continued on following page

Table 4. Continued

Country	Product	Economic Value	Sources
	Acacia decurrens (SSA)	1.3 times more income than sole Teff, 11 times more income than sole fodder	(Mekonnen et al., 2023)
India	Eucalyptus + Wheat	Gross income: Rs. 67,733.50/ha	(Singh et al., 2022)
	Various Agroforestry Products	High potential for economic benefits and carbon absorption	(Ahmad et al., 2022)
	Rice-wheat-Eucalyptus agroforestry	NPV of 4.8k€/ha, carbon sequestration increases NPV by 0.1-1%	(Wicke et al., 2013)
	Poplar	Average annual net returns of USD 346.36 for Poplar-based agrisilviculture	(Joshi et al., 2024)
	Ficus	Higher financial net present value (NPV) in rainfed systems	(Dhanya et al., 2016)
	Arecanut, Rubber, Jackfruit, Pineapple	Notably higher annual farm incomes for households practicing agroforestry	(Mathur & Bhattacharya, 2024)
	Various species in homestead agroforestry	Significant shifts in livelihood significance index (LSI) due to socio-economic changes	(Das et al., 2025)
	Acacia nilotica, Eucalyptus teriticornis, Leptochloa fusca	Highest NPV observed in Gudha due to high economic value of these species	(Stille et al., 2011)
Pakistan	Acacia nilotica plantation	NPV of 2.8k€/ha, carbon sequestration increases NPV by 2-24%	(Wicke et al., 2013)
Bangladesh	Rice-Eucalyptus agroforestry	NPV of 1.1k€/ha, carbon sequestration increases NPV by 1-12%	(Wicke et al., 2013)
Mexico	Nature-based Tourism	USD 7.7 million annually	(Revollo-Fernández et al., 2024)
Peru (Guatemala)	Non-coffee products (fuelwood, construction materials)	20-33% of total agroforestry system value	(Rice, 2008)
Madagascar	Clove, vanilla, coffee, cocoa	27.4% of exports, income for 500,000 farmers	(Andriatsitohaina et al., 2024)
Brazil	Fruit trees	NPV confirms recovery of costs within 20 years	(Tremblay et al., 2015)
	Coffee	Mean NPV of US$ 50,585.91 and mean IRR of 89.93%	(Goncalves et al., 2021)
Lao PDR	Bamboo products	Handcraft model generates 1.12-2.17 times higher income/labor hour	(Lee et al., 2021)
Indonesia	Rattan, Rubber	Contributes to economic development in Central Kalimantan	(Tata, 2019)

continued on following page

Table 4. Continued

Country	Product	Economic Value	Sources
	Nuts, fruits, green leafy vegetables	Supports food security and ecosystem services	(Putra et al., 2022)
	Coffee	Potential profit due to rising market demand and environmental benefits	(Ulya et al., 2023)
	Durian, Ornamental Plants, Laying Hens	Potential profit of IDR 143,857,559 (USD 9,652)/hectare/month	(Nurrochmat et al., 2023)
	Jackfruit, Ornamental Plants, Laying Hens	Potential profit of IDR 146,908,379 (USD 9,789)/hectare/month.	(Nurrochmat et al., 2023)
	Nutmeg, Ornamental Plants	Potential profit of IDR 36,814,280 (USD 2,470)/hectare/month	(Nurrochmat et al., 2023)
Spain	Cork, Timber, Firewood, Nuts	Economic activities valued in 2010	(Campos et al., 2020)
	Pine nuts	Small but positive contribution to GDP and employment	(Campoy-Muñoz et al., 2022)
Denmark	Arable crops, livestock, biomass trees	Gross margin of 112 €/ha/year	(Lehmann et al., 2020)
Germany	Alley cropping agroforestry systems	Farmers expect a yearly subsidy of €512 (~$606)/hectare	(Otter & Deutsch, 2023)
United Kingdom	Arable crops, livestock, biomass trees	Gross margin of 5083 €/ha/year	(Lehmann et al., 2020)
Europe (Mediterranean)	Biomass Production	Greater financial value compared to agricultural systems	(Kay et al., 2019)

In temperate agroforestry systems in Europe, the sustainability trade-offs and synergies are influenced by factors such as social capital, animal health and welfare management, and fossil fuel use (Rice, 2008). The economic viability of diverse agroforestry systems in Europe varies, with different systems demonstrating varying levels of agronomic productivity and economic viability (Campoy-Muñoz et al., 2022). Gross margin and economic viability of agroforestry systems have been found to vary across different countries, with differences in returns from crop and tree components (Campoy-Muñoz et al., 2022).

In tropical regions, agroforestry systems in multi-layered ecosystem often incorporate with a variety of tree species alongside crops and livestock. Agroforestry has been identified as a way to protect food access, increase peasant revenue, and provide various ecosystem services, ultimately improving nutrition and probiotic ecosystems (Belfrid et al., 2022). Policy conditions, internalizing environmental externalities, soil and site characteristics, prices, and the profitability of specific system components are some of the elements that affect the economic performance of agroforestry systems (Wicke et al., 2013). It has been demonstrated that strict management and a greater comprehension of agroforestry systems improve their competitiveness and economic results (Wicke et al., 2013). One notable example is the forest-garden system practiced in Southeast Asia, where farmers cultivate a mix of fruit trees, vegetables, and medicinal plants within their homesteads (Roshetko & Bertomeu, 2015).

In Bengkulu, Indonesia, the economic valuation of these benefits indicated that biodiversity-friendly practices could increase farm income by up to 20%. Coffee agroforestry systems are financially viable, with studies showing a positive Net Present Value (NPV), a Benefit-Cost Ratio (BCR) exceeding 1, and an Internal Rate of Return (IRR) of 33% (Oktoyoki et al., 2024). In Central Java, agroforestry contributes significantly to household incomes, with a 38% contribution to Pesanggem households, equating to approximately IDR 11,000,000/household annually (Santosa & Padmaningrum, 2024). In East Java, the "Tumpangsari" system, which combines trees with crops like coffee and vegetables, is a primary income source for many households, with earnings ranging from Rp3.33M to Rp4.00M annually (Jambut et al., 2022). Non-Wood Forest Products (NWFPs) in Gempolan Village, such as durian and ginger contribute significantly to the local economy, with a total economic value of Rp.1,908,394,000/year. Durian alone accounts for 45.9% of this value (Indratna et al., 2023). In West Sumatra, the Parak agroforestry system supports both direct and indirect economic benefits, with a total income of IDR 70,502,200/year from various commodities (Helmayuni & Mardianto, 2021).

The Colombian coffee industry stands as a pivotal economic engine within the nation, encompassing an extensive area of approximately 844,000 hectares that are specifically allocated for the purpose of coffee cultivation, which not only leads to the creation of around 730,000 direct employment opportunities but also provides essential support for an estimated 2 million individuals reliant on this vital sector for their livelihoods (Rendón-Sandoval et al., 2023). The coffee sector holds substantial importance in the economic framework of Colombia (Parente-Laverde, 2020). Accounting for roughly 10% of the nation total exports, while simultaneously representing approximately 11.2% of the agricultural Gross Domestic Product (GDP) of Colombia (Parente-Laverde, 2020). Coffee agroforestry practices play a vital role in promoting the socio-economic of small-scale farmers. A significant 70% of the total coffee producers in Colombia livelihoods (Rendón-Sandoval et al., 2023), highlighting the importance of this sector in fostering sustainable livelihoods for these individuals.

Ghana's cocoa agroforestry systems, which integrate shade trees, can improve both biodiversity and farmer livelihoods while also reducing labour costs (Boadi et al., 2023). This diversification of income sources enhances the farmers economic resilience and long-term productivity of cocoa. For instance, hybrid cocoa grown under moderate shade in Ghana's Eastern region demonstrated the highest value of production (VOP) at 669 USD/ha/year, outperforming other systems like full-sun cocoa (Hawkins et al., 2024). Trees in cocoa farms offer non-market ecosystem services such as habitat for pollinators and nutrient cycling, valued at approximately USD 164/hectare/year (Obeng et al., 2020). In carbon sequestration, Cocoa agroforestry systems sequester significant amounts of carbon, which can potentially be monetized through carbon trading. Organic systems, in particular, sequester 30% more carbon than conventional systems, enhancing their economic attractiveness (Agbotui et al., 2023).

Agroforestry with silvopastoral systems type in Brazil that combine livestock grazing with tree planting have demonstrated substantial economic and ecological benefits. A case study in the Brazilian agroforestry revealed that farms utilizing silvopastoral practices higher profitability due to improved pasture quality and increased livestock productivity. Moreover, these systems supported greater biodiversity, including native plant species and wildlife, which contributed to ecosystem stability and resilience. In southern Brazil, the Caíva silvopastoral system has shown a 400% increase in animal production through improved management practices, enhancing income while maintaining forest cover (Hanisch, 2024). In the Descoberto Basin-Brazil, agroforestry systems have proven financially viable, with a net present value (NPV) of USD 63,097.49 and an internal rate of return (IRR) of 71%, especially when combined with payments for environmental services (Dominicis et al., 2023). In Pará, agroforestry systems combining acai, cocoa, and black pepper have demonstrated a competitive advantage over monocultures, with an NPV of US$ 6,508.94/ha and an IRR of 13.93% (Santana et al., 2023). Agroforestry systems with coffee agroforestry systems in Brazil contribute to carbon sequestration and economic impact of carbon revenue (Goncalves et al., 2021). In the Eastern Amazon, agroforestry systems are valued for their ecosystem services, with community willingness to pay for conservation estimated at R$ 5,011.19/ha, reflecting their perceived socioeconomic and environmental benefits (Oliveira et al., 2020).

CONCLUSION AND FUTURE CHALLENGES

Based on 2,564 documents from 119 countries around the world that draw up the implemented agroforestry systems, 65 countries have shown the importance of agroforestry systems to economic value. Investing in biodiversity through agroforestry is crucial for building a sustainable bioeconomy. Agroforestry has enhanced ecosystem services, promoted biodiversity, resilient and sustainable agriculture, supported circular bioeconomy, climate adaptation and mitigation, and economic and social benefits. In enhancing ecosystem services, agroforestry has contributed to soil health improvement, water regulation and carbon sequestration. Trees in agroforestry systems help reduce water runoff, increase infiltration, soil fertility and regulate local hydrology. Trees and shrubs act as carbon sinks, mitigating climate change while creating carbon credits for bioeconomy markets. Diverse habitats for plants, animals, and microorganisms, providing species richness and ecosystem resilience in agroforestry systems.

Agroforestry built renewable biomass resources that can feed into bioproduct industries (biofuels, bioplastics, and pharmaceuticals) and promote sustainable production systems. Mixed-species systems in agroforestry can withstand extreme weather better than monocultures, reducing crop losses due to climate change. The sequestration of greenhouse gases in agroforestry systems directly contributes to global climate goals. Agroforestry provides rural communities with diverse sources of income, reduces poverty and increases livelihood security. It fosters local entrepreneurship in value-added products like medicinal plants, biochar, and sustainably harvested timber.

Strategic agroforestry interventions can improve landscape connectivity and ecological stability, addressing broader-scale landscape vulnerabilities. Conducting comprehensive multi-scale analyses helps in understanding the ecological contributions of agroforestry at both local and broader landscape levels, enabling more effective interventions. Integrating knowledge from agronomy, ecology, and forestry can optimize agroforestry design and implementation, maximizing benefits for biodiversity and crop production. Agroforestry's potential is sometimes hindered by knowledge gaps and policy recognition issues. By capturing lessons learned through both post-mortem and continuous improvement approaches, continuous learning and adaptation, as well as the dissemination of successful practice in agroforestry can be refined and its benefits maximized.

REFERENCES

Agbotui, D. K., Ingold, M., Wiehle, M., & Buerkert, A. (2023). Can carbon payments improve profitability of traditional conventional and organic cocoa agroforests? A case study in the Eastern Region of Ghana. *Agroforestry Systems*, *97*(5), 813–831. DOI: 10.1007/s10457-023-00828-0

Ahmad, F., Talukdar, N. R., Goparaju, L., Biradar, C., Dhyani, S. K., & Rizvi, J. (2022). GIS-based assessment of land-agroforestry potentiality of Jharkhand State, India. *Regional Sustainability*, *3*(3), 254–268. https://www.scopus.com/inward/record.uri?eid=2-s2.0-85141875839&doi=10.1016%2Fj.regsus.2022.10.004&partnerID=40&md5=70195ae0f51802cf59dceb27996ef053

Alifa, A. N., Isnawati, H., Syamsu, R. M. &, & Imam, M. (2024). Socioeconomic and ecological sustainability of agroforestry. *Biodiversity and Bioeconomy: Status Quo, Challenges, and Opportunities*, 281–294. DOI: 10.1016/B978-0-323-95482-2.00013-4

Amare, D., & Darr, D. (2023). Profitability of Smallholder Agroforestry Woodlot Innovations. *Human Ecology: an Interdisciplinary Journal*, *51*(5), 1009–1019. DOI: 10.1007/s10745-023-00446-5

Andriatsitohaina, R. N. N., Laby, P., Llopis, J. C., & Martin, D. A. (2024). Agroforestry in Madagascar: Past, present, and future. *Agroforestry Systems*, *98*(6), 1659–1680. DOI: 10.1007/s10457-024-00975-y PMID: 39086741

Anthelme, G., Junior, P. F. P., Boué, V. B. B. N., Larissa, G., Marie-Solange, T., & Yves, A. Y. C. (2024). Evaluating carbon sequestration and biodiversity in Cocoa agroforestry Systems: A case study from the Man Region, Côte d'Ivoire. *International Journal of Environment and Climate Change*, *14*(10), 691–700. DOI: 10.9734/ijecc/2024/v14i104517

Arshad, M. A., Rouf, S., Abbas, R. N., Shahbaz, Z., Aleem, K., Shahbaz, H., Pervaiz, R., Sarwar, A., & Rehman, H. (2024). Navigating synergies: A comprehensive review of agroforestry system and agronomy crops. *Haya: The Saudi Journal of Life Sciences*, *9*(04), 97–113. DOI: 10.36348/sjls.2024.v09i04.003

Asare, R., Afari-Sefa, V., Osei-Owusu, Y., & Pabi, O. (2014). Cocoa agroforestry for increasing forest connectivity in a fragmented landscape in Ghana. *Agroforestry Systems*, *88*(6), 1143–1156. DOI: 10.1007/s10457-014-9688-3

Awazi, N. P., Tientcheu-Avana, M.-L., Temgoua, L. F., Tsufac, A. R., Forje, G. W., Cedric, C. D., Nfornkah, B. N., & Tchamba, M. N. (2023). Agroforestry-based bioeconomy transition in Cameroon: realities and the way forward. In *Biodiversity and Bioeconomy*. Status Quo, Challenges, and Opportunities., DOI: 10.1016/B978-0-323-95482-2.00014-6

Barrios, E., Sileshi, G. W., Shepherd, K., & Sinclair, F. (2012). Agroforestry and soil health: linking trees, soil biota, and ecosystem services. In *Soil Ecology and Ecosystem Services* (pp. 315–330). Oxford University Press., DOI: 10.1093/acprof:oso/9780199575923.003.0028

Belfrid, D. M., Arcadius, D., N'tcha Yekanbessoun, M., Mariano, S., Fidèle, S., Jean, A., & Pepin, A. M. (2022). Inventory of agroecosystem services and perceptions potential implications due to climate change: A case study from Benin in West Africa. *Environmental Impact Assessment Review*, *95*, •••. https://www.scopus.com/inward/record.uri?eid=2-s2.0-85128162100&doi=10.1016%2Fj.eiar.2022.106792&partnerID=40&md5=ec9437a14903ffcb1192bc7c9822efb4

Boadi, S. A., Bosselmann, A. S., Owusu, K., Asare, R., & Olwig, M. F. (2023). Household economics of cocoa agroforestry: costs and benefits. In *Agroforestry as Climate Change Adaptation* (pp. 121–145). Springer International Publishing., DOI: 10.1007/978-3-031-45635-0_5

Boeri, P., Piñuel, L., Dalzotto, D., & Sharry, S. (2020). Native biodiversity: A strategic resource to accelerate bioeconomy development in Latin America and the Caribbean. In *Agricultural, Forestry and Bioindustry Biotechnology and Biodiscovery* (pp. 163–174). Springer International Publishing. DOI: 10.1007/978-3-030-51358-0_10

Bucheli, V. J. P., & Bokelmann, W. (2017). Agroforestry systems for biodiversity and ecosystem services: The case of the sibundoy valley in the colombian province of putumayo. *The International Journal of Biodiversity Science, Ecosystem Services & Management*, *13*(1), 380–397. DOI: 10.1080/21513732.2017.1391879

Campos, P., Álvarez, A., Oviedo, J. L., Mesa, B., Caparrós, A., & Ovando, P. (2020). Environmental incomes: Refined standard and extended accounts applied to cork oak open woodlands in Andalusia, Spain. *Ecological Indicators*, *117*, 106551. https://www.scopus.com/inward/record.uri?eid=2-s2.0-85085523786&doi=10.1016%2Fj.ecolind.2020.106551&partnerID=40&md5=aac1770e738494d1ad2cf29c7e232ab4. DOI: 10.1016/j.ecolind.2020.106551

Campoy-Muñoz, P., Cardenete, M. A., Pérez-Hernández, P. P., Martín-Lozano, J. M., & Martínez-Rueda, J. (2022). A SAM Approach for the Analysis of Small-Scale Forestry-Based Activities: A Case Study of a Landowner Cooperative for Non-wooded Products from Pinus pinea L. in Southern Spain. *Small-scale Forestry, 21*(2), 249–273. DOI: 10.1007/s11842-021-09497-2

Dagar, J. C. (2017). Agroforestry : Four decades of research development. *Indian J. Agrofor, January 1999*, 0–32. https://www.researchgate.net/profile/J-Dagar-2/publication/353259684_Agroforestry_Four_decades_of_research_development/links/60efdb639541032c6d3ec0aa/Agroforestry-Four-decades-of-research-development.pdf

Das, B., Bhardwaj, P. K., Chaudhary, S. K., Pathaw, N., Singh, H. K., Tampha, S., Singh, K. K., Sharma, N., & Mukherjee, P. K. (2024). Bioeconomy and ethnopharmacology – Translational perspective and sustainability of the bioresources of northeast region of India. *Journal of Ethnopharmacology, 330*(April), 118203. DOI: 10.1016/j.jep.2024.118203 PMID: 38641075

Das, B. R., Bhuyan, M. J., Deka, N., & Bhattacharya, P. (2025). Assessing homestead agroforestry-based livelihoods in Sivasagar District of Assam, India, using livelihood significance index (LSI). *Forest Policy and Economics, 170*, 103388. Advance online publication. DOI: 10.1016/j.forpol.2024.103388

de Queiroz-Stein, G., & Siegel, K. M. (2023). Possibilities for mainstreaming biodiversity? Two perspectives on the concept of bioeconomy. *Earth System Governance, 17*(May), 100181. DOI: 10.1016/j.esg.2023.100181

Dhanya, B., Purushothaman, S., & Viswanath, S. (2016). Economic rationale of traditional agroforestry systems: A case-study of Ficus trees in semiarid agroecosystems of Karnataka, southern India. *Forests, Trees and Livelihoods, 25*(4), 267–281. DOI: 10.1080/14728028.2016.1218800

Dominicis, L. F., Lima, M. F. B., Souza, Á. N., Joaquim, M. S., Araújo, J. B. C. N., Coelho, L. M.Junior, Ribeiro, J. F., & do Santos, P. P. (2023). Payment for Environmental Services and the Financial Viability of Agroforestry Systems: An Integrated Analysis of Socio-Environmental Projects in the Descoberto Basin—Federal District. *Forests, 14*(10), 2110. Advance online publication. DOI: 10.3390/f14102110

Dondina, O., Saura, S., Bani, L., & Mateo-Sánchez, M. C. (2018). Enhancing connectivity in agroecosystems: Focus on the best existing corridors or on new pathways? *Landscape Ecology, 33*(10), 1741–1756. DOI: 10.1007/s10980-018-0698-9

Edo, M., Entling, M. H., & Rösch, V. (2023). Agroforestry supports high bird diversity in European farmland. *Agronomy for Sustainable Development*, *44*(1), 1. Advance online publication. DOI: 10.1007/s13593-023-00936-2

Estreguil, C., Caudullo, G., Rega, C., & Paracchini, M. L. (2016). *Enhancing connectivity improving green enhancing connectivity improving green infrastructure. Cost-benefit solutions for forest and agri-environment A pilot study in Lombardy.* Issue October., DOI: 10.2788/170924

Eyasu, G., & Gebrewahid, Y. (2024). Potential of agroforestry practices on woody species diversity and composition: Implication for conservation of indigenous trees in selected districts of Tigray, Northern Ethiopia. *Asian Journal of Research in Agriculture and Forestry*, *10*(4), 88–100. DOI: 10.9734/ajraf/2024/v10i4319

Fahad, S., Chavan, S. B., Chichaghare, A. R., Uthappa, A. R., Kumar, M., Kakade, V., Pradhan, A., Jinger, D., Rawale, G., Yadav, D. K., Kumar, V., Farooq, T. H., Ali, B., Sawant, A. V., Saud, S., Chen, S., & Poczai, P. (2022). Agroforestry systems for soil health improvement and maintenance. *Sustainability (Basel)*, *14*(22), 14877. Advance online publication. DOI: 10.3390/su142214877

Fikry, M. Y., & Sarjan, M. (2024). Peran agroforestri dalam mendukung pengelolaan sumberdaya alam berkelanjutan. *LAMBDA : Jurnal Ilmiah Pendidikan MIPA Dan Aplikasinya*, *4*(1), 16–22. DOI: 10.58218/lambda.v4i1.846

Global Bioeconomy Summit. (2020). *Expanding the Sustainable Bioeconomy-Vision and Way Forward* (Communiqu'e of the Global Bioeconomy Summit (ed.)). Communiqu'e of the Global Bioeconomy Summit.

Goncalves, N., Andrade, D., Batista, A., Cullen, L.Jr, Souza, A., Gomes, H., & Uezu, A. (2021). Potential economic impact of carbon sequestration in coffee agroforestry systems. *Agroforestry Systems*, *95*(2), 419–430. DOI: 10.1007/s10457-020-00569-4

Hanisch, A. L. (2024). Improving an agroforestry system with livestock in southern Brazil. *Tropical Forest Issues*, *62*(62), 2–5. DOI: 10.55515/JCWC8581

Hawkins, J. W., Gallagher, E. J., van der Haar, S., Sevor, M. K. E., Weng, X., Rufino, M. C., & Schoneveld, G. C. (2024). Low-emissions and profitable cocoa through moderate-shade agroforestry: Insights from Ghana. *Agriculture, Ecosystems & Environment*, *367*(January), 108961. DOI: 10.1016/j.agee.2024.108961

Helmayuni, & Mardianto. (2021). Economic valuation of "Parak" traditional agroforestry west sumatera (case study in Nagari Paninggahan, Solok regency. *INFLUENCE International Journal of Science Review, 3*(3), 34–42.

Hidalgo, P. J., Hernández, H., Sánchez-Almendro, A. J., López-Tirado, J., Vessella, F., & Porras, R. (2021). Fragmentation and connectivity of island forests in agricultural mediterranean environments: A comparative study between the guadalquivir valley (Spain) and the apulia region (Italy). *Forests*, *12*(9), 1201. Advance online publication. DOI: 10.3390/f12091201

Holyoak, M. (2008). Connectance and connectivity. *Encyclopedia of Ecology, Five-Volume Set, 1–5*, 737–743. DOI: 10.1016/B978-008045405-4.00095-1

Indratna, A. D., Putri, A. P. I. K. S., Azzahra, E. O., Oktavia, M. S., Widodo, T., Agustina, A., & Pertiwi, Y. A. B. (2023). Potential economic value of non-wood forest products in agroforestry system at Gempolan Village, Kerjo District, Karanganyar Regency: An economic perspective. *IOP Conference Series. Earth and Environmental Science*, *1220*(1), 012014. Advance online publication. DOI: 10.1088/1755-1315/1220/1/012014

Jambut, A., Hassan Zaki, P., & Hasmadi Ismail, M. (2022). Profitability of agroforestry in tumpangsari-gunung forest reserve arjuna-lalijiwo, East Java, Indonesia. *Вестник Атырауского Университета Имени Халела Досмухамедова*, *64*(1), 63–73. DOI: 10.47649/vau.2022.v64.i1.07

Jitendra, A. (2024). Linking ecological restoration and biodiversity conservation with bioeconomy. *Biodiversity and Bioeconomy: Status Quo, Challenges, and Opportunities*, 61–79. DOI: 10.1016/B978-0-323-95482-2.00004-3

Jose, S. (2009). Agroforestry for ecosystem services and environmental benefits: An overview. *Agroforestry Systems*, *76*(1), 1–10. https://www.scopus.com/inward/record.uri?eid=2-s2.0-63649140240&doi=10.1007%2Fs10457-009-9229-7&partnerID=40&md5=c297b224f129b06436d775ac90490152. DOI: 10.1007/s10457-009-9229-7

Jose, S., & Udawatta, R. P. (2021). Agroforestry for ecosystem services: An introduction. In *Agroforestry and Ecosystem Services* (pp. 1–17). Springer International Publishing., DOI: 10.1007/978-3-030-80060-4_1

Joshi, P. (2023). Silvopasture: A sustainable agro-system practice for rural farmers. *Agriculture Extension in Developing Countries*, *1*(May), 27–30. DOI: 10.26480/aedc.02.2023.46.48

Joshi, R., Sharma, B., Singh, H., Dhakal, N., Ayer, S., & Maraseni, T. (2024). Poplar Plantation as an Agroforestry Approach: Economic Benefits and Its Role in Carbon Sequestration in North India. *Journal of Resources and Ecology*, *15*(4), 880–888. DOI: 10.5814/j.issn.1674-764x.2024.04.009

Kag, S., Kukreti, N., Kumar, P., & Kataria, R. (2024). Bioeconomy for Sustainable Bioenergy and Biofuel Generation. In *Bioeconomy for Sustainability* (pp. 83–105). Springer Nature Singapore., DOI: 10.1007/978-981-97-1837-5_4

Kaur, A., & Monga, R. (2024). Recent trends in agroforestry research and development. In *Futuristic Trends in Agriculture Engineering & Food Sciences: Vol. 3 Book 14* (pp. 183–188). Iterative International Publisher, Selfypage Developers Pvt Ltd. DOI: 10.58532/V3BCAG14P5CH2

Kay, S., Graves, A., Palma, J. H. N., Moreno, G., Roces-Díaz, J. V., Aviron, S., Chouvardas, D., Crous-Duran, J., Ferreiro-Domínguez, N., García de Jalón, S., Măcicăşan, V., Mosquera-Losada, M. R., Pantera, A., Santiago-Freijanes, J. J., Szerencsits, E., Torralba, M., Burgess, P. J., & Herzog, F. (2019). Agroforestry is paying off – Economic evaluation of ecosystem services in European landscapes with and without agroforestry systems. *Ecosystem Services, 36*, 100896. https://www.scopus.com/inward/record.uri?eid=2-s2.0-85060894709&doi=10.1016%2Fj.ecoser.2019.100896&partnerID=40&md5=d17aa01b3e132884047dedd94fd84244. DOI: 10.1016/j.ecoser.2019.100896

Khan, N., Jhariya, M. K., Raj, A., Banerjee, A., Meena, R. S., Bargali, S. S., Yadav, S. K., & Kumawat, A. (2021). Agroforestry and its services for soil management and sustainability. In *Sustainable Intensification for Agroecosystem Services and Management* (pp. 353–377). Springer Singapore., DOI: 10.1007/978-981-16-3207-5_11

Kremer, R. J. (2021). Soil health ecosystem services of agroforestry. In *Agroforestry and Ecosystem Services* (pp. 49–63). Springer International Publishing., DOI: 10.1007/978-3-030-80060-4_3

Kriesch, L., & Losacker, S. (2024). Bioeconomy firms and where to find them. *Region, 11*(1), 55–78. DOI: 10.18335/region.v11i1.523

Kuyah, S., Öborn, I., & Jonsson, M. (2017). Regulating ecosystem services delivered in agroforestry systems. In *Agroforestry: Anecdotal to Modern Science* (pp. 797–815). Springer Singapore. DOI: 10.1007/978-981-10-7650-3_33

Lasmar, D. J., Ferreira, I. S., & Silva, M. L. A. (2024). Bioeconomia: perspectiva de desenvolvimento de cadeias produtivas da biodiversidade do amazonas. In *A multidisciplinaridade em foco: ensino, pesquisa e extensão* (Vol. 3, pp. 42–66). DOI: 10.37885/241017841

Lee, B., Rhee, H., Kim, S., Lee, J.-W., Koo, S., Lee, S.-J., Alounsavath, P., & Kim, Y.-S. (2021). Assessing sustainable bamboo-based income generation using a value chain approach: Case study of nongboua village in Lao PDR. *Forests, 12*(2), 153. Advance online publication. DOI: 10.3390/f12020153

Lehmann, L. M., Smith, J., Westaway, S., Pisanelli, A., Russo, G., Borek, R., Sandor, M., Gliga, A., Smith, L., & Ghaley, B. B. (2020). Productivity and economic evaluation of agroforestry systems for sustainable production of food and non-food products. *Sustainability (Basel), 12*(13), 5429. Advance online publication. DOI: 10.3390/su12135429

Maciejczak, M. (2023). Nature based innovations in the development of bioeconomy. *Annals of the Polish Association of Agricultural and Agribusiness Economists, 25*(2), 97–108. DOI: 10.5604/01.3001.0053.6818

Magry, M. A., Cahill, D., Rookes, J., & Narula, S. A. (2024). Non-timber forest products: Evolution, development and research. *Biodiversity (Nepean), 25*(2), 120–141. DOI: 10.1080/14888386.2024.2336216

Mathur, I., & Bhattacharya, P. (2024). Perception of agroforestry practices and factors influencing adoption among shifting cultivators in Tripura, India. *Forests, Trees and Livelihoods, 33*(1), 23–41. DOI: 10.1080/14728028.2023.2286022

Mekonnen, M., Worku, T., Yitaferu, B., Cerdà, A., & Keesstra, S. (2023). Economics of agroforestry land use system, Upper Blue Nile Basin, northwest Ethiopia. *Agroforestry Systems, 97*(3), 305–317. DOI: 10.1007/s10457-021-00612-y

Mlambo, D., Sebata, A., Chichinye, A., & Mabidi, A. (2024). Agroforestry and biodiversity conservation. *Agroforestry for Carbon and Ecosystem Management*, 63–78. DOI: 10.1016/B978-0-323-95393-1.00008-7

Molua, E. L. (2005). The economics of tropical agroforestry systems: The case of agroforestry farms in Cameroon. *Forest Policy and Economics, 7*(2), 199–211. DOI: 10.1016/S1389-9341(03)00032-7

Mondal, S., Angon, P. B., & Roy, A. R. (2023). Ecological Advancements and Developments of Agroforestry. *Turkish Journal of Agriculture - Food Science and Technology, 11*(12), 2476–2480. DOI: 10.24925/turjaf.v11i12.2476-2480.6172

Mony, C., Landais, Q., Georges, R., Butet, A., Burel, F., Jambon, O., Gouesbet, V., & Ernoult, A. (2022). Effects of connectivity on seed dispersal patterns in hedgerows. *Journal of Vegetation Science, 33*(1), e13113. Advance online publication. DOI: 10.1111/jvs.13113

Nair, P. K. R. (1998). Directions in tropical agroforestry research: past, present, and future. *Directions in Tropical Agroforestry Research: Adapted from Selected Papers Presented to a Symposium on Tropical Agroforestry Organized in Connection with the Annual Meetings of the American Society of Agronomy,* 5 November 1996, Indianapolis, Indiana, USA, 223–245.

Nair, P. K. R., Mohan Kumar, B., & Nair, V. D. (2009). Agroforestry as a strategy for carbon sequestration. *Journal of Plant Nutrition and Soil Science*, *172*(1), 10–23. DOI: 10.1002/jpln.200800030

Nurrochmat, N. A., Purnomo, H., & Erbaugh, J. T. (2023). Developing sustainable models of agroforest-landscape in Bogor, Indonesia. *Biodiversitas (Surakarta)*, *24*(8), 4324–4333. DOI: 10.13057/biodiv/d240814

Obeng, E. A., Obiri, B. D., Oduro, K. A., Pentsil, S., Anglaaere, L. C., Foli, E. G., & Ofori, D. A. (2020). Economic value of non-market ecosystem services derived from trees on cocoa farms. *Current Research in Environmental Sustainability*, *2*, 100019. DOI: 10.1016/j.crsust.2020.100019

Oktoyoki, H., Pratama, B., Putra, Y. P., Novrian, W., Douni, J. B., & Ansiska, P. (2024). Financial feasibility study and development of technical specifications for coffee agroforestry for community forest farmers with revolving fund support for productivity optimization. *Jendela Pengetahuan*, *17*(2), 239–251. DOI: 10.30598/jp17iss2pp239-251

Oliveira, G. M. T., Santana, A. C., Oliveira, E. S., Silva, R. J., Santos, W. A. S., Santana, Á. L., & Costa, V. C. N. (2020). The value of agroforestry ecosystem services provided in rural communities in the Eastern Amazon (Tomé-Açu – PA, Brazil). *Journal of Agricultural Studies*, *8*(4), 202. DOI: 10.5296/jas.v8i4.17338

Otter, V., & Deutsch, M. (2023). Did policy lose sight of the wood for the trees? An UTAUT-based partial least squares estimation of farmers acceptance of innovative sustainable land use systems. *Land Use Policy*, *126*, 106467. Advance online publication. DOI: 10.1016/j.landusepol.2022.106467

Pa, C. M., Salimath, S. K., Hegde, R., v, M., Nl, D. D., & Gooli, M. (2024). Agrobiodiversity vis-à-vis agroforestry: Cultivating a sustainable future. *International Journal of Research in Agronomy*, *7*(3S), 176–186. DOI: 10.33545/2618060X.2024.v7.i3Sc.428

Pancholi, R., Yadav, R., Gupta, H., Vasure, N., Choudhary, S., Singh, M. N., & Rastogi, M. (2023). The Role of Agroforestry Systems in Enhancing Climate Resilience and Sustainability- A Review. *International Journal of Environment and Climate Change*, *13*(11), 4342–4353. DOI: 10.9734/ijecc/2023/v13i113615

Pantera, M.-L., R., M., Herzog, F., & Herder, M. (2021). Agroforestry and the environment. *Agroforestry Systems*, *95*(ue 5), 767–774. DOI: 10.1007/s10457-021-00640-8

Parente-Laverde, A. M. (2020). Value chain and economic development: The case of the Colombian coffee industry. *Organizations and Markets in Emerging Economies, 11*(1), 173–188. DOI: 10.15388/omee.2020.11.29

Peri, P. L., Dube, F., & Varella, A. C. (2016). Silvopastoral systems in the subtropical and temperate zones of South America: an overview. *Silvopastoral Systems in Southern South America*, 1–8.

Prasad, M. N. V. (2024). Bioremediation, bioeconomy, circular economy, and circular bioeconomy—Strategies for sustainability. *Bioremediation and Bioeconomy: A Circular Economy Approach, Second Edition*, 3–32. DOI: 10.1016/B978-0-443-16120-9.00025-X

Putra, P. B., Cahyono, S. A., Agus, C., Susanti, P. D., & Indrajaya, Y. (2022). The Role of Integrated Organic Cycle Farming in Tropical Agroforestry Systems for Sustainable Food Production. In *World Sustainability Series* (pp. 171–182). DOI: 10.1007/978-3-030-98617-9_10

Rahman, S. A., Samsudin, Y. B., Bhatta, K. P., Aryal, A., Hayati, D., Cahya, M., Trihadmojo, B., Husain, I., Andini, S., Narulita, S., Rahman, M. A., Lu, X., Emang, D., Abimbola, A. I., Kustanti, A., & Larjavaara, M. (2023). The Role of Agroforestry Systems for Enhancing Biodiversity and Provision of Ecosystem Services in Agricultural Landscapes in Southeast Asia. In *In Agroforestry for Sustainable Intensification of Agriculture in Asia and Africa* (pp. 303–319). Issue April., DOI: 10.1007/978-981-19-4602-8_10

Raj, A., Jhariya, M. K., Banerjee, A., Meena, R. S., Kumar, S., Devi, A., & Poonam. (2024). Agroforestry and ecosystem services. *Agroforestry for Carbon and Ecosystem Management*, 205–221. DOI: 10.1016/B978-0-323-95393-1.00027-0

Raj, A., Jhariya, M. K., Banerjee, A., Meena, R. S., Nema, S., Khan, N., Yadav, S. K., & Pradhan, G. (2022). Agroforestry a model for ecological sustainability. *Natural Resources Conservation and Advances for Sustainability*, 289–307. DOI: 10.1016/B978-0-12-822976-7.00002-8

Rendón-Sandoval, F. J., Casas, A., Moreno-Calles, A. I., Torres-García, I., & García-Frapolli, E. (2023). Traditional agroforestry systems and conservation of native plant diversity of seasonally dry tropical forests. *Sustainability (Switzerland, 12*(11). DOI: 10.3390/su12114600

Revollo-Fernández, D. A., Lithgow, D., Von Thaden, J. J., Salazar-Vargas, M. D. P., & Rodríguez de los Santos, A. (2024). Unlocking Local and Regional Development through Nature-Based Tourism: Exploring the Potential of Agroforestry and Regenerative Livestock Farming in Mexico. *Economies, 12*(6). https://www.scopus.com/inward/record.uri?eid=2-s2.0-85196781687&doi=10.3390%2Feconomies12060137&partnerID=40&md5=7fbd5dcbab6623865b01eacfa7cc7b4a

Rice, R. A. (2008). Agricultural intensification within agroforestry: The case of coffee and wood products. *Agriculture, Ecosystems & Environment, 128*(4), 212–218. DOI: 10.1016/j.agee.2008.06.007

Roshetko, J. M., & Bertomeu, M. (2015). Multi-species and multifunctional smallholder tree farming systems in Southeast Asia: Timber, NTFPs, plus environmental benefits. *Annals of Silvicultural Research, 39*(ue 2), 62–69. DOI: 10.12899/ASR-1125

Salve, A., Tiwari, C., & Baghele, L. (2022). Impact of Agroforestry Systems: A Review. *Asian Journal of Microbiology, Biotechnology & Environmental Sciences : AJMBES*, (June), 214–223. DOI: 10.53550/AJMBES.2022.v24i02.002

Santana, A. C., Santana, Á. L., Santana, Á. L., Oliveira, G. M. T., & Dos Santos, M. A. S. (2023). Bioeconomic evaluation of an agroforestry system and the potential to recover degraded areas and capitalize producers in the state of Pará, Brazilian Amazon. *Desenvolvimento e Meio Ambiente, 61*, 439–455. DOI: 10.5380/dma.v61i0.80516

Santos, M., Cajaiba, R. L., Bastos, R., Gonzalez, D., Petrescu Bakış, A. L., Ferreira, D., Leote, P., Silva, W., Cabral, J. A., Gonçalves, B., & Mosquera-Losada, M. R. (2022). Why Do Agroforestry Systems Enhance Biodiversity? Evidence From Habitat Amount Hypothesis Predictions. *Frontiers in Ecology and Evolution, 9*, 630151. Advance online publication. DOI: 10.3389/fevo.2021.630151

Santosa, F. J., & Padmaningrum, D. (2024). The economic impact of agroforestry practice in production forest areas, Central Java province, Indonesia. *Scientific Horizons, 27*(4), 141–153. DOI: 10.48077/scihor4.2024.141

Sarkar, P. K., Dhakar, M. K., Das, B., Das, B., Sarkar, P., & Bhatt, B. P. (2024). Medicinal Plant-Based Agroforestry Systems. In *Bioprospecting of Ethnomedicinal Plant Resources* (pp. 35–54). Apple Academic Press., DOI: 10.1201/9781003451488-3

Satish, P., Madiwalar, A. F., Lallawmkimi, M. C., Reddy, K. J., Parveen, S. P. A., Laxman, T., Kiruba, M., & Anand, G. (2024). Agroforestry: Multifunctional Benefits and Implementation Strategies. *Journal of Geography. Environment and Earth Science International, 28*(10), 1–12. DOI: 10.9734/jgeesi/2024/v28i10821

Singh, K., & Mishra, N. (2024). Biodiversity and bioeconomy: are these two faces of a single coin? *Biodiversity and Bioeconomy: Status Quo, Challenges, and Opportunities*, 3–23. DOI: 10.1016/B978-0-323-95482-2.00001-8

Singh, V., Raj, A., Jhariya, M. K., & Thakur, S. (2022). Economic evaluation of agroforestry and non-agroforestry systems in Eastern Uttar Pradesh, India. *Vegetos*, *35*(3), 810–815. https://www.scopus.com/inward/record.uri?eid=2-s2.0-85124725622&doi=10.1007%2Fs42535-022-00348-9&partnerID=40&md5=97494bb75c00eca7c8ffa64575726e4b. DOI: 10.1007/s42535-022-00348-9

Stille, L., Smeets, E., Wicke, B., Singh, R., & Singh, G. (2011). The economic performance of four (agro-) forestry systems on alkaline soils in the state of Haryana in India. *Energy for Sustainable Development*, *15*(4), 388–397. DOI: 10.1016/j.esd.2011.08.002

Tata, H. L. (2019). Is smallholder farmer maintaining biodiversity in rattan agroforest? *IOP Conference Series. Earth and Environmental Science*, *298*(1), 012034. https://www.scopus.com/inward/record.uri?eid=2-s2.0-85070640436&doi=10.1088%2F1755-1315%2F298%2F1%2F012034&partnerID=40&md5=7641b3ac6f556c86d823d1ffa9f59b41. DOI: 10.1088/1755-1315/298/1/012034

Titova, E. S., Shishkin, S. S., & Shtykhno, D. A. (2023). Bioeconomy as One of the Ways To Sustainable Development of Russian Regions. *Federalism*, *28*(1), 56–79. DOI: 10.21686/2073-1051-2023-1-56-79

Toplicean, I. M., & Datcu, A. D. (2024). An Overview on Bioeconomy in Agricultural Sector, Biomass Production, Recycling Methods, and Circular Economy Considerations. *Agriculture (Switzerland*, *14*(7). DOI: 10.3390/agriculture14071143

Torquebiau, E. (2024). *The agroforestry-biodiversity-climate change nexus*. 21–27.

Tremblay, S., Lucotte, M., Revéret, J.-P., Davidson, R., Mertens, F., Passos, C. J. S., & Romaña, C. A. (2015). Agroforestry systems as a profitable alternative to slash and burn practices in small-scale agriculture of the Brazilian Amazon. *Agroforestry Systems*, *89*(2), 193–204. DOI: 10.1007/s10457-014-9753-y

Udawatta, P., R., R., L., & Jose, S. (. (2019). Agroforestry and biodiversity. *Sustainability*, *11*(10), 2879. DOI: 10.3390/su11102879

Ulya, N. A., Harianja, A. H., Sayekti, A. L., Yulianti, A., Djaenudin, D., Martin, E., Hariyadi, H., Witjaksono, J., Malau, L. R. E., Mudhofir, M. R. T., Mudhofir, M. R. T., & Astana, S. (2023). Coffee agroforestry as an alternative to the implementation of green economy practices in Indonesia: A systematic review. *AIMS Agriculture and Food*, *8*(3), 762–788. DOI: 10.3934/agrfood.2023041

Verma, K., Sharma, P., Bhardwaj, D. R., Sharma, V., & Thakur, P. (2024). Towards a greener future: Scaling up agroforestry for global sustainability. In *Green and Low-Carbon Economy*. Green and Low-Carbon Economy., DOI: 10.47852/bonviewGLCE42022481

Vostriakova, V. I. (2023). Bioeconomy as a complex system of ensuring the country's sustainable development. *Bìznes Ìnform, 1*(540), 167–177. DOI: 10.32983/2222-4459-2023-1-167-177

Wartman, P., Acker, R., & Martin, R. C. (2018). Temperate agroforestry: How forest garden systems combined with people-based ethics can transform culture. *Sustainability (Switzerland, 10*(ue 7)). DOI: 10.3390/su10072246

Wicke, B., Smeets, E. M. W., Akanda, R., Stille, L., Singh, R. K., Awan, A. R., Mahmood, K., & Faaij, A. P. C. (2013). Biomass production in agroforestry and forestry systems on salt-affected soils in South Asia: Exploration of the GHG balance and economic performance of three case studies. *Journal of Environmental Management, 127*, 324–334. DOI: 10.1016/j.jenvman.2013.05.060 PMID: 23810966

Woźniak-Gientka, E., & Tyczewska, A. (2023). Genome editing in plants as a key technology in sustainable bioeconomy. *EFB Bioeconomy Journal, 3*(February), 100057. DOI: 10.1016/j.bioeco.2023.100057

Wright, D. R., Gordon, A., Bennett, R. E., Selinske, M. J., Lentini, P. E., Garrard, G. E., Rodewald, A. D., & Bekessy, S. A. (2024). Biodiverse coffee plantations provide co-benefits without compromising yield. *Journal of Sustainable Agriculture and Environment, 3*(3), sae2.70005. Advance online publication. DOI: 10.1002/sae2.70005

Yashmita-Ulman, , Singh, M., Kumar, A., & Sharma, M.Yashmita-Ulman. (2021). Conservation of plant diversity in agroforestry systems in a biodiversity Hotspot Region of Northeast India. *Agricultural Research, 10*(4), 569–581. DOI: 10.1007/s40003-020-00525-9

Zhu, X., Liu, W., Chen, J., Bruijnzeel, L. A., Mao, Z., Yang, X., Cardinael, R., Meng, F. R., Sidle, R. C., Seitz, S., Nair, V. D., Nanko, K., Zou, X., Chen, C., & Jiang, X. J. (2020). Reductions in water, soil and nutrient losses and pesticide pollution in agroforestry practices: A review of evidence and processes. *Plant and Soil, 453*(1–2), 45–86. DOI: 10.1007/s11104-019-04377-3

Chapter 9
Integrating Millet and Tribal Agroforestry for Sustainable Agriculture in North Karnataka, India

Guneeta Kaur Gill
https://orcid.org/0000-0002-8534-5723
St. Claret College, Bangalore, India

Sridip Basu
https://orcid.org/0009-0009-4265-2959
St. Claret College, Bangalore, India

Geethu Varghese
https://orcid.org/0000-0002-3330-9843
St. Claret College, Bangalore, India

Zoolu Tanwar
https://orcid.org/0009-0008-7578-5830
St. Claret College, Bangalore, India

R. Tarun
https://orcid.org/0009-0007-5137-2911
St. Claret College, Bangalore, India

Nishchiththa N. Halale
https://orcid.org/0009-0002-3363-255X
St. Claret College, Bangalore, India

ABSTRACT

This chapter explores sustainable farming methods that effectively incorporate biodiversity into food production systems, specifically highlighting tribal agroforestry and millet farming in North Karnataka, India. Utilizing a mixed-methods approach, the research analyses how traditional agroforestry practices maintain ecological functions while boosting agricultural productivity. The study assesses the combined potential of indigenous knowledge and contemporary techniques such as organic farming and agroecology in optimizing crop production within biodiversity-rich environments. Results reveal significant positive relationships between the integration of biodiversity and enhancements in ecosystem resilience and the socioeconomic well-being of tribal communities. The chapter provides evidence-based insights

DOI: 10.4018/979-8-3373-2003-8.ch009

for agricultural policy and practice, demonstrating how farming systems centered on biodiversity significantly contribute to various Sustainable Development Goals, including zero hunger, responsible consumption, and climate action.

INTRODUCTION

Tribal communities in North Karnataka, India, are stewards of sustainable agricultural practices that enhance biodiversity. They utilize agroforestry systems that integrate food crops with trees, fostering diverse and resilient landscapes. Central to these practices is millet, a nutritious and climate-resilient crop essential for food security and cultural heritage. Millets thrive in the region's arid climate and fit well within traditional agroforestry, emphasizing soil health, water conservation, and ecosystem preservation.

The increasing global demand for millet is driving interest in scaling up its production through modern agricultural methods, which may impact the biodiversity-friendly practices of tribal communities. This chapter explores how traditional agroforestry systems can be integrated with contemporary techniques to promote the commercial viability of millets while preserving ecological balance and the cultural integrity of these communities. The aim is to create pathways that satisfy market demands while enhancing biodiversity and sustainability in food production.

The traditional agricultural system used by tribes in northern Karnataka is both ecologically balanced and productive. This sustainable agroforestry practice combines trees, crops, and sometimes livestock, enhancing biodiversity, soil fertility, and climatic conditions while promoting livelihoods. By mimicking natural ecosystems, agroforestry fosters long-term ecological resilience and is essential for food security, medicine, fodder, housing, and tools, deeply connecting to the social and cultural fabric of tribal life.

The agroforestry practices of the tribals involve a wide range of crops such as millet, different seeds, vegetables and fruits to be grown parallel to the local species of trees such as Ficus, Neem and Tamarindus indica, etc. Millets, particularly, have been a staple crop, valued for their nutritional richness and adaptability to arid conditions. However, with the increasing requirement of commercialisation, the government has tried to club agriculture and commerce - for example, the millet mission, which required farmers to escalate millet production using advanced and modern techniques and technologies. No doubt it benefits economically, but at the same time, it threatens traditional practices. Moreover, rapid land-use changes, deforestation, and policy shifts risk undermining the indigenous knowledge systems that have sustained these landscapes for generations.

This study aims to assess how Agroforestry techniques used in the northern Karnataka region contribute to biodiversity conservation, climate resilience, and sustainable agricultural practices. It seeks to highlight the interplay between traditional knowledge and contemporary requirements while proposing paths that benefit tribal cultural heritage regarding environmental sustainability and economic viability. Moreover, the study explores the challenges and opportunities of integrating these time-honoured practices with modern agricultural strategies to maintain ecological functions and preserve tribal cultural heritage.

The study majorly talks about the agricultural practices performed by the Siddi tribe, also known as Habshi or Sheedi, which is an ethnic group in India of African descent, primarily found in the states of Karnataka (especially Uttara Kannada district), Gujarat, and parts of Maharashtra. P. Bhattacharya (2023) underscored the significance of agriculture by exploring the livelihood systems of the Siddi, Kunabi, and Gond tribes in Uttara Kannada and investigating their reliance on wage labour and forest resources. The tribes' traditional knowledge highlights the importance of preserving Indigenous practices for sustainable development, particularly in light of contemporary challenges and socioeconomic marginalisation (Bhattacharya, 2023). Furthermore, Siddis are unaffected by this globalised age's widespread progress and modernisation (Altieri, 2018). Increased access to education has enhanced awareness among tribal communities about their cultural heritage and helped them maintain connections with dispersed community members. Improving the economy enhances their prosperity and helps them consolidate their identity and a good life (Siddi & Dalawai, 2024). The governmental and non-governmental organisations have been working for their upliftment, and there have been significant improvements in the condition of these tribes (Karmwar & Smita, 2022).

Agroforestry, the integration of trees, crops, and sometimes livestock, has historically been a cornerstone of sustainable farming practices among tribal communities in Karnataka, India. These practices are beneficial because they promote biodiversity, maintain soil fertility, and improve ecosystem services, promoting livelihood and ecological sustainability (Nair, 2011).

Many studies have emphasised the versatility of the Agroforestry system (AFS). Furthermore, by providing a variety of habitats for plants and animals, AFS guarantees food security and plays a vital part in biodiversity conservation. Likewise, Jose (2009) highlighted that AFS can reduce land degradation and help promote ecological resilience, a significant requirement in semi-arid regions like North Karnataka.

Role of Traditional Agroforestry in Biodiversity Conservation

In traditional tribal agroforestry systems, crops such as millets (e.g., finger millet and pearl millet), pulses, oilseeds, and indigenous vegetables are intercropped with native trees like Ficus species, neem, tamarind, and acacia. These systems contribute to ecological stability and cultural identity (Patil et al., 2018). They further argued for long-term sustainability by focusing on the fact that combining traditional knowledge with tribal farming practices can help maintain ecological balance and Agroforestry.

The role of native tree species in supporting ecosystem services in AFS is well-documented. According to Rahman et al. (2023), local species significantly improve soil quality and organic matter. Crucially, they offer a habitat for organisms, which boosts the farming system's resilience to climate stress. Similarly, Bhagwat et al. (2008) demonstrated that traditional agroforestry systems in the Western Ghats contribute substantially to biodiversity conservation and can be considered as important tools for landscape-level conservation strategies and according to one of his research projects based in Kodagu, Karnataka, highlights how coffee-based agroforestry systems retain significant native biodiversity, even under economic pressures to intensify production. Moreover, decisions in most traditional AFS favour the immediate production of crops instead of trying to maintain the ecosystem services (Sanchez, 1995).

Recent studies have emphasised the need for farmers to be aware of the long-term benefits of ecosystem services provided by trees, highlighting the importance of rural extension services to educate farmers. They have also demonstrated that Indigenous agroforestry practices can simultaneously support agricultural yields and biodiversity, underlining the importance of maintaining these landscapes. Recognising the escalating dangers of climate change and land degradation, local species in agroforestry systems bolster the ecosystem's resilience.

Challenges of Modern Agricultural Intensification

With the increase in demand for millet and other commercial crops, there came the need for initiatives, as done by the Indian government's millet mission, which aimed at scaling up the production of the crops. However, agricultural intensification often results in monocultures, chemical dependency, and biodiversity loss (Pretty et al. 2009). Patil et al. (2018) argued that replacing traditional mixed farming with single-crop farming threatens the ecological balance and the cultural identities of tribal communities.

Incorporating very high varieties and artificial fertilisers disturbs traditional forestry practices, especially in dry-land regions and Mandya. While commercially attractive, Rahman et al. (2023) showed that exotic tree species offer fewer ecosystem benefits than native trees.

Using modern technologies to intensify crop productivity is good, but raises concerns about the natural aspects. For example, it has raised concerns about the erosion of biodiversity and traditional systems, which eventually trade off the need to combine modern technologies with Indigenous practices. Modern agricultural intensification aimed at increasing millet production has raised concerns about the erosion of biodiversity in traditional systems. The Millet mission launched by the Indian government aimed at increasing the millet yield, which supported the concept of single crop cultivation, and it eventually reduced the diversity farming system and, as a result, compromised ecological sustainability (Islam & Manaloor, 2021).

Effects of Agricultural Commercialisation

Commercialising agriculture, especially millets, presents opportunities and challenges for tribal communities. As per (Altieri, 2018), profitable agriculture compared to traditional AFS often leads to monoculture crops, which compromises tradition and becomes heavily dependent upon the external inputs, unlike traditional forestry systems, present in the pathway to improve and maintain the diversity of crops, enhance soil fertility, and promote conservation of water.

The social effects that agricultural commercialisation has on tribal communities are still undiscovered. However, according to Malhotra et al. (2021), it has been found that money-driven agriculture often leads to the loss of cultural identity, community resilience, and diminishing of traditional knowledge systems. Therefore, integrating sustainable practices into commercialisation strategies is vital for protecting tribal livelihoods.

Climate Resilience through Agroforestry

Climate resilience is another critical arena to focus on. Agroforestry systems have been recognised for their ability to buffer farms against climate variability. Lin (2007) stated that, compared to monocultures and single cropping systems, diversified farms are more flexible to extreme climate changes, whether droughts or floods. In the context of North Karnataka, traditional agroforestry with drought-tolerant crops like millets and deep-rooted trees offers a viable strategy for climate adaptation.

Policy Gaps and Challenges in Tribal Agroforestry in Karnataka

Karnataka faces significant challenges with the misuse of funds for Scheduled Castes (SCSP) and Tribal Sub-Plans (TSP). In 2025–26, 11,896 crore was diverted to state schemes, violating the Karnataka SCSP/TSP Act, 2013 (The Hindu Bureau, 2025; The News Minute, 2025). Many departments struggle to utilize these funds due to poor design and lack of engagement.

While the National Agroforestry Policy encourages tree planting, restrictive regulations hinder tribal harvesting, and permit requirements lead to delays and exploitation. The unimplemented National Transit Pass System limits timber sales in remote areas (Singh et al., 2016).

Current agricultural schemes, including the Pradhan Mantri Krishi Sinchayee Yojana (PMKSY), prioritize monoculture irrigation over agroforestry benefits. Delays in land titling under the Forest Rights Act leave tribes without legal planting rights. Furthermore, women tribal farmers are often excluded from decision-making, limiting their access to resources. The Voluntary Tribal Relocation Program inadequately supports community-led agroforestry and provides poor training for resettled farmers (The News Minute, 2025). An integrated, community-focused governance approach is essential to address these issues.

Policy Support and Future Directions

Agroforestry systems provide many benefits for the environment and crop productivity. However, it remains unnoticed when policy support in future directions is considered. There is an urgent need to integrate biodiversity conservation into agricultural development frameworks. Government initiatives should promote native species planting, agroecological farming, and community seed banks to safeguard agricultural biodiversity.

Therefore, the research should primarily focus on considering the tribal knowledge of their ecosystem services and calculating the socio-economic outcomes of the Agroforestry system. Cross-disciplinary approaches involving ecologists, agronomists, and anthropologists are needed to understand and support these systems holistically.

The economic and policy dimensions of sustaining traditional agroforestry systems have also been discussed extensively. Rawat & Kumar (2020) suggested that to get sustainably produced goods, the farmers practising agroforestry must be facilitated with incentives that include payments for ecosystem services and subsidy reforms that are environmentally friendly. A study conducted in Kodagu and Mandya districts by Nair et al. (2021) showed that native trees in traditional agroforestry systems offer higher ecosystem service benefits than exotic species, regardless of system type and management differences.

Additionally, payments for ecosystem services (PES) have been proposed to incentivise biodiversity conservation in agricultural landscapes. Wunder (2005) argues that PES schemes, if well-designed, can align economic incentives with environmental goals. In tribal agroforestry, such mechanisms could support farmers who maintain diverse, sustainable farming systems.

Certification schemes like organic or biodiversity-friendly certifications can also help tribal farmers access premium markets while encouraging sustainable practices (Raynolds, 2000). However, the literature also points to the need for support systems to help smallholder farmers navigate certification processes.

Research Gaps

The research gaps emerging from the literature review are listed below:

- Insufficient studies explore combining traditional agroforestry methods with modern sustainable farming practices.
- Limited investigation into the commercial cultivation of millet and traditional crops while preserving biodiversity.
- Insufficient understanding of agricultural commercialisation's socioeconomic and cultural effects on tribal communities.
- A shortage of comprehensive studies examining how agroforestry biodiversity contributes to climate resilience.
- Poor alignment of state-level agricultural initiatives, such as the Millet Mission, with biodiversity conservation objectives.

One of the unaddressed issues in previous research is the incorporation of policy initiatives with ground realities. While the millet mission focuses mainly on millet cultivation, there is little focus on how such policies can shape biodiversity-friendly practices among the tribal people. (Priyadarshini & Abhilash, 2020) have raised concerns about incorporating policies that favour local ecological balance and the cultural practices of the tribal areas.

Limited research also exists on the potential collaborations between traditional agroforestry systems and contemporary sustainable farming innovations. As Francis et al. (2003), discussed, agroecological principles suggest that integrating traditional knowledge with scientific innovations can enhance productivity and ecological integrity.

Future research should focus on creating frameworks for the coexistence of traditional ecological knowledge and contemporary sustainable agricultural practices. This would ensure that tribal communities' socio-ecological integrity is preserved while meeting the demands of modern markets.

Literature indicates that traditional agroforestry systems of tribal communities in Karnataka offer a sustainable agriculture model that promotes biodiversity, food security, and climate resilience. However, the commercialization of agriculture, especially millet production, creates challenges. Integrated approaches that combine traditional knowledge with modern innovations and supportive policies are essential.

The agroforestry practices of tribal communities in North Karnataka showcase how indigenous knowledge can provide sustainable solutions to agricultural and environmental challenges. By integrating trees, crops, and livestock, these practices enhance biodiversity, improve soil health, conserve water, and support food security. They represent a sophisticated form of environmental management that has sustained communities for generations. In an era of climate uncertainty and declining soil quality, valuing these traditional systems is essential for bridging the gaps in environmental degradation while preserving culture and heritage.

Thoughtfully integrating these practices with modern agricultural techniques can enhance productivity and resilience, provided that cultural and spiritual values are respected. Supporting traditional agroforestry is not only about cultural preservation but also a strategic approach toward long-term ecological sustainability and equitable food distribution.

Objectives

This study seeks to scrutinise ways to improve tribal agroforestry methods in North Karnataka to boost crop yields while protecting and incorporating biodiversity into farming systems. The specific objectives of the study are:

- Examine the Significance of Traditional Agroforestry Systems in Biodiversity Conservation: This study will investigate how traditional agroforestry methods employed by tribal communities support local biodiversity, enhance soil health, and provide ecosystem services, including water retention and carbon sequestration.
- Assess the Feasibility of Sustainable Millet Production: The study will analyse how millets can be incorporated into these biodiversity-supportive agroforestry systems to improve commercial and subsistence production without jeopardising environmental sustainability.
- Explore the Role of Biodiversity in Contemporary Agricultural Practices: This study aims to evaluate how modern agricultural methods, such as organic farming, precision agriculture, and agroecology, can be aligned with traditional agroforestry systems to enhance productivity while upholding biodiversity.
- Examine Socioeconomic and Environmental Impacts: The study will assess the effects of merging biodiversity-rich agroforestry systems with millet production on food security, ecosystem health, and the socioeconomic prosperity of tribal communities.
- Contribute to Advancing Sustainable Development Goals (SDGS): This chapter will discuss how integrating biodiversity into millet production systems aids in achieving essential SDGS.

METHODOLOGY AND PARTICIPANT PROFILE

The study used mixed methods research, a systematic approach combining qualitative and quantitative data collection and analysis in one study, to better understand the research objectives.

Quantitative research recognises causality in the objective world, maintaining that it exists separately from human perception; qualitative research best suits exploratory, interpretative, and complex issues (Zou & Xu, 2023). A primary reason for using mixed methods research is the need for diverse data sources and varied collection and analysis techniques to address research challenges that are too complex for a single approach (Zou & Xu, 2023). Mixed methods research is grounded in problem-centred, practical pragmatism (Zou & Xu, 2023).

The research utilised a simultaneous mixed-methods approach, incorporating a qualitative aspect to contextualise the quantitative findings. The integration of both sections occurred during the results interpretation phase. The data was collected through a structured questionnaire, and semi-structured open-ended questions allowed for a qualitative analysis.

A structured questionnaire was developed based on the research objectives and administered to 34 households to gather data on agroforestry practices, land use, millet production, biodiversity indicators, and environmental outcomes like water use and soil fertility. The collected data was organized in Excel for analysis, where descriptive statistics provided an overview. Findings were presented using tables and graphs created with Python, and correlation maps were generated to identify key relationships between variables.

Thematic analysis is among the most commonly used techniques for examining qualitative data, which provides a systematic but adaptable framework for identifying, analysing, and interpreting meaning patterns within datasets (Ahmed et al., 2025). Not restricted to any theoretical framework, thematic analysis is a versatile tool appropriate for inductive and deductive data analysis methods (Ahmed et al., 2025). The study employed thematic analysis to iteratively map the most significant themes in the data (Braun & Clarke, 2006). Through an extensive literature review and the research objectives, an interview protocol (Appendix A) was developed and utilised with participants during the iterative and analytical process to emphasise each participant's unique meanings, context, and interpretations.

Data collection was carried out in March 2025, targeting tribal and non-tribal communities in seven villages of Haliyal Taluka, Uttara Kannada. The tribal group consists of the Siddi people, identifying as Christians or Muslims. Haliyal's economy is mainly agricultural. A demographic overview of the villages is provided in Table 1.

Table 1. Demographic overview of selected villages

Village Name	Population
Addigera	289
Agasalkatta	273
Bidrolli	856
Bukkankoppa	525
Gadagera	169
Gardoli	662
Kurigadde	279

The researchers interacted with 34 households for the quantitative surveys during the study. The snowball sampling method was applied to select a purposive sample of convenience for both the quantitative survey and the qualitative interviews. Twelve participants were recruited from 34 households for in-depth semi-structured interviews. Participants were informed that their involvement in the study was voluntary and that they could withdraw at any time. To ensure anonymity, participants were assigned pseudonyms.

The average age of the participants was 49.6 years, ranging between 30 and 65 years. Most of the farmers surveyed were in the 41–50 age range (28.6%), while not a single respondent was under 30, suggesting that younger generations may be less involved in farming. A noticeable gender gap was present, with nearly 80% of male respondents. Regarding education, a large portion (44.12%) had only completed primary school, and about a quarter had no formal education. Income levels varied, but around one-third earned between 1–2 lakh per year, indicating moderate economic stability for many households. Table 2 gives a demographic and socio-economic profile of the participants.

Table 2. Demographic & socio-economic profile

Variable	Categories	Frequency (n)	Percentage (%)
Age	Below 30	0	0
	30-40	9	22.9%
	41-50	11	28.6%
	51-60	4	20.0%
	Above 60	10	14.3%
Gender	Male	27	79.41
	Female	07	20.58

continued on following page

Table 2. Continued

Variable	Categories	Frequency (n)	Percentage (%)
Education	No formal education	08	23.52%
	Primary school	15	44.12%
	High school	08	21.59%
	College or higher	03	8.82%
Household Income ( )	Below 50,000	07	20.58%
	50,000 - 1,00,000	08	21.59%
	1,00,000 - 2,00,000	11	32.35%
	Above 2,00,000	08	21.59%

RESULTS AND DISCUSSIONS

Quantitative Analysis

This section presents the quantitative information obtained from participants regarding their farm size, yields, earnings, agricultural practices, agroforestry methods adopted, and corresponding socio-economic variables. To attain the research objectives, these details were collected through a survey guided by a structured questionnaire. According to the participants' ease, the interviews were conducted using local languages like Kannada, Hindi, and Marathi. Their details were recorded carefully and analysed. This subsection will emphasise the statistical analysis of the data obtained from 34 farmers in Haliyal Taluk, Uttara Kannada district of North Karnataka.

Theme 1: Agroforestry vs. Non-Agroforestry Agriculture: Exploring the Differences in Yield, Sustainability, and Ecology

Table 3 shows that medium-sized landholders (2–5 acres) have the highest number of agroforestry farmers (14) compared to non-agroforestry farmers (8). This suggests that farmers with 2–5 acres are likely to adopt agroforestry because they seek to optimise limited land, improve soil health, and diversify income through trees + crops. Smallholders (<2 acres) may lack the space or resources to integrate trees effectively. Large landowners (>10 acres) may prefer monoculture (like sugarcane or wheat) for mechanisation and market efficiency, avoiding the complexity of managing mixed systems. Agroforestry requires planning, knowledge, and time, which may not align with large-scale commercial farming goals.

Table 3. Distribution of landholding sizes agroforestry and non-agroforestry farmers

Landholding Size (Acres)	Agroforestry Farmers (n)	Non-Agroforestry Farmers (n)
Less than 2	02	04
2-5	14	08
6-10	02	01
More than 10	03	00

Figure 1 shows the presence of pollinators on both agroforestry and non-agroforestry lands. The data clearly shows that pollinators are more attracted to agroforestry lands and the presence of pollinators is nearly doubled when it comes to agroforestry lands as the trees planted in the agroforestry lands are mainly fruit-bearing trees, which attract a lot of small fauna.

Figure 1. Comparison of pollinator presence between agroforestry and non-agroforestry farming systems

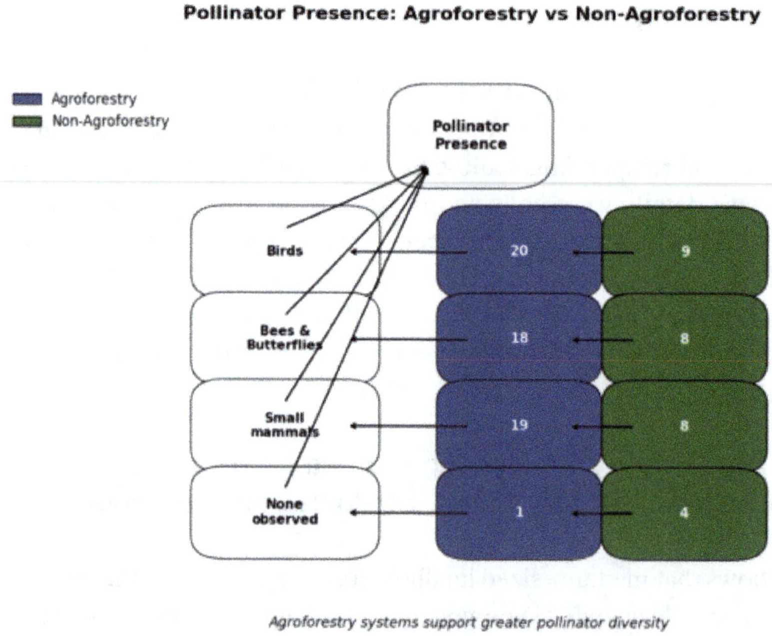

Figure 2 presents a violin plot comparing the annual income (in Rupees) of agroforestry and non-agroforestry farmers. The plot reveals a notable difference in income distribution between the two groups. Farmers practising agroforestry exhibit a wider spread of income values, with a higher median than non-agroforestry farmers. This suggests that agroforestry adoption may be associated with improved financial outcomes. Additionally, the violin's shape for agroforestry farmers indicates greater variability in income, potentially reflecting diverse agroforestry systems and varying levels of market integration. In contrast, non-agroforestry farmers show a narrower income range and a lower central tendency, indicating relatively lower and less variable earnings. These findings imply that agroforestry could enhance economic resilience among farming households.

Figure 2. Violin plot comparing annual income (in rupees) between farmers who practice agroforestry and farmers who do not practice agroforestry

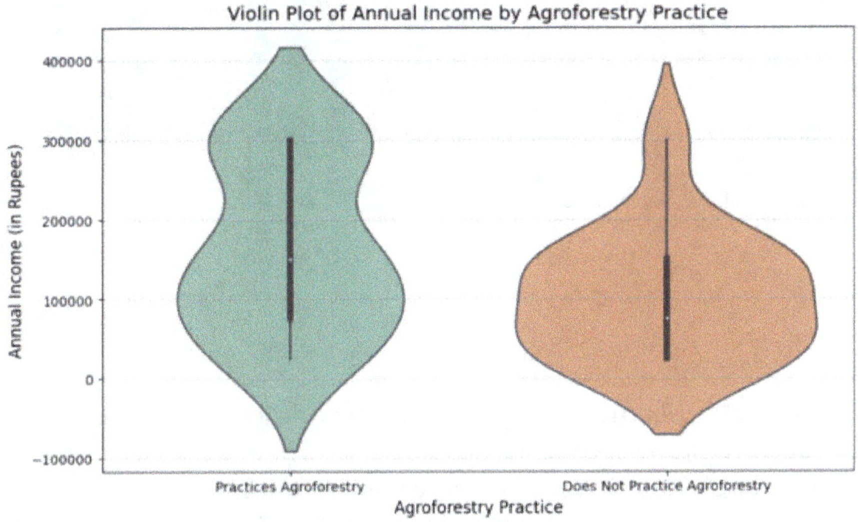

Theme 2: Agroforestry as a Multifunctional Land-Use Practice – Linking Land Management, Ecology, and Sustainability

21 farmers out of 34 interviewed practice agroforestry. Figure 3 presents the percentage of land under agroforestry of the total land owned among farmers who practice agroforestry. The data shows that 9 farmers have their entire land under agroforestry. Most farmers have above 50% of their land under agroforestry, meaning most of the farmers employ indigenous/traditional practices. A notable population segment has allocated less than 25% of their land to agroforestry practices.

Figure 3. Comparison of total land and land under agroforestry for among farmers who practice agroforestry

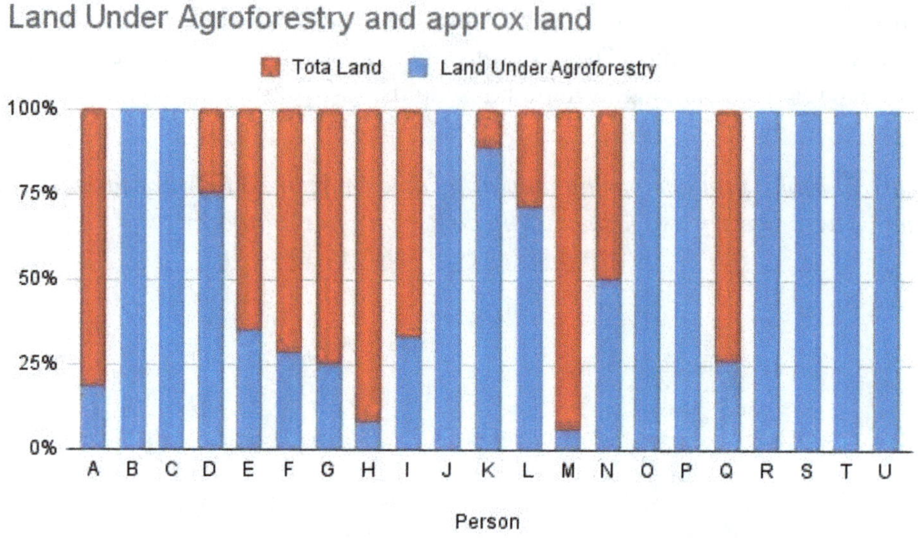

Table 4 shows that agrihorticulture (growing crops with horticultural plants like fruits or vegetables) is the most common agroforestry practice, adopted by 85.71% of farmers. Agrihorticulture dominates because it involves combining agricultural crops with fruit trees, vegetables, or shrubs, which offer quick economic returns, are suitable for small landholdings, and require less specialised knowledge or long-term investment compared to other systems. Agrisilvihorticulture (9.52%) includes a mix of agriculture, trees, and horticulture, making it more complex and resource-intensive, so fewer farmers adopt it unless they have specific training or market access. Silvopasture (4.76%) integrates trees with livestock grazing. Its low adoption rate may be due to land use conflict, lack of technical know-how, or limited livestock integration in the study area. Agrisilviculture (0%), growing crops with trees only, is absent, possibly because tree-crop combinations take longer to yield returns and may not align with immediate income needs. The absence of home gardens (0%) suggests limited interest or awareness in that specific system among the surveyed farmers.

Table 4. Frequency and percentage distribution of different types of agroforestry practiced among farmers

Agroforestry Type	Frequency (n)	Percentage (%)
Agrisilviculture	00	0.00%
Agrihorticulture	18	85.71%
Silvopasture	01	4.76%
Agrisilvihorticulture	02	9.52%
Home gardens	00	0.00%

Table 5 shows that soil quality saw the highest improvement after adopting agroforestry, with nearly half of the farmers (47.62%) reporting significant improvement, and none experiencing a decline. This highlights the positive impact of agroforestry on soil health through organic matter addition and reduced erosion. In contrast, water retention had more varied responses, with only 23.81% each for significant and slight improvement, while over 47% reported no change, possibly due to slow infiltration or unsuitable species. The positive effect of agroforestry in increasing soil quality than water retention suggests that agroforestry benefits the soil faster than hydrological properties.

Table 5. Percentage distribution of changes in environmental factors (soil quality and water retention) observed in agroforestry systems

Environmental Factor	Significant Improvement (%)	Slight Improvement (%)	No Change (%)	Worsened (%)
Soil Quality	47.62%	09.53%	42.85%	0%
Water Retention	23.81%	23.81%	47.62%	4.76%

Table 6 shows that farmers largely believe agroforestry helps prevent soil erosion (15 Yes vs. 6 No) and climate change impacts (15 Yes vs. 6 No). However, its perceived effectiveness is lower against water scarcity (9 Yes vs. 12 No) and least against pest and disease outbreaks (3 Yes vs. 18 No). This indicates that agroforestry is viewed as a robust method for protecting soil and climate. However, it appears to be less effective or requires more knowledge for managing water availability and challenges related to pest or diseases.

Soil Erosion & Climate Change (High Agreement): Trees in agroforestry systems stabilise soil with roots, reduce runoff, and improve carbon sequestration, benefits that are often visible and immediate, leading to positive farmer perception.

Water Scarcity (Mixed Views): While trees can improve groundwater recharge and moisture retention over time, some farmers may experience short-term competition for water between trees and crops, especially if species are not selected carefully.

Pest and Disease (Low Agreement): Agroforestry can enhance biodiversity, sometimes reducing pest pressure, but it can also create habitats for pests or pathogens if not well-managed. Lack of knowledge on integrated pest management in agroforestry likely contributes to negative perceptions.

Table 6. Frequency of responses indicating whether agroforestry helped to prevent specific environmental challenges

Did Agroforestry help in preventing	Yes	No
Soil Erosion	15	06
Water Scarcity	09	12
Pest and Disease Outbreak	03	18
Climate Changes	15	06

The correlation map between agroforestry practice and selected environmental indicators is presented in figure 4 and pairwise correlations among agroforestry and environmental variables is presented in table 7. The analysis of the correlation map is given below.

(i) Agroforestry Practice (Years)

The duration of agroforestry practice exhibited varying degrees of correlation with key ecological indicators, as follows:

High correlation: Climate Resilience Improvement (r = 0.70)

Moderate correlation: Soil Quality Improvement (r = 0.56)

Moderate correlation: Pollinators' Presence (r = 0.41)

Low correlation: Water Retention Improvement (r = 0.37)

The integration of trees within agricultural systems has been strongly associated with improvements in climate resilience *(r = 0.70)*. Trees function as natural buffers by reducing the impact of extreme weather conditions, stabilising microclimatic conditions, and promoting overall agroecosystem health. In addition to climate benefits, extended agroforestry practices contribute to gradual improvements in soil quality *(r = 0.56)*. Leaf litter from trees adds organic matter to the soil, enhancing nutrient cycling and reducing erosion, which supports long-term land productivity. Agroforestry also positively influences pollinator presence *(r = 0.41)*, as the increased biodiversity provides habitat for species such as bees and butterflies. Nonetheless, the continued use of agrochemicals can limit pollinator abundance despite these habitat improvements. Furthermore, tree root systems have been shown to enhance soil structure and porosity, leading to improved water retention *(r = 0.37)*. However, the effectiveness of this benefit depends on selecting tree species that are well-suited to the local environmental conditions.

(ii) Soil Quality Improvement

Moderate Correlation: Climate Resilience Improvement (r = 0.54)
Moderate Correlation: Water Retention Improvement (r = 0.43)
Low Correlation: Pollinators' Presence (r = 0.32)

Climate Resilience Improvement (r = 0.54): Improved soil quality contributes significantly to climate resilience. Soil with high organic matter and good structure act like sponges, retaining moisture during dry periods and improving drainage during heavy rainfall, enhancing plant survival and crop stability under variable climatic conditions.

Water Retention Improvement (r = 0.43): Soils rich in organic matter exhibit enhanced water infiltration and retention capacity. This allows them to absorb rainfall more effectively and maintain soil moisture for extended periods, supporting plant growth during dry spells.

Pollinators' Presence (r = 0.32): Healthier soils support the growth of more vigorous and diverse vegetation, increasing floral resources for pollinators such as bees. However, other factors—such as pesticide use and habitat availability—also play important roles in determining pollinator abundance and diversity.

(iii) Water Retention Improvement

Low Correlation: Climate resilience improvement (r = 0.34)
Very Low Correlation: Pollinators' presence (r = 0.18)

Climate Resilience (r=0.34): Soil's capacity to retain water enhances crop resilience during arid conditions, although factors such as shading also contribute to this effect.

Pollinators' presence (r=0.18): Bees are indifferent primarily to soil moisture levels, suggesting minimal impact on their foraging behaviour.

(iv) Pollinators' Presence

Low Correlations: Climate resilience improvement (r = 0.28)
Soil quality improvement (r = 0.32)
Very Low Correlation: Water retention improvement (r = 0.18)

The correlations discussed above are weak because bees and other pollinators require specific plant species for their survival and can be adversely affected by pesticide exposure. Furthermore, while optimal soil and water conditions are beneficial, these factors alone cannot support pollinator health and resilience.

(v) Climate Resilience Improvement

High Correlation: Agroforestry Practice (r = 0.70)
Moderate Correlation: Soil quality improvement (r = 0.54)
Low Correlations: Water retention improvement (r = 0.34)
Pollinators' presence (r = 0.28)

Agroforestry practice (r=0.70): The integration of trees with agricultural crops constitutes an effective strategy for enhancing system resilience. This approach fosters a balanced ecosystem capable of effectively mitigating the impacts of extreme weather conditions.

Soil quality improvement (r=0.54): Quality soil plays a crucial role in sustaining plant life during challenging environmental conditions; however, the influence of trees extends this support significantly, thereby enhancing overall ecological resilience.

Water retention improvement & pollinators' presence (r= 0.34 & r= 0.28): Supportive yet non-essential water management enhances resilience, whereas pollinators, while beneficial, are not critical for climate adaptation.

Figure 4. Correlation map between agroforestry practice and selected environmental indicators

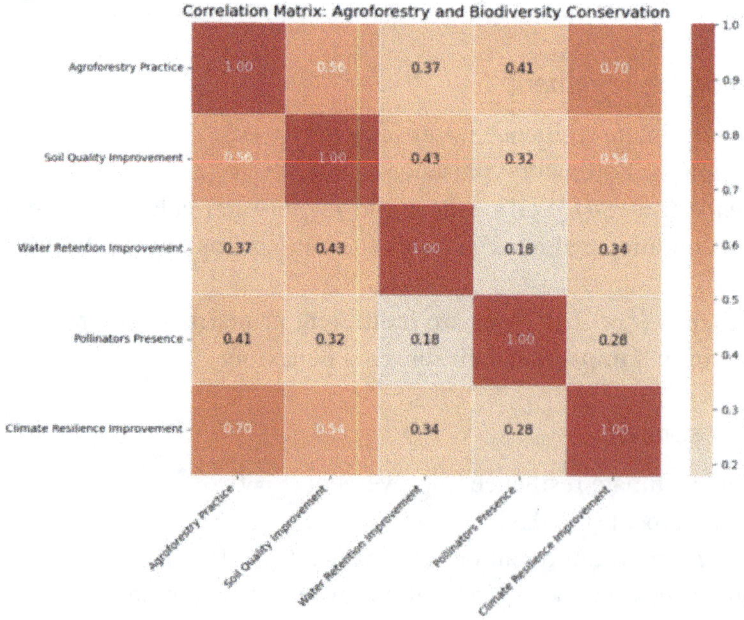

Table 7. Pairwise correlations among agroforestry and environmental variables

Variable 1	Variable 2	Correlation Coefficient (r)
Agroforestry Practice	Soil Quality Improvement	0.56
Agroforestry Practice	Water Retention Improvement	0.37
Agroforestry Practice	Pollinators Presence	0.41
Agroforestry Practice	Climate Resilience Improvement	0.70
Soil Quality Improvement	Water Retention Improvement	0.43
Soil Quality Improvement	Pollinators Presence	0.32
Soil Quality Improvement	Climate Resilience Improvement	0.54
Water Retention Improvement	Pollinators Presence	0.18
Water Retention Improvement	Climate Resilience Improvement	0.34
Pollinators Presence	Climate Resilience Improvement	0.28

Theme 3: Fertiliser Dependency Patterns in Agroforestry: A Land Size Perspective

The data in Figure 5 reveal how fertiliser use patterns and sugarcane yields vary across landholding sizes. Farmers with less than 2 acres predominantly use chemical fertilisers, yet show low yields, possibly due to poor soil quality or inefficient application. Medium-sized farms (2–5 acres) exhibit the most diversity in input use, indicating a transition toward integrated nutrient management. Larger holdings (6–10 and >10 acres) increasingly adopt mixed or organic-dominant practices, suggesting better access to resources and knowledge. This shift implies that with more land and investment capacity, farmers are more willing to explore sustainable methods that boost long-term productivity.

Smallholders (<2 acres): With limited land and financial resources, these farmers often rely on chemical fertilisers for quick results. However, low yields suggest overuse without proper agronomic practices, leading to diminishing returns and soil degradation.

Mid-Sized Farms (2–5 acres): These farmers are experimenting with a mix of inputs, especially using chemicals more than organics, likely aiming to balance yield optimisation with cost and sustainability. The presence of significant mid-range yields supports this transitional behaviour.

Larger Farms (6–10 & >10 acres): With greater resources, these farmers can afford to invest in organic practices, which improve soil health over time. Some even achieve high yields with organic-dominant or organic-only use, pointing to better farm management and possible access to premium markets or certification systems.

Yield vs. Input Relationship: While chemical fertilisers provide immediate nutrients, their effectiveness diminishes without balanced soil management. Organic and mixed systems appear to enable stable yields, especially when land size and farmer capacity increases.

Figure 5. Comparative analysis of fertiliser uses and yield by land size in agroforestry contexts

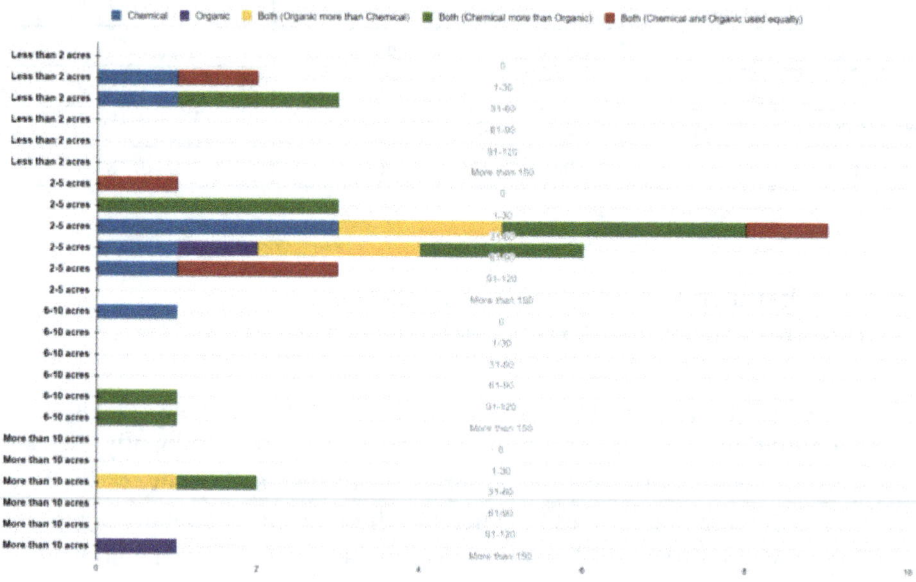

Qualitative Analysis

The study aimed to combine quantitative and qualitative research methods, which is appropriate when research aims for a holistic understanding of a phenomenon where existing studies are fragmented, inconclusive, or equivocal (Venkatesh et al., 2016).

This subsection focuses on the participants' experiences and interpretations of traditional agroforestry systems in tribal (Siddi community) and non-tribal communities based on the research objectives and the Interview protocol (Appendix A) in the seven villages in Haliyal Taluka of the Uttara Kannada district in North Karnataka. The interviews were audio recorded and conducted in three regional languages: Kannada, Hindi, and Marathi. They were translated and transcribed into English for further analysis and coding. The participants were made aware of the varying levels of time commitment required and were assured that their involvement would not exceed two hours. Participants are coded as P1 to P12 to maintain their confidentiality and anonymity.

This subsection will emphasise each participant's distinct meanings, contexts, and interpretations.

Theme 1: Agroforestry and Crop Diversification: A Double-Edged Sword of Risk and Reward

Most participants practice agroforestry, with only one out of twelve interviewed not involved in these practices. P4 (Siddi male) states, "We have grown trees nowadays.... like cashews in our land." P12 (Siddi male), the sole individual interviewed not engaged in agroforestry, mentions that "(he) cut all the mango and other fruit trees he had...they were too much work and he works alone on his land...they (the fruit trees) create a mess."

When asked about the years of practising agroforestry, P1 (non-tribal male) states, "It has been many years. Since my grandfather's generation, this is what our family has been doing. During my grandfather's time, they used to walk all the way to Haliyal. There was no cycle or anything at that time. Now there are vehicles."

The agroforestry being practised includes mixed orchard farming (for example, mango, coconut, and fruit trees alongside sugarcane, corn and lobia intercrops) and silvo-arable systems (combining trees with field crops). P5 (Siddi female) notes that "Yes, there are (they grow) mango trees on one side." She further states her reasons for practising agroforestry: "We get the mango fruits from it...We sell and make money out of it. And it feels cool to sit under the tree.... the cool breeze is nice. The leaves fall and become compost, which we also use on the farm." P8 (Siddi female) explains agroforestry's advantages by stating, "From the trees, the soil fertility is good, and it does not get washed away. The shade is more, and the climate is cool. The trees help for foraging and medicinal purposes."

P7 (Siddi male) states that "(reason why he started growing trees) One is because of the fruits we get to eat and the other is the climate like rain, wind etc (to protect against) ... and also since the trees are being cut, I thought of growing them in my field."

While millet was an important crop ten years ago, participants noted that it is now grown only in restricted areas, primarily for local cattle feed. As P1 (non-tribal male), P2 (Siddi male), and P11 (Siddi male) remarked, "No market for millets now" exists. P1 notes, "My grandfather used to grow ragi.... but we do not grow it anymore. We are not used to eating that here. We do not have the knowledge on how to grow it either. I am currently growing lobia, Bengal gram, horse gram, and coriander to use at home."

Participants noted that the drop in millet farming is due to low market demand in their Taluka and inadequate government backing for millet agriculture. They also mentioned that small land sizes and the lack of processing facilities make millet production less viable. Even though they recognise millet's nutritional and environmental advantages, they favour cultivating more profitable crops such as sugarcane.

P1 indicates that "Yes (millets have more nutritious value compared to rice), but we are not used to eating ragi that much (anymore). We mainly eat rice that we grow. We make roti (flatbread) out of rice and eat that." Moreover, "Yes (land suitable for growing millets) … we can grow, but I do not grow it." This is mainly due to the zero-market feasibility of millet in the Taluka. Indeed, P2 and P6 (non-tribal males) noted that, similar to sugarcane grown in their Taluka, 100 kilometres away, the people exclusively cultivated and sold millet due to reliable market conditions and government assistance. P12 noted that "(he) would grow millet if the government gave a price for it…and taught them about the best ways to grow them (millets)".

Sugarcane, followed by rice and corn, is the dominant crop grown currently due to its market reliability. The local sugar mill is a fixed buyer for sugarcane. Furthermore, since it was set up, all the farmers in the Taluka shifted to growing sugarcane. All the participants corroborated this. P1 states, "Growing sugarcane is profitable compared to growing other crops due to the demand in the market. So, we have reduced the growth of other crops." Government-set prices (MSP) and existing irrigation infrastructure favour sugarcane over millets. Multiple participants noted that "There is no market for millets now." Sugarcane offers rapid income with minimal labour, whereas fruit trees take years to yield returns, and crops like millet are susceptible to animal damage and demand more labour. In this case, the ecological advantages of agroforestry and crop diversity outweigh the necessity for profits. Nevertheless, along with sugarcane, all participants engage in crop diversification by cultivating rice, corn, lobia, Bengal gram, horse gram, and vegetables for personal consumption.

Participants observed bees and butterflies around fruit trees, but only a few recognised the significance of pollinators. For many, these insects were merely an incidental result of having trees. Notably, despite trees' ability to attract bees, little effort was made to support pollinators, like planting wildflower strips, and no proactive steps were taken to improve pollination for fruit crops.

Agroforestry marginally improves drought adaptation (e.g., shade-tolerant lobia) and climate resilience, but sugarcane's water demands exacerbate vulnerabilities, with many participants stressing water scarcity. Once climate-resilient staples, millets are now non-viable due to market neglect and land fragmentation.

Theme 2: Systemic Barriers to Sustainable Practices

A. Chemical Dependency vs. Ecological Knowledge

Farmers depend on synthetic chemical fertilisers like DAP and Urea for sugarcane because of sugar factory quotas and market demand, even though they acknowledge the benefits of compost and cow dung, which are limited by a lack of subsidies and access. All participants, except for one (P9, a non-tribal male), primarily depend on chemical fertilisers and pesticides. P9 mentions that "We use more organic fertiliser," yet regarding insects and pests, he states, "(We use) Chemicals." P10 (Siddi male) similarly mentions that he manages pests in his crops using pesticides by stating, "Yes, when we find it (pests), we use chemicals". This application of pesticides occurs even though the potential harm to birds and other mammals is recognised. The participants felt pressured by the necessity for high yields and profits.

Interestingly, the female participants interviewed expressed more concern about balancing the use of organic and chemical fertilisers, mainly due to health worries regarding their children and the long-term health of the soil. P5 (Siddi female) states, "Yes, I do feel that way," when asked if organic manure is better for your crops than chemicals. She further stressed, "I am using both organic manure and chemicals. I try to limit the use of chemicals for my crops." She expressed worries about her community's health status now compared to before, attributing changes to chemical dependence by stating that "Health now has gotten worse. Earlier, we used to visit the hospitals a lot less; now, the hospital visits have increased. I do not feel like we are getting nutritious food anymore. Now I buy rice and other groceries for my house from the ration shop, but when I was younger, we used to eat the rice we grew on our farm itself."

P6 (Siddi female) only grows sugarcane and has mango trees between her crops. She relies heavily on chemical fertilisers on her 1-acre land, leased by the Forest Department. She states, "Using (organic) manure is good, but we do not get more yield". P6 understands the negative impacts of increased chemical fertiliser use on food quality, soil fertility, and health status. She states, "The chemicals that we use are the main reason". Further, "Because other people grow more than us, we use more chemicals to grow more. People accuse us of not growing anything. If we use less fertiliser, we will not get that much. So, they use it, and we too use it because they grow more than we do." She states, "We get more when we use chemicals, but it is injurious to our health. We know that! But the other people grow more than us, so we use them (chemical fertilisers)." She emphasises the government's insufficient role in promoting organic fertilisers by stating, "So the government should give the (organic) fertiliser; we farmers cannot get proper fertilisers. In old times, we used good fertilisers, but now they sell only these harmful fertilisers, so we are using them. We would check and buy good ones. The less we use chemicals, the less yield we get; others say they use more chemicals to get more yield and question us: Why do we use less? We even borrowed money and purchased manure. .... So, others get a good yield; when we use less (chemicals), we get less yield."

Participants observed that the government, especially the Agricultural Department, failed to support drought-resistant seeds (millets/lobia) or subsidies for agroforestry, particularly concerning organic fertilisers. All the participants were highly critical of the Agriculture Department, with many noting that "(they) are not getting anything (from government schemes), and nobody comes to educate them about planting techniques or crops." Moreover, most participants stated that "they (the agriculture department) only give a subsidy on Urea and DAP". They noted that representatives from the Agriculture Department rarely come to meet them and only serve their own self-interest.

B. Human-Animal Conflict as Systemic Failure

Shortcomings in compensation policies worsen wildlife damage to crops and cattle. The Forest Department's strict "conservation bias" overlooks the livelihood needs, traditional knowledge, and rights of tribal communities regarding forest land. All the villages covered by the study are surrounded by thick forests under the department's control. This leads to numerous problems, especially for the tribal population regarding access to foraging. Many tribal participants reported being treated like "criminals" by the Forest Department.

P12 (Siddi male) is highly antagonistic towards the Forest Department. He recounts losing many cattle without receiving any compensation from the Forest Department, "They (Forest Department) brought and left a cheetah (leopard) in the forest here a few years back from the wildlife sanctuary and did not listen to us.... This is not an area where it (the leopard) will survive.... It has eaten so many cattle, but they (the Forest Department) do not care and give no money for the loss. The elephants and the pigs (wild boar) also come and destroy everything....... we cannot kill them anymore. There is so much loss."

All participants reported frequent crop damage caused by wild animals, including elephants, various deer species, wild boars, and rabbits. Additionally, they pointed out the threat posed by larger carnivorous animals such as leopards, tigers, and black bears. They highlighted the absence of assistance from the Forest Department. The participants shared that they camped overnight in their fields, using crackers, stones and drums to frighten off wild animals and protect their crops. P1 (non-tribal male) states, "Elephants come and eat rice and millet, so we do not grow... they eat sugarcane, but we have to grow (sugarcane) because of the factory." He further narrates, "A bear came a few days back and attacked a villager in his field...his eyes were injured and his face was destroyed."

P11 (Siddi male) notes that the Forest Department has recently involved the tribal community by offering contracts to harvest and sell foraged products such as honey. However, there is resistance, which he has attributed to his tribal community's limited information and financial literacy.

Theme 3: The Paradox of Marginalisation and Untapped Potential

The small size of landholdings and the absence of processing units render millets and non-commercial crops unfeasible. Existing systems facilitate sugarcane production via irrigation and the sugar factory, neglecting agroforestry and millet value chains. Most participants report using drip or sprinkler irrigation systems in their fields, praising their water-saving potential, especially given the region's water scarcity. P9 (non-tribal male) states, "We have done sprinkler (irrigation) for our land". Numerous participants also possess borewells. This is in stark contrast to the natural scarcity of water in the area.

Meanwhile, some depend entirely on rainfall, affecting the crops they cultivate. P3 (Siddi male) states, "We do not have any collection of water; only if it rains, then water is supplied to the crops; there is no other water collection for the crops." He further states, "We will only grow millet more because of water scarcity; we do not grow any other crops in the land." He added, "No, after we grow only millet in that land, we will not do any crop rotation due to lack of water." P3 mentioned a lack of government support to deal with water scarcity by stating, "They (the government) said that if we are doing agriculture they will give money or if there is a barren land they will provide work for us.....we also applied for it but there was no use or any benefits from the government ......they also said that for water scarcity to build a well and for that where will we find labourers......we did not have money we did not go for it but at the end of the day there was no use at all."

Although some participants have adapted by cultivating shade-tolerant crops, they are stuck with water-intensive sugarcane due to systemic dependencies. Land fragmentation and labour shortages compromise the possibility of agroforestry.

The human-animal conflict highlights institutional negligence, with compensation schemes regularly ignoring diverse crops and animal loss. The participants recognise substantial challenges in the forest department's implementation of these schemes on the ground level. Furthermore, the absence of governmental support for more organic options hampers farmers' ecological knowledge of composting.

Discussion: Qualitative Analysis

The study reveals that traditional agroforestry systems in the tribal and non-tribal communities play a dual role in biodiversity conservation and socio-economic resilience, although with considerable constraints. The agroforestry practices by the participants prove inherent biodiversity benefits, including pollinator attraction and soil erosion control. However, the lack of planned pollinator support (e.g., wildflower strips) underlines the unexploited potential to enhance ecosystem services. These systems also influence water retention and carbon sequestration, yet their sustainability is compromised by competing pressures, such as sugarcane's dominance due to market incentives and irrigation infrastructure.

Notwithstanding its historical role as a climate-resilient crop, the decline of millet cultivation highlights systemic barriers to sustainable millet production. Participants identified market neglect, land fragmentation, and absent processing units as key challenges, yet their sporadic cultivation of millets for fodder suggests latent potential. Integrating millets into agroforestry systems could revive their commercial viability while leveraging shade tolerance (e.g., lobia's success under trees). However, this requires addressing institutional gaps, such as drought-resistant seed access and value-chain development, to align with SDG 2 (Zero Hunger) and SDG 13 (Climate Action).

The contradiction of modern-technique integration emerges in farmers' reliance on chemical inputs for sugarcane (driven by sugar factory quotas) alongside their awareness of organic alternatives (compost, cow dung). This reflects a disconnect between ecological knowledge and practice, exacerbated by policy blind spots (e.g., no subsidies for agroforestry, organic manure). Harmonising modern agroecology with traditional systems would require systemic support, such as precision irrigation to reduce water competition between sugarcane and agroforestry crops.

Finally, the socioeconomic and environmental outcomes reveal a cycle of marginalisation. Smallholders' adoption of agroforestry is hindered by labour intensity and land fragmentation, while wildlife conflicts and exclusionary compensation schemes further erode livelihoods. Nevertheless, their adaptive strategies (e.g., shade-tolerant intercrops) demonstrate resilience. Strengthening these systems could advance SDG 12 (Responsible Consumption) by diversifying production and SDG 3 (Health) through nutrient-rich millets.

LIMITATIONS

The study's findings are based on a small sample size (n=34 for Quantitative and n=12 for Qualitative), which may limit their applicability to broader tribal or agroforestry-dependent communities. The participant gender ratio was uneven, with fewer women, which may overlook important gender-specific perspectives on labour roles and access to resources. Future research should intentionally include more female farmers.

The qualitative analysis heavily relied on self-reported practices, which can be influenced by recall and social desirability biases, but these were mitigated by adding quantitative data. Language and cultural barriers also hindered deeper connections with farmers, even though some team members spoke Kannada. Dialect differences restricted understanding of the socio-cultural dynamics important to agroforestry. While structured questionnaires were used for quantitative data, they limited the richness of insights. Involving local translators and cultural mediators could improve the quality of qualitative inquiries and enhance data collection in future studies.

CONCLUSION

This study highlights traditional agroforestry systems as a valuable yet often overlooked approach to biodiversity conservation and sustainability. It emphasizes the need for policy interventions that link ecological benefits with market access, modern techniques, and equitable practices to achieve the SDGs.

Agroforestry offers significant economic, ecological, and agronomical advantages over conventional farming, particularly for medium-sized landowners, who experience improved financial resilience. Ecologically, agroforestry enhances biodiversity, increasing pollinator populations and soil quality, though water retention improvements are slower. Farmers recognize its benefits in combating soil erosion and climate change, but mixed views on water scarcity and pest management suggest a need for better technical guidance. Agrihorticulture presents practical benefits for smallholders, while more complex systems like silvopasture are underutilized. To boost adoption, targeted extension services and educational initiatives are essential, urging policymakers to promote agroforestry for its socioeconomic and ecological benefits.

ACKNOWLEDGEMENTS

This study was funded by St. Claret College, Autonomous in Bengaluru. The authors thank Rev. Dr. Thomas Thennadiyil CMF, Dr. Chinmaya Dash, Mr. Jayanth H., Dr. Siddalingappa Poojari, and Ms. Zuliana for their support. They also appreciate Yeshaswini B., Rakshitha G., Suresh S., and Harshitha S. for assistance with Kannada translations, and express gratitude to the participants for sharing their time and stories.

REFERENCES

Ahmed, S. K., Mohammed, R. A., Nashwan, A. J., Ibrahim, R. H., Abdalla, A. Q., Ameen, B. M. M., & Khdhir, R. M. (2025). Using thematic analysis in qualitative research. *Journal of Medicine, Surgery, and Public Health*, *6*, 100198. DOI: 10.1016/j.glmedi.2025.100198

Altieri, M. A. (2018). *Agroecology: The science of sustainable agriculture* (2nd ed.). CRC Press., DOI: 10.1201/9780429495465

Bhagwat, S. A., Willis, K. J., Birks, H. J. B., & Whittaker, R. J. (2008). Agroforestry: A refuge for tropical biodiversity? *Trends in Ecology & Evolution*, *23*(5), 261–267. DOI: 10.1016/j.tree.2008.01.005 PMID: 18359125

Bhattacharya, P. (2023). *Indigenous knowledge and resilience among Indian tribes: Cultural practices in the face of modernity.*

Braun, V., & Clarke, V. (2006). Using thematic analysis in psychology. *Qualitative Research in Psychology*, *3*(2), 77–101. DOI: 10.1191/1478088706qp063oa

Francis, C. A., Lieblein, G., Gliessman, S. R., Breland, T. A., Creamer, N., Harwood, R., Salomonsson, L., Helenius, J., Rickerl, D., Salvador, R., Wiedenhoeft, M., Simmons, S., Allen, P., Altieri, M., Flora, C., & Poincelot, R. (2003). Agroecology: The ecology of food systems. *Journal of Sustainable Agriculture*, *22*(3), 99–118. DOI: 10.1300/J064v22n03_10

Islam, S., & Manaloor, V. (2021). Millets for food and nutrition security in India: Determinants and policy implications. *Journal of Nutrition and Food Security*, *6*(2), 185–198. DOI: 10.18502/jnfs.v6i2.6074

Jose, S. (2009). Agroforestry for ecosystem services and environmental benefits: An overview. *Agroforestry Systems*, *76*(1), 1–10. DOI: 10.1007/s10457-009-9229-7

Karmwar, M., & Smita. (2022). Tribal development in India: A case of the Siddi community. *Mekal Insight, 6*(1–2), 45–56. https://www.igntu.ac.in/MekalInsight/Vol6-Issue1-2-year-2022/7.pdf

Lin, B. B. (2007). Agroforestry management as an adaptive strategy against potential climate change impacts. *Mitigation and Adaptation Strategies for Global Change, 12*(5), 835–853. DOI: 10.1007/s11027-007-9101-x

Malhotra, A., Nandigama, S., & Bhattacharya, K. S. (2021). Food, fields and forage: A socio-ecological account of cultural transitions among the Gaddis of Himachal Pradesh in India. *Heliyon, 7*(7), e07569. DOI: 10.1016/j.heliyon.2021.e07569 PMID: 34355085

Nair, P. R. (2011). Agroforestry systems and environmental quality: Introduction. *Journal of Environmental Quality, 40*(3), 784–790. DOI: 10.2134/jeq2011.0076 PMID: 21546663

Patil, D., Chittapur, B., & Murthy, M. M. (2018). Traditional agroforestry systems and biodiversity conservation. *Bangladesh Journal of Botany, 47*(4), 785–794. https://www.researchgate.net/publication/330479881_Traditional_agroforestry_systems_and_biodiversity_conservation

Pretty, J., Adams, B., Berkes, F., De Athayde, S. F., Dudley, N., Hunn, E., & Pilgrim, S. (2009). The intersections of biological diversity and cultural diversity: Towards integration. *Conservation & Society, 7*(2), 100–112. DOI: 10.4103/0972-4923.58642

Priyadarshini, P., & Abhilash, P. C. (2020). Policy recommendations for enabling transition towards sustainable agriculture in India. *Land Use Policy, 96*, 104718. DOI: 10.1016/j.landusepol.2020.104718

Rahman, S. A., Samsudin, Y. B., Bhatta, K. P., & Aryal, A. (2023). The role of agroforestry systems for enhancing biodiversity and provision of ecosystem services in agricultural landscapes in Southeast Asia. In *Agroforestry for sustainable intensification of agriculture in Asia and Africa* (pp. 215–240). Springer., DOI: 10.1007/978-981-19-4602-8_10

Rawat, N., & Kumar, R. (2020). Agroforestry: A sustainable land use system. *ResearchGate*. https://www.researchgate.net/publication/382001712_AGROFORESTRY_A_Sustainable_Land_Use_System

Raynolds, L. T. (2000). Re-embedding global agriculture: The international organic and fair trade movements. *Agriculture and Human Values, 17*(3), 297–309. DOI: 10.1023/A:1007608805843

Sanchez, P. A. (1995). Science in agroforestry. *Agroforestry Systems*, *30*(1–2), 5–55. DOI: 10.1007/BF00708912

Siddi, A. K., & Dalawai, R. S. (2024). A cultural drawbridge: A study of the Siddi tribe. *International Journal of Social Sciences and Humanities Review*, *11*(2), 12–20. https://www.ijsp.in/admin/mvc/upload/Vol%2011_02_03.pdf

Singh, V. P., Sinha, R. B., Nayak, D., Neufeldt, H., van Noordwijk, M., & Rizvi, J. (2016). *The national agroforestry policy of India: Experiential learning in development and delivery phases* (ICRAF Working Paper No. 240). World Agroforestry Centre (ICRAF). https://doi.org/DOI: 10.5716/WP16143.PDF

The Hindu Bureau. (2025). *Deputy Commissioner stresses full and proper implementation of SCSP, TSP schemes*. The Hindu. https://www.thehindu.com/news/national/karnataka/deputy-commissioner-stresses-full-and-proper-implementation-of-scsp-tsp-schemes/article69827779.ece

The News Minute. (2025). *BJP accuses Karnataka govt of plans to use SCs/STs funds for guarantee schemes*. https://www.thenewsminute.com/karnataka/bjp-accuses-karnataka-govt-of-plans-to-use-scssts-funds-for-guarantee-schemes

Venkatesh, V., Brown, S. A., & Sullivan, Y. W. (2016). Guidelines for conducting mixed-methods research: An extension and illustration. *Journal of the Association for Information Systems*, *17*(7), 435–495. DOI: 10.17705/1jais.00433

Wunder, S. (2005). *Payments for environmental services: Some nuts and bolts* (CIFOR Occasional Paper No. 42). Center for International Forestry Research. https://doi.org/DOI: 10.17528/cifor/001760

Zou, P. X. W., & Xu, X. (2023). Mixed methods research. In *Research methodology and strategy* (pp. 89–112). Wiley., DOI: 10.1002/9781394190256.ch4

ADDITIONAL READINGS

Altieri, M., & Rosset, P. (2017). Agroecology: science and politics.

Bandyopadhyay, U., & Patnaik, A. (2025). From Green Revolution to Millet Revolution: Understanding India's transition through agri-food policies. *Food, Culture, & Society*, •••, 1–21.

Joshi, P. K., Birthal, P. S., & Minot, N. (2006). Sources of agricultural growth in India: Role of diversification towards high-value crops.

Nair, P. R., Kumar, B. M., & Nair, V. D. (2021). *An introduction to agroforestry: four decades of scientific developments*. Springer. DOI: 10.1007/978-3-030-75358-0

Perfecto, I., Vandermeer, J., & Wright, A. (2019). *Nature's matrix: linking agriculture, biodiversity conservation and food sovereignty*. Routledge. DOI: 10.4324/9780429028557

Sundar, N. (2016). *The scheduled tribes and their India: Politics, identities, policies, and work*. Oxford University Press.

KEY TERMS AND DEFINITIONS

Agrihorticulture: Agrihorticulture is a complete land-use approach that combines agriculture and horticulture with tree-based systems to enhance productivity, sustainability, and biodiversity.

Agrisilviculture: Nurturing various seasonal crops alongside various mixed tree species, creating a vibrant and harmonious ecosystem.

Agrisilvihorticulture: Cultivating crops harmoniously with vibrant horticultural plants and towering forest trees creates a diverse and thriving ecosystem.

Agroecology: Agroecology is a combined approach that applies ecological and social concepts to designing and managing food systems. It improves interactions between plants, animals, humans, and the environment while addressing social aspects for a sustainable and equitable food system.

Agroforestry: Agroforestry combines trees and shrubs with crops and livestock in sustainable systems. It improves ecological diversity and productivity and grants environmental, economic, and social benefits through combined agriculture and forestry technologies.

Biodiversity: Biodiversity refers to the diversity of life on Earth at every level, from genes to ecosystems.

Climate-Resilient Crop: These crops are resilient to climate change and engineered to withstand drought, flooding, and salinity. They are developed using conventional breeding and genetic modification methods to expand their resistance to environmental stress.

Conservation: Conservation is the investigation of the decline in Earth's biological diversity and strategies for its prevention. Conservation efforts aim to protect the variety of life at all biological organisation levels.

Crop Diversification: Crop diversification includes adding new crops or systems to existing farmland based on investment returns. It shifts from less profitable to more profitable crops. Diversification may also include changing cropping patterns or expanding into non-farming activities like poultry or animal husbandry, helping farmers boost production and income.

Ecological Balance: Ecological balance refers to a vibrant equilibrium among organisms where genetic, species, and ecosystem diversity remain stable, permitting slow changes through natural succession.

Indigenous Knowledge Systems: Indigenous Knowledge Systems represent the collective knowledge, practices, and beliefs that Indigenous peoples have developed over generations through their relationship with the environment.

Mixed Methods Research: Mixed methods research mixes quantitative and qualitative methods to challenge the research question. This method provides a more systematic understanding than relying solely on quantitative or qualitative studies, as it combines the benefits of both research types.

Pollinators: Pollinators, like bees, butterflies, and hummingbirds, help plants reproduce by transferring pollen. Bees are the most numerous and effective pollinators, aiding crops and wild plants, providing one-third of our food and supporting the natural food web.

Silvopasture: Silvopasture incorporates trees, livestock, and forage within a single piece of land.

Sustainable Agriculture: Sustainable agriculture protects the environment, enhances natural resources, and most efficiently uses non-renewable resources.

Sustainable Development Goals: The United Nations recognised 17 global Sustainable Development Goals (SDGS) in the year 2015 to tackle social, economic, and environmental challenges for peace and prosperity by 2030. These goals include ending poverty, ensuring quality education, and addressing climate change.

Violin Plot: A violin plot demonstrates numerical data distributions across one or more groups through density curves. The width of each curve indicates the estimated frequency of data points in different areas. Density curves are often complemented by an additional chart type, like a box plot, for more context.

APPENDIX

Interview Protocol

Interview Guide:

What crops are cultivated, and what agroforestry practices are employed?
What are the primary agroforestry methods used in millet cultivation, if any?
What traditions and folklore do these practices reference?
What is their perception of biodiversity conservation and the preservation of ecological services?
How do these practices contribute to biodiversity conservation and soil health?
How does the integration of biodiversity in crop production systems improve food security for your community?
What role do diverse crops and agroforestry species play in ensuring a stable food supply?
How does millet production, if any, contribute to dietary diversity and nutrition in your community?
How does the consumption of millets and other crops from biodiversity-rich systems impact the health of your community?
Are there any traditional medicinal plants or herbs grown alongside crops contributing to community well-being?
How does using organic or low-chemical inputs in these systems affect health outcomes?
How do traditional agroforestry practices promote sustainable land use and resource management?
What measures are taken to ensure crop production does not deplete soil fertility or water resources?
How does the integration of biodiversity reduce the need for chemical inputs (e.g., fertilisers, pesticides)?
How do agroforestry practices enhance climate resilience (e.g., drought resistance, flood mitigation)?
How do trees and diverse crops affect carbon sequestration and reduce greenhouse gas emissions?
How does the integration of biodiversity help your community adapt to climate change impacts?
What modern agricultural practices are being adopted, if any?
Is there any visible integration of modern and traditional techniques, and how do these practices impact traditional systems and biodiversity?

What opportunities exist for scaling up millet production sustainably?
How have state policies influenced farming practices?
What policy changes are needed to support biodiversity-friendly agriculture?

Chapter 10
Harnessing Fungal Diversity for Sustainable Food Systems:
Bridging Tradition and Innovation

Sharmistha Sharma
https://orcid.org/0009-0006-1101-504X
Assam University, Silchar, India

Anjali Roy
https://orcid.org/0009-0001-2759-7201
Assam University, Silchar, India

Himanshu Kishore Prasad
https://orcid.org/0000-0003-0215-1323
Assam University, Silchar, India

ABSTRACT

As global food security challenges intensify due to climate change, population growth, and resource constraints, there is an urgent need to develop sustainable and nutritious alternative food sources. Fungi represent a vastly underutilized yet promising domain for food innovation, offering unique advantages in nutrition, sustainability, and economic viability. This chapter explores the intersection of fungal biodiversity, food technology, and bioeconomy, examining both traditional knowledge and cutting-edge innovations in fungal-based food systems. The chapter comprehensively analyzes how fungal biodiversity can be used to address global food security challenges while supporting conservation efforts. Secondly, it examines integrating traditional knowledge with modern biotechnology in fungal food production, offering insights into the sustainable scaling of these practices.

DOI: 10.4018/979-8-3373-2003-8.ch010

The emerging production technologies for the future of food systems, followed by practical frameworks for assessing and implementing fungal-based food solutions in various contexts, are discussed.

1. INTRODUCTION

The global food system is at an inflection point: it must feed an ever-growing population while facing dwindling resources and a changing climate. The numbers tell the story: we expect protein demand to increase by 52% by 2050 (Food and Agriculture Organization of the United Nations, 2023), but crop yields could be reduced by up to 15–30% under the effects of climate change in that same time-frame (Intergovernmental Panel on Climate Change, 2022). Further complicating this challenge is the environmental impact of conventional agriculture. As a fundamental component of our current food supply, it accounts for about 24% of global greenhouse gas (GHG) emissions and relies on increasingly unsustainable practices (Crippa et al., 2021). These pressures have created a call to action that requires new approaches to food production that are resilient and fit among the environmental, social, and economic pillars of sustainability (Hellwig et al., 2022). The kingdom of Fungi is perhaps one of the most promising solutions in this context because it is the most metabolically diverse group of organisms on Earth, which have the least resource requirements, and are estimated to include 2.2-3.8 million species (Pérez-Moreno et al., 2021), a number that makes Fungi one of the most diverse groups of organisms in existence (Stamets, 2006). Their evolutionary history over at least 600 million years has provided them with an advanced biochemical armamentarium, including secretion of an enormous variety of exogenous enzymes for degradation of complex organic matter as food sources (Pérez-Moreno et al., 2021; Stamets, 2006). For centuries, humans have used their capacity to ferment and leaven foods: *Saccharomyces* yeasts in baking bread and making alcoholic drinks, and *Aspergillus* molds to make basic East Asia fermented foods, including miso, soy sauce, and koji (Figure 1) (Nout & Aidoo, 2010). Although these traditional applications reflect the long-standing relationship between humans and fungi, current biotechnology is developing their potential on an unprecedented scale, as they are leading a new food revolution to produce high-quality protein, bioactive compounds, and value-added ingredients with significantly less environmental impact than conventional animal agriculture but are facing many systemic hurdles that need to be critically evaluated (Shahid et al., 2024). Fungi are at the forefront of this new food revolution, providing pathways for producing high-quality protein, bioactive compounds, and value-added ingredients with less environmental footprint than conventional animal agriculture (Shahid et al., 2024). Yet, as we will discuss in this chapter, moving from ancient tradition to laboratory innovation, to global mainstream adoption is not without significant systemic barriers that need critical evaluation and overcoming. In this chapter, we present a comprehensive overview of how fungi can be used to construct sustainable food systems beginning with the cultural and historical roots of fungal foods including indigenous and traditional practices, followed by modern biotechnological innovations that are scaling up production through commercialization pathways and examining the major challenges such as economic viability, technological scalability, regulatory frameworks across diverse global regions, and consumer acceptance. It then integrates this analysis into broader policy and sustainability frameworks to evaluate how mycological systems have the potential to support the Sustainable Development Goals of the United Nations and circular bioeconomy, assess current market trends, and conclude with an outline of future research gaps and a roadmap for the future of fungal foods.

Figure 1. An illustration showcasing various methods of mycological food product fermentation

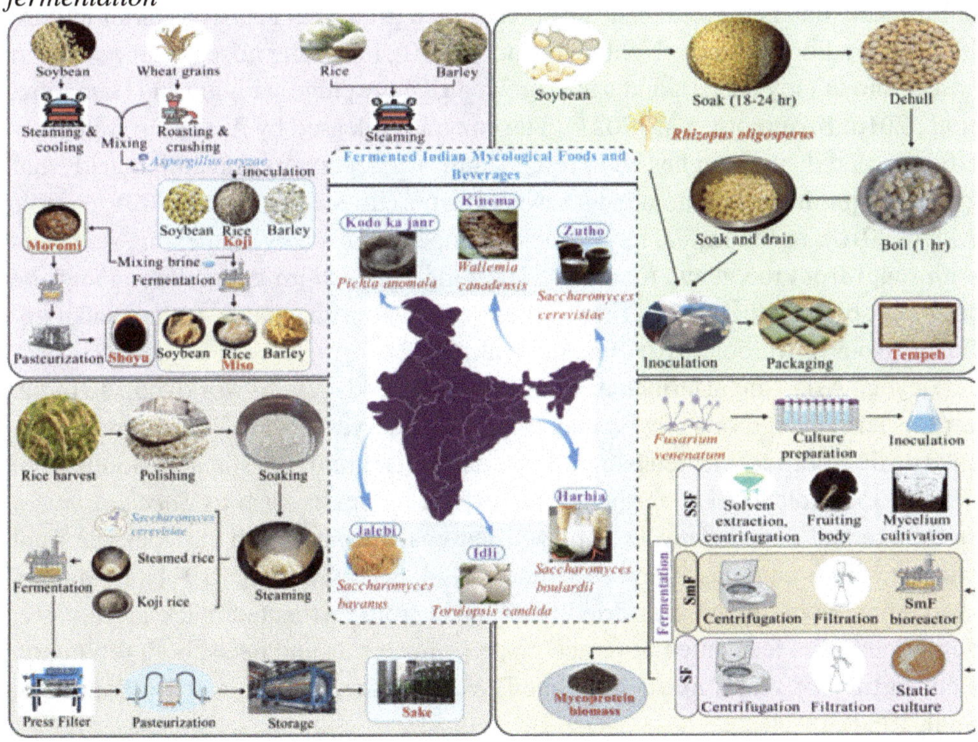

Fermented products like koji, moromi, shoyu, sake, tempeh, and mycoprotein productions are emphasized. The central section of the diagram highlights fermented food products commonly produced in different regions of India. SSF, solid state fermentation; SmF, submerged fermentation; SF, surface fermentation (Finnigan et al., 2024a; Finnigan et al., 2024b; Jiang et al., 2025; Nout & Aidoo, 2010; Zhang et al., 2020). (Created with BioGDP)

2. CULTURAL AND HISTORICAL FOUNDATIONS OF FUNGAL FOOD SYSTEMS

Fungi have been used for centuries by indigenous and ethnic communities around the world to convert and preserve food through fermentation, which is a technique based on the ability of local yeasts and molds to metabolize substrates with specific flavor profiles, improved digestibility, and increased shelf life (Amara & El-Baky, 2023; Tamang et al., 2022; Shapiro & Cumbelich, 2025). This process often relies on traditional techniques developed over generations and not easily reproduced through industrial standardization. A mold, *Aspergillus oryzae*, is the basis of a food culture in East Asia that is used to make koji, fermented grain that acts as a starter culture for many traditional foods (Nout and Aidoo, 2010), including miso (savory soybean paste), shoyu (soy sauce), and sake (brewed rice beverage) (Figure 1) (Hamajima et al., 2016; Kusumoto et al., 2021). The enzymes released by *A. oryzae* are highly effective and digest starches and proteins into simpler sugars and amino acids that give these products their umami and sweet flavors (Ito & Matsuyama, 2021; Nout & Aidoo, 2010). In Indonesia, for example, 'tempeh' is made by fermenting soybeans with fungi from the genus *Rhizopus*. The fungal mycelium grows throughout the soybeans, binding them into a firm, protein-rich cake with a meaty texture and nutty flavor (Nout & Aidoo, 2010; Tamang et al., 2022).

Beyond Asia, fungal fermentation is integral to food systems worldwide. In India, various fermented foods rely on naturally occurring yeasts. 'Idli,' a steamed savory cake made from a fermented batter of rice and black gram, is a staple in South India (Figure 1) (Table 1). The fermentation, driven by yeasts such as *Saccharomyces cerevisiae* and *Debaryomyces hansenii*, leavens the batter and enriches the final product with B vitamins (Batra & Millner, 1974; Soni & Sandhu, 1991). In Punjab, 'wari' consists of spiced, sun-dried balls made from fermented black gram paste, while 'kanji' is a fermented beverage made from carrots and beets, both relying on a consortium of yeasts for their unique flavors (Batra & Millner, 1974; Sandhu & Soni, 1989).

In addition to fermented foods, the direct consumption of wild and cultivated mushrooms has deep cultural roots. For millennia, mushrooms have been foraged for their nutritional value and medicinal properties. In traditional Chinese medicine, for instance, mushrooms like 'Reishi' (*Ganoderma lucidum*) are revered for their ability to enhance vitality and support the immune system (Sarkar et al., 2025). A global assessment of traditional fungal-fermented foods from various regions, highlighting their microbial agents and functional roles, is presented in Table 1, while nutritional and therapeutic attributes of various edible mushrooms are detailed in Table 2. This traditional knowledge, encompassing specific fungal species and fermentation techniques, forms the invaluable foundation for modern fungal biotechnology.

Table 1. Region-specific global assessment of fungal foods in traditional practices

Species	Consumption Mode	Region	Functional Role	References
Saccharomyces bayanus	Jalebi	India, Nepal, and Pakistan	Confectionary product	(Batra & Millner, 1974)
Saccharomyces sp., *Lactobacillus* sp., *Candida* sp., *Leuconostoc* sp.	Leben or Lben	Africa	Fermentable dairy product	(Tamang et al., 2016)
Inonotus obliquus (Ach. Ex. Pers)	Chaga tea	China	Medicinal tonic against infection, diabetes, or as an immune booster	(Cheung et al., 2023)
Torulopsis candida and *Trichosporon pullulans*	Idli	South India	Staple diet	(Batra & Millner, 1974)
Rhizopus sp., *Saccharomyces cerevisiae*	Kodo ko janr	Darjeeling and Sikkim, India	Mild alcoholic beverage	(Thapa & Tamang, 2006)
Parmeliaceae, Ascomycota Fungi (*Cetraria islandica* (L.) Ach	Condiments, soups, salads, or bread additives	Estonia, Finland, Iceland, Italy, and Norway	Therapeutic uses	(Thakur et al., 2023)
Candida humilis, Cryptococcus diffluens, Bandoniozyma tunnelae, Kazachstania spp., and *Hanseniaspora thailandica.*	Soibum	Manipur, India	Non-salted fermented bamboo shoot product	(Tamang, 2020)
Lactic acid bacteria (LAB) and *Saccharomyces cerevisiae*	Tarhana	Turkey	Protein-enriched product	(Daglio lu, 2000)
Saccharomyces cerevisiae, Candida tropicalis, Saccharomyces boulardii, Candida nitratophila, Zygosaccharomyces cidri, Pichia sp., *Candida musae, Saccharomycopsis* sp., and *Issatchenkia*	Haria	West Bengal, India	Rice-based alcoholic beverage	(Tamang et al., 2012)
Rhizopus chinensis, Rhizopus stolonifera var. *lyococcus, Hansenula anomala, Mucor cicinelloides, Saccharomycopsis fibuligera, Saccharomyces cerevisiae.*	Rokshi or Jhar	Sikkim, India	Beverage product	(Sekar & Mariappan, 2007)

continued on following page

Table 1. Continued

Species	Consumption Mode	Region	Functional Role	References
Saccharomyces fibuligera, Saccharomycopsis burtonii, Saccharomyces cerevisiae	Xaj	Assam, India	Rice-based alcoholic beverage	(Tamang, 2020)
Saccharomyces, Schizosaccharomyces spp., and *Geotrichum*	Toddy	Uttar Pradesh, India	Fermented alcoholic drink	(Batra & Millner, 1974)

3. MODERN TECHNOLOGIES IN FUNGAL BIOTECHNOLOGY

Building upon the rich foundation of traditional knowledge, modern biotechnology has developed sophisticated methods to harness and optimize the metabolic capabilities of fungi for food production (Ahmed et al., 2024; Pouris et al., 2024). These technological advancements aim to improve efficiency, consistency, and scalability, transforming fungi into highly productive "cell factories" for alternative proteins and other value-added products. Key innovations span fermentation techniques and advanced strain engineering.

3.1 Techniques of Fermentation

The application of fermentation in the alternative protein sector can be broadly categorized into three main approaches: traditional fermentation, biomass fermentation, and precision fermentation (Majumder et al., 2023; Finnigan et al., 2024b). As discussed previously, traditional fermentation uses live microorganisms to process ingredients, as seen in tempeh production. Biomass fermentation leverages the rapid growth of microorganisms to produce large quantities of protein-rich biomass, which becomes the primary food ingredient; mycoprotein production is a prime example (Finnigan et al., 2024a; Finnigan et al., 2024b). Precision fermentation employs microorganisms, often genetically engineered, to produce specific, high-value molecules like proteins, fats, or flavor compounds (Linder, 2024). The industrial implementation of these approaches relies on several core fermentation technologies.

Solid-State Fermentation (SSF) is a process that involves the growth of microorganisms on a solid or semi-solid substrate (like agro-waste) with minimal free water (Cedeno et al., 2025; Soccol et al., 2017). This method closely mimics the natural growth conditions for many filamentous fungi, such as *Aspergillus oryzae*, allowing for the development of aerial and penetrative hyphae (Soccol et al., 2017; Tamang et al., 2022). SSF is particularly advantageous for its sustainability, as it requires less water and can effectively utilize abundant and often underutilized agro-industrial wastes as substrates (Soccol et al., 2017). It is a cornerstone technology for producing traditional foods like koji and tempeh, and its low barrier to entry makes it accessible for smaller-scale producers (Tamang et al., 2022).

Submerged Fermentation (SmF) entails cultivating microorganisms in a liquid nutrient medium within a bioreactor (Majumder et al., 2023; Finnian et al., 2024b). A key advantage of SmF is its capacity to precisely regulate process variables like pH, temperature, dissolved oxygen, and nutrient concentration, which is crucial for optimizing yield and consistency at an industrial scale (Majumder et al., 2023). SmF simplifies the processes of mixing, aeration, and downstream separation of biomass, making it highly suitable for mass synthesis of single-cell proteins and other microbial products (Finnigan et al., 2024b).

Surface Fermentation (SF) is another method where microorganisms grow on a substrate that is either liquid or solid. This technique is used in the production of 'Fy protein' by the fungus *Fusarium strain flavolapsis*, where a semi-solid fungal biomat is formed on the surface of the fermentation medium (Furey et al., 2022; Gnaim et al., 2025; Ng et al., 2024). The biomass is then harvested, deactivated with heat, and processed into a final product (Furey et al., 2022).

These fermentation techniques have enabled significant innovations in meat alternatives. Marlow Foods pioneered SmF in large-scale airlift pressure cycle fermenters to produce mycoprotein from *Fusarium venenatum*, marketed as Quorn™ (Finnigan et al., 2019; Finnigan et al., 2024b; Gnaim et al., 2025). These fermenters, which can be up to 30 meters tall with a capacity of 150,000 liters, are designed to optimize hyphal growth and produce approximately two metric tons of mycoprotein per hour, highlighting the immense scalability of fungal fermentation (Finnigan et al., 2019; Finnigan et al., 2024a; Risner et al., 2023).

3.2 Strain Engineering

In addition to hardware improvements in fermentation equipment, the biological software of fermentation, or the fungal strains themselves, is also an important area of innovation. Strain engineering includes both classical and modern genetic techniques to increase the efficiency, yield, and quality of products produced by fungi (Zhang et al., 2020). For example, *Aspergillus oryzae* produces approximately 50 enzymes, including amylases that are essential for hydrolyzing starch in rice; genes like *amyA*, *amyB*, and *amyC* are important targets for genetic modification to increase fermentation efficiency (Zhang et al., 2020). Likewise, an isoenzyme XynG2 isolated from *A. oryzae* that continues to function at high ethanol concentrations was added to sake mash and produced a 4.4% increase in yield (Sato et al., 2010). In the yeast *Saccharomyces cerevisiae* used for making sake, a strain resistant to the inhibitor cerulenin of fatty-acid synthase was generated, resulting in an increased five-fold production of ethyl caproate, which gives sake its desirable fruity aroma (Ichikawa et al., 1991). However, in recent years, powerful tools to modify fungal genomes, such as CRISPR/Cas9, have opened the door to much more precise alterations (Li et al., 2025). A prominent example of this is the engineering of an *Agaricus bisporus* (button mushroom) that browns less due to a loss-of-function mutation in one gene encoding polyphenol oxidase (PPO), reducing its activity by 30% and extending shelf life, all accomplished with CRISPR/Cas9 (Waltz, 2016). Importantly, this gene-edited mushroom was not regulated as a genetically modified food product by the U.S. Department of Agriculture because it did not contain any foreign DNA; however, this decision has opened another potential pathway for the regulation of other foods containing engineered DNA (Waltz, 2016). CRISPR/Cas9 can also be used in basic research to characterize biosynthetic pathways; for instance, CRISPR/Cas9 knockout of the key gene *SbaPKS* led to reduced hypocrellin production in the fungus *Shiraia bambusicola* and identified it as a potential target to overexpress for increased yield (Deng et al., 2017). These molecular tools will be critical for realizing the full potential of fungi through the development of strains with enhanced nutritional profiles, flavors, and process efficiency (Li et al., 2025).

4. NUTRITIONAL AND HEALTH IMPACTS

Fungal foods, from traditionally fermented products to modern mycoprotein, provide numerous nutritional and health benefits, the fermentation process itself usually improving the quality of nutrients available for absorption (bioavailability), and even supplementing some of the vitamins that are often not present in sufficient quantities in plant-based diets (such as B12, D, and E) (Amara & El-Baky, 2023; Finnigan et al., 2024b; Sharma et al., 2025). Due to their high-quality protein, dietary fiber, and bioactive compounds, fungi are part of a healthy and sustainable diet (Table 2).

4.1 Protein Quality and Enzymatic Contribution

Mushrooms also provide high-quality protein: edible mushrooms and mycoproteins are often used as meat substitutes in vegetarian or vegan diets (Wang & Zhao, 2022). For example, mycoprotein from *Fusarium venenatum* possesses a PDCAAS (Protein Digestibility-Corrected Amino Acid Score) of 0.91 (equivalent to beef and soy), contains all essential amino acids, and has low phytic acid content, which minimizes the potential for mineral absorption inhibitors like iron or zinc (Finnigan et al., 2024b; Gnaim et al., 2025; Linder, 2024). The fibrous structure of fungi gives them a meat-like mouthfeel; this is an advantage over many plant-based alternatives that require extensive processing to attain the same texture (Finnigan et al., 2024b). Fungi also provide nutrition indirectly by producing food-useful enzymes like lipases, proteases, amylases, and xylanases that serve as highly effective biological catalysts (Dai et al., 2021). Fungal proteases from *Aspergillus niger* are employed in brewing and baking to digest proteins, enhancing the crumb structure of baked goods; pectinases help clarify fruit juices, and lipases can enhance flavor and texture in dairy products (Li et al., 2020). Enzymes for food improvement also support efficiency and sustainability by enabling the conversion of food waste into simple sugars for fermentation to fuel a circular food system (Amara & El-Baky, 2023; Dai et al., 2021).

Table 2. A detailed account of edible mushrooms' nutritional and medicinal properties (Das et al., 2021)

Species	Vernacular name	Crude protein (g/Kg)		Carbohydrates (g/Kg)	Energy (Kcal/ Kg)	Medicinal properties	References
Pleurotus ostreatus	Oyster mushroom	132.3	35.8	751.1	330	Beta-glucans and antioxidants like flavonoids, ergothioneine, phenolics, tocopherols, and ascorbic acid exhibit anticancerous and anti-inflammatory properties.	(Akata et al., 2012)
Pleurotus eryngii	King oyster mushroom	110.0	14.5	813.7	421	Its polysaccharides reduce oxidative and inflammatory stress, with potential anti-obesity, antidiabetic, and neuroprotective benefits.	(Reis et al., 2012)
Agaricus bisporus	Button mushroom	140	21.8	740	325	Source of Vitamin D, antioxidants such as ergothioneine and selenium.	(Reis et al., 2012)
Agaricus brasiliensis	Almond mushroom	267.4	26.2	638.3	-	The source of phenolics such as pyrogallol, gallic acid, and other secondary metabolites demonstrates antioxidant, therapeutic, and immunomodulatory properties.	(Tsai et al., 2008)
Cordyceps sp.	Murrill's Agaricus	398.4	18.2	594.2	379	Triggers cancer cells, causing apoptosis and anti-angiogenesis.	(Hsu et al., 2002)
Lentinula edodes	Shiitake	44.0	17.3	871.4	772	Lentinan (beta-glucan) possesses anti-cancerous effects, preventing lung metastasis.	(Vetter, 2019)
Flammulina velutipes	Golden needle mushroom	38.7	28.9	859.9	467	Its bioactive compounds strengthen the immune system, reduce inflammation and free radicals, and maintain blood sugar levels, regulating type 2 diabetes.	(Reis et al., 2012)
Volvariella volvacea	Straw mushroom	200-250	20-40	400-570	300-400	Anti-tumor, anti-diabetic, anti-inflammatory, and a good source of antioxidants.	(Das et al., 2021)
Pleurotus sajor-caju	Grey oyster mushroom	374	10	553	2750	Regulates hyperglycemic conditions.	(Akyüz & Kirbağ, 2010)
Tricholoma giganteum	Matsutake	329	9.1	101	636	It inhibits angiotensin-converting enzymes, thereby regulating blood pressure.	(Mau et al., 2001)

4.2 Bioactive Compounds and Functional Benefits

While fungal foods are rich in basic nutrients, they also contain a range of bioactive compounds that may provide specific health benefits and can be categorized as macromolecules (polysaccharides) or lower molecular weight compounds (phenolics and peptides) (Rousta et al., 2024; Wang & Zhao, 2022).

Bioactive Peptides: Fungal proteases digest large proteins into smaller fragments called bioactive peptides, inactive when enclosed in the parent protein but can cause physiological effects when released during fermentation (Rousta et al., 2024). For instance, tripeptides V-P-P and I-P-P produced from miso made by fermenting casein have been reported to have anti-hypertensive activity due to inhibition of ACE (angiotensin-converting enzyme) (Inoue et al., 2009), while peptides isolated from *Aspergillus* species-fermented meat sauce showed potent antioxidant activity (Ohata et al., 2016).

Dietary Fibers (Chitin and β-glucans): The primary structural element of fungal cell walls and a source of special dietary fibers is chitin. Chitin is a non-digestible fiber that prevents constipation, and it is being researched as a potential management tool in colorectal disorders (Chiu et al., 2019). Its derivative chitosan has been proven effective in clinical studies in blocking fat absorption; LDL ("bad") cholesterol can be reduced while HDL ("good") cholesterol increases (Figure 2) (Moraru et al., 2018).

Another class of soluble dietary fibers with well-established beneficial effects on metabolic health is the β-glucans (Edo et al., 2025). The mechanisms of action are complex, but as a viscous fiber it slows gastric emptying and intestinal absorption of glucose, which aids blood sugar control (which can be important for type 2 diabetics), also, Glucagon-like peptide-1 (GLP-1) and peptide YY (PYY) are two satiety hormones that can be released when the gut microbiota ferments short-chain fatty acids and prevents the liver from synthesizing cholesterol (Edo et al., 2025). In fact, β-glucans are fermented by gut microbiota to produce SCFAs (Edo et al., 2025), which reduce cholesterol synthesis in the liver and stimulate the release of PYY and GLP-1, reducing cholesterol and stimulating satiety (Figure 1) (Sima et al., 2018). These examples illustrate that fungal-based foods are functional foods capable of being proactive for health maintenance and disease prevention.

Figure 2. Diagrammatic representation of the positive nutritional advantages associated with consuming fungal foods. This visual representation features four sections emphasizing mycoprotein, mushrooms, and fungal-derived fermented foods and beverages. HDL-C, high density lipoprotein cholesterol; LDL-C, low density lipoprotein cholesterol; VLDL, very low density lipoprotein; TC, total cholesterol; IL-6, interleukin-6; TNF-α, tumour necrosis factor; GLP-1, glucagon-like peptide-1; BCAA, branched chain amino acid; LPS, lipo polysaccharide; TLR, toll-like receptor; RTK, receptor tyrosine kinase; PI3K, phosphoinositide 3-kinase; AKT, protein kinase B; ARE, antioxidant response element; Cat, catalases; GPx, glutathione peroxidase; GR, glutathione reductase; SOD, superoxide dismutase; IFAs, isoflavone aglycones; IFGs, isoflavone glycosides; Keap1, Kelch-like ECH-associated protein 1; Nrf2, nuclear factor erythroid 2-related factor (Azizi et al., 2021; Ahmad et al., 2022; Jiang et al., 2025; Paul et al., 2023; Surya et al., 2024; Wang & Zhao, 2022)

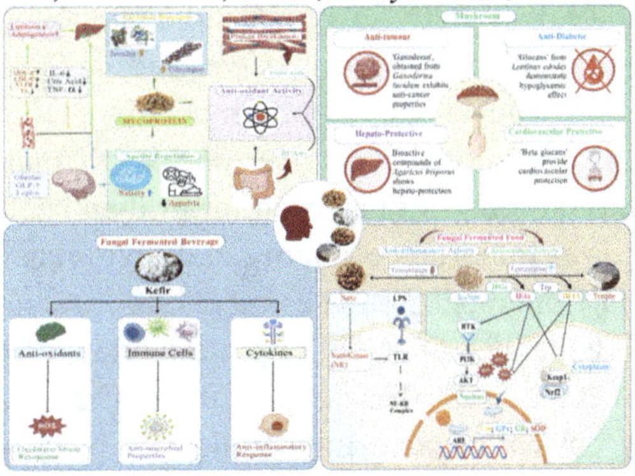

5. SYSTEMIC CHALLENGES AND COMMERCIALIZATION RISKS

Even though fungi have enormous potential as a sustainable food source, there are multiple challenges from lab innovation to widespread commercial success (Awasthi et al., 2022). Many of them form a self-reinforcing negative feedback loop that is holding back the entire sector: high production costs lead to premium pricing, which discourages mainstream consumer adoption; lack of broad market demand makes it difficult for companies to secure the large capital investment needed to build large-scale facilities, so they remain small, which means their production costs stay high; slow, complex, and expensive regulatory approval processes that can drain a start-up's financial runway long before a product comes to market (Awasthi et al., 2022;

De Cianni et al., 2024; Pouris et al., 2024). This section examines these systemic barriers that collectively represent the biggest risk to the fungal food revolution.

5.1 Economic and Scalability Hurdles

One significant and complex challenge is the economic viability of fungal-based proteins. Although touted as potentially less expensive than meat, technoeconomic analyses have shown that mycoprotein can compete with high-priced meats like beef and lamb but not commodity meats like chicken (Risner et al., 2023). According to a thorough techno-economic analysis, the cost of producing mycoprotein from lignocellulosic agricultural residues was $5.04 per kilogram for the crude paste. This is significantly more expensive than the cost of chicken per protein (Upcraft et al., 2021). One major hurdle is this price discrepancy, as cost is one of the main drivers of consumer food choice, especially in an inflationary economy (Deloitte, 2023). The production process is also energy-intensive, and large-scale fermentation in bioreactors requires significant energy inputs for aeration, temperature control, and mixing, contributing to operational costs (Cerrone & O'Connor, 2025). Although mycoprotein's total carbon footprint has been lower than beef's, direct energy use during manufacturing is an important economic and environmental factor (Shahid et al., 2024; Cerrone & O'Connor, 2025). These economic pressures are compounded by the technological challenges of scaling up production, often called the "ghost of scale-up" (Awasthi et al., 2022; Teng et al., 2021). The transition from a highly controlled laboratory environment to industrial-scale bioreactors is not a linear process and is fraught with complexity (Teng et al., 2021). A critical factor for filamentous fungi is managing their morphology; whether the fungus grows as dispersed mycelia or forms dense pellets can dramatically affect process viscosity, nutrient uptake, and product yield (Veiter et al., 2018). Maintaining process consistency, ensuring adequate oxygen delivery, and achieving effective mixing in fermenters with volumes exceeding 100,000 liters are non-trivial engineering challenges (Veiter et al., 2018). Overcoming these hurdles requires substantial capital expenditure to build and operate state-of-the-art fermentation facilities and downstream processing equipment, creating a high barrier to entry for new companies (Teng et al., 2021).

5.2 Navigating the Global Regulatory Maze: FDA and EFSA Frameworks

For any novel food ingredient, navigating the regulatory landscape is a critical step for market access. However, the stark divergence between the frameworks in major markets like the United States and the European Union creates more than just a procedural hurdle; it poses a significant strategic challenge that is a major barrier to global commercialization. This "regulatory asynchrony" effectively fragments the global market, increases compliance costs, and dictates where innovation and investment are prioritized. The result, as some analysts contend, is a system that often leaves European consumers and companies at a competitive disadvantage.

Under the generally recognized as safe (GRAS) framework, the Food and Drug Administration (FDA) primarily regulates novel food ingredients in the United States. This system offers two primary options: a business can voluntarily submit a GRAS notification to the FDA or assemble its own group of experts to "self-affirm" GRAS status (Table 3; Food and Drug Administration, 2023). If the FDA has no objections to the submitted safety data, it issues a "no questions" letter, making the notice publicly available (Table 3). This process can be remarkably swift; for example, Marlow Foods' GRAS notice for Quorn mycoprotein was filed in December 2001 and received a "no questions" letter in January 2002 (Food and Drug Administration, 2002). This relative speed allows companies to enter the US market quickly, begin generating revenue, and attract further investment. Recent examples of successful notifications include The Better Meat Co. for its Rhiza mycoprotein derived from *Neurospora crassa* and MycoTechnology for its pea protein fermented with shiitake mycelia (Bartholomai et al., 2022; Food and Drug Administration, 2023; Shapiro & Cumbelich, 2025).

In stark contrast, the European Union employs a much more stringent, centralized, and lengthy process under its Novel Food Regulation (EU) 2015/2283 (European Union, 2015; Ververis et al., 2020). According to the European Union (2015), any food that was not consumed widely in the EU prior to May 1997 is deemed novel and needs pre-market authorization (European Union, 2015). The process is mandatory and involves multiple stages of review, first by the European Commission (EC) and then a scientific risk assessment by the European Food Safety Authority (EFSA) (European Food Safety Authority, 2025). Recent analysis of this process reveals a brutal reality for innovators: the average time from applying to receiving a final EFSA opinion is 2.56 years, with some cases taking as long as six years (Bloch et al., 2025; Ververis et al., 2020). EFSA's requests for more data "almost always extend" this timeline and can cause the review clock to pause for months at a time (European Food Safety Authority, 2025; Ververis et al., 2020).

The practical implication of this divergence is profound. A company can launch a product and establish a market presence in the US while its identical application in the EU still navigates the initial administrative checks (Bloch et al., 2025; Ververis et al., 2020; Table 3). This long and unpredictable waiting period in the EU can be a "death sentence" for startups, draining their limited financial resources and delaying market traction (Bloch et al., 2025). This regulatory lag stifles innovation within the EU and creates a significant barrier for companies with global ambitions, forcing them to adopt market-by-market strategies and navigate two fundamentally different and costly regulatory paradigms. The following table provides a comparative summary of these two key regulatory pathways.

Table 3. Comparative analysis of regulatory pathways for novel fungal foods: US (FDA GRAS) vs. EU (EFSA novel food)

Feature	US (FDA GRAS Notification)	EU (EFSA Novel Food Regulation)
Governing Regulation	Federal Food, Drug, and Cosmetic Act (FD&C Act)	Regulation (EU) 2015/2283
Process Type	Self-affirmed and FDA-notified GRAS.	Mandatory, centralized pre-market authorization (European Union, 2015; Ververis et al., 2020)
Average Timeline to Market	Rapid; can be a few months to a year.	Extremely lengthy; average of 2.56 years to receive EFSA opinion, can extend to 6 years (Bloch et al., 2025; Ververis et al., 2020)
Key Data Requirements	Evidence demonstrating safety through scientific procedures, including manufacturing processes, specifications, dietary exposure, and toxicological data (Food and Drug Administration,2023)	Comprehensive dossier covering compositional, nutritional, toxicological, and allergenic properties, production process, and history of use (European Food Safety Authority, 2025; Ververis et al., 2020)
Regulatory Outcome	"No questions" letter from FDA, which is not a formal approval but an indication of no objection. (Food and Drug Administration,2002; 2023)	Formal authorization by the European Commission, published in the Union list of novel foods (European Union, 2015)
Public Transparency	GRAS notices and FDA response letters are publicly available on the FDA website (Food and Drug Administration,2002; 2023)	Applications and EFSA scientific opinions are made public, with provisions for confidentiality (European Food Safety Authority, 2025)
Key Challenge for Applicants	Ensuring scientific evidence is robust and convincing enough to withstand FDA scrutiny and potential public challenge.	The extreme length, unpredictability, and high cost of the process, particularly the delays caused by requests for additional data (Bloch et al., 2025).

5.3 Consumer Acceptance: Overcoming Sensory Barriers and Food Neophobia

The consumer determines the success of any food product. For fungal-based foods, achieving widespread consumer acceptance requires overcoming significant sensory and psychological barriers. When it comes to food purchases, taste and price are always the most important considerations, and at present, many alternative protein products are perceived as lagging conventional meat on both fronts (Maseko et al., 2025). The sensory challenges are substantial (Table 4). While mycoprotein is often praised for its naturally fibrous, meat-like texture, many products in the broader alternative protein category struggle with undesirable sensory attributes (De

Cianni et al., 2024; Finnigan et al., 2024a; Gamarra-Castillo et al., 2022). These can include off-flavors often described as “beany,” “green,” or “bitter,” which arise from compounds like lipid oxidation products in plant-based ingredients (Michel et al., 2021). Poor mouthfeel, a lack of juiciness, and an unappealing appearance or color that does not change appropriately during cooking are also common complaints that deter repeat purchases (Table 4) (Michel et al., 2021).

Beyond these tangible sensory issues lie deeper psychological hurdles. Food neophobia, a natural reluctance to eat novel or unfamiliar foods, is a major barrier for products derived from microorganisms (Chezan et al., 2022). The association of fungal proteins with “mold” can evoke feelings of disgust or concern about safety, even if the strains used are perfectly safe (Chezan et al., 2022). Studies have shown that a consumer's comfort level with this association is a significant predictor of their willingness to purchase mycoprotein products (Chezan et al., 2022). Additionally, though healthiness is a commonly reported driver behind alternative protein trials (Deloitte, 2023), initial interest in the nutritional aspects of plant-based foods has shifted into more skepticism about levels of processing and sodium content in some meat alternatives (De Cianni et al., 2024). These barriers are technological, which improve sensory qualities, and marketing and communications strategies, which build familiarity and trust.

Table 4. Sensory challenges and mitigation strategies in fungal-based meat alternatives

Sensory Challenge	Underlying Cause	Potential Mitigation Strategy (with Fungal Application)	References
Off-Flavors (e.g., beany, bitter, grassy)	Lipid oxidation products (e.g., n-hexanal) and non-volatiles (e.g., saponins, phenolic compounds) from plant-based ingredients.	Fermentation with specific fungal strains (e.g., *Aspergillus oryzae; Rhizopus oligosporus*) can break down off-flavor compounds or produce desirable masking volatiles and umami flavor precursors.	(Gamarra-Castillo et al., 2022; Massa et al., 2025)
Lack of Meaty Flavor	Absence of key flavor precursors (e.g., sulfur-containing amino acids, reducing sugars) that form meaty notes via the Maillard reaction during cooking.	Fungal fermentation can generate these precursors. Precision fermentation can be used to produce key meaty flavor compounds like heme protein.	(De Cianni et al., 2024 ; Hilgendorf et al., 2024; Maseko et al., 2025)
Poor Texture (e.g., spongy, not fibrous, lack of chewiness)	Weak protein networks in plant-based formulations; lack of structure mimicking animal muscle fibers.	Utilize the natural filamentous structure of mycoprotein (*Fusarium venenatum*) as a whole-food ingredient. Fungal hydrocolloids can act as binders to improve texture.	(Gnaim et al., 2025)

continued on following page

Table 4. Continued

Sensory Challenge	Underlying Cause	Potential Mitigation Strategy (with Fungal Application)	References
Lack of Juiciness / Dry Mouthfeel	Poor water-holding and fat-binding capacity of the protein matrix.	Fungal fermentation can modify protein structures to improve water and oil-holding capacity. Hybrid products can incorporate cultivated animal fat produced via fermentation.	(Maseko et al., 2025; Zhang et al., 2025)
Unappealing Color	Lack of natural meat-like pigments; use of artificial colors or plant pigments (e.g., beetroot) that do not change color appropriately during cooking.	Use of natural, food-grade pigments produced by fungi (e.g., *Monascus purpureus*) or other microorganisms to provide stable, appealing colors.	(Chaudhary et al., 2021)

5.4 Ethical Bioprospecting and Equitable Benefit Sharing

The search for new strains of fungi with desirable properties for food production (whether harvested from different environments or discovered in traditional foods) is a type of bioprospecting (Kaur et al., 2024), which comes with both ethical obligations and real hazards. Biopiracy occurs when the knowledge and biological resources of indigenous peoples and local communities are accessed without their consent or just compensation, potentially resulting in reputational harm, legal disputes, and loss of trust (Hamilton, 2006).

For instance, the patenting of *Rhizopus oligosporus* for tempeh production has been cited as an example of where Indonesian communities who have stewarded this fermentation knowledge for centuries were denied access to commercial benefits associated with their cultural heritage (Marden et al., 2021). In response, international law has developed in the form of two treaties: the Convention on Biological Diversity (CBD) and its supplementary Nagoya Protocol that imposes a legal duty upon users of genetic materials and related traditional knowledge (TK) to adhere to Access and Benefit Sharing (ABS) principles (Nagoya Protocol, 2011).

The core tenets of ABS require that researchers and companies acquire Prior Informed Consent (PIC) from the providers of the resources (be it a national government or an indigenous community) before access is granted. Furthermore, they must negotiate Mutually Agreed Terms (MAT) that outline how the advantages that result from using these resources will be distributed in a way that is just and equal (Nagoya Protocol, 2011).

Benefit sharing is not limited to simple monetary royalties. It encompasses a wide range of arrangements that can be tailored to the needs and desires of the community. Monetary benefits can include upfront payments, access fees per sample, contributions to community trust funds, salaries for local collaborators, and joint ownership of intellectual property (Tsioumani, 2018). Equally important are non-monetary benefits, which can foster long-term capacity and empowerment. These may include sharing research and development results, providing education and training, collaborating on scientific programs, transferring relevant technologies, and participating directly in product development (Tsioumani, 2018). By embedding these principles of ethical partnership into the R&D process, the fungal food industry can ensure that its growth is sustainable and just, honoring the contributions of traditional knowledge holders who paved the way for modern innovation.

6. SUSTAINABILITY, POLICY, AND THE CIRCULAR BIOECONOMY

The potential of fungi to help create a food system that is more sustainable extends far beyond simply offering a low-impact protein alternative. Fungi are uniquely positioned to act as keystone enablers of a circular bioeconomy, capable of transforming waste into value and aligning with global sustainability policies. A comprehensive assessment of their role requires an explicit connection to frameworks like the UN Sustainable Development Goals (SDGs) and a critical analysis of their entire life cycle (United Nations Department of Economic and Social Affairs, 2023).

6.1 Contribution to UN Sustainable Development Goals (SDGs)

The cultivation and utilization of fungi in food systems can directly or indirectly contribute to achieving numerous SDGs. This systemic impact underscores their potential as a multifaceted solution to pressing global challenges.

- **SDG 2 (Zero Hunger)** (United Nations Department of Economic and Social Affairs, 2023): Fungi directly address food security. Mycoprotein and edible mushrooms provide a high-quality, nutrient-dense protein source (Awasthi et al., 2022; Pérez-Moreno et al., 2021). Moreover, low-technology mushroom cultivation techniques can be deployed in food-insecure regions, using local agricultural waste to produce nutritious food, thereby increasing food availability and resilience at the community level (Pérez-Moreno et al., 2021; Violo, 2024). Solid-state fermentation of staple crops can also enhance their nutritional value and bioavailability, transforming low-cost materials into better food (Tamang et al., 2022).
- **SDG 1 (No Poverty) and SDG 5 (Gender Equality)** (United Nations Department of Economic and Social Affairs, 2023): The cultivation and trade

of mushrooms can be a significant source of income, particularly in rural communities. This economic activity can help alleviate poverty. In many cultures, the trade of foraged or cultivated mushrooms is an enterprise led by women, providing them with economic empowerment and contributing to gender equality (Pérez-Moreno et al., 2021; Violo, 2024).

- **SDG 3 (Good Health and Well-being)** (United Nations Department of Economic and Social Affairs, 2023): As detailed in Section 4 Fungal foods promote good health by providing nutritionally dense foods with bioactive compounds such as β-glucans that are beneficial for cholesterol management and blood sugar control (Finnigan et al., 2024b; Pérez-Moreno et al., 2021; Sima et al., 2018).
- **SDG 12 (Responsible Consumption and Production) and SDG 13 (Climate Action)** (United Nations Department of Economic and Social Affairs, 2023; Yee et al., 2025): This is where fungi have one of their most profound impacts. Mycoprotein production has a much smaller environmental footprint than conventional livestock; life cycle assessments demonstrate that its carbon footprint can be more than 90% lower than beef, with greatly reduced land and water use (Finnigan et al., 2024a; Linder, 2024), fungi sequester carbon in forest ecosystems to contribute to climate action, and a circular economy helps reduce waste and GHG emissions (Ahlborn et al., 2020; Pérez-Moreno et al., 2021; Violo, 2024).

6.2 Fungi in the Circular Bioeconomy: Valorizing Agro-Industrial Waste

Fungi are not just a component of the circular economy; they facilitate closing resource loops by breaking down recalcitrant lignocellulosic waste produced annually from global agriculture (Cedeno et al., 2025). The breakdown of cellulose and lignin-rich byproducts, including straw, husks, bagasse, and spent grains, is a major challenge in the food and agriculture industries; it treats the waste as a disposal problem. Fungi provide a solution to this through their powerful enzymatic machinery. In particular, filamentous fungi produce a cocktail of enzymes such as cellulases and xylanases that efficiently degrade these complex polymers into simpler, more readily used molecules; in this way, they serve an important technological function by converting the vast potential value that is locked up in low-value or even negative-value waste streams (Ahlborn et al., 2020; Cedeno et al., 2025; Dai et al., 2021). Many of these resources would otherwise remain unavailable if not for this biochemical "key" and could instead pollute the environment. Solid-state fermentation (SSF) is a good technology for conducting this process because it is low in water, inexpensive, and replicates the natural growth habit of fungi (Ahlborn et al., 2020; Cedeno et al., 2025; Soccol et al., 2017).

We are starting to see concrete examples of this valorization: for example:

- Fungi grown on substrates such as brewer's spent grain, grape bagasse, and surplus bread have been used to produce mycoprotein with an edible protein content (Ahlborn et al., 2020; Ng et al., 2024; Stoffel et al., 2019).
- Rice husks and cocoa shells are examples of agro-industrial waste that can be converted to valuable industrial enzymes that are then applied in other biotechnological processes (Cedeno et al., 2025).
- SSF is also being used to convert waste products such as cassava bagasse into protein-enriched animal feed, solving a waste management problem while creating an agricultural input at the same time (Soccol et al., 2017).

In all these applications, fungi act as the bioengineers of a circular system that converts waste to food, feed, and other high-value bioproducts to reduce pollution, conserve resources, and generate economic value from materials that would otherwise be wasted (Cedeno et al., 2025; Dai et al., 2021).

6.3 Life Cycle Assessments (LCA) and Climate Mitigation Strategies

Life cycle assessments (LCAs) quantify the environmental impact of fungal food production from cradle to grave, and multiple LCAs consistently show the substantial environmental benefits of mycoprotein over conventional meat, particularly beef (Neville et al., 2017). A recent study concluded that the carbon footprint of Quorn mince has a ten-fold lower protein content than beef and a four-fold lower protein content than chicken (Finnigan et al., 2024a; Finnigan et al., 2024b). Another study using mycoprotein produced from lignocellulosic residues estimated greenhouse gas emissions to be below 14% of those associated with beef production (Upcraft et al., 2021).

However, a critical analysis of these LCAs reveals important nuances. The dramatic advantage in land use is unequivocal. However, other impact categories depend heavily on system boundaries and production methods. For example, the water consumption for mycoprotein can be high if the feedstock is an irrigated crop like rice, as the water used for agriculture is allocated to the straw byproduct (Upcraft et al., 2021). The environmental profile improves dramatically if the feedstock is considered a true "waste" product, to which zero upstream agricultural burden is allocated (Upcraft et al., 2021). Furthermore, the LCAs highlight that a major contributor to the carbon footprint of mycoprotein is the energy consumed during the fermentation process itself (Shahid et al., 2024; Cerrone & O'Connor, 2025). This indicates that the overall climate benefit of fungal foods is intrinsically linked to the decarbonization of the energy grid. To maximize their climate mitigation potential, mycoprotein production facilities must be powered by renewable energy sources (Awasthi et al., 2022; Finnigan et al., 2024a). Therefore, while fungi offer a clear pathway to reducing the climate impact of our protein supply, realizing this potential requires a holistic approach that integrates sustainable feedstock sourcing with clean energy.

7. MARKET TRENDS AND CONSUMER INSIGHTS

The commercial landscape for fungal-based foods is dynamic and rapidly evolving, characterized by significant investment, emerging product categories, and a complex interplay of market forces and consumer psychology. While market projections are optimistic, long-term growth hinges on the industry's ability to navigate consumer perceptions and develop products that can successfully compete with conventional options on the key metrics of taste, price, and convenience (Allied Market Research, 2021; Gnaim et al., 2025).

7.1 Global Market Analysis and Economic Projections

The market for fungal-derived products is expanding significantly, attracting substantial capital investment that signals strong confidence in its future growth. From its 2020 valuation of $156.6 million, the fungal protein market is projected to grow at a compound annual growth rate (CAGR) of 9.5 percent to reach $386.6 million by 2030 (Allied Market Research, 2021; Gnaim et al., 2025).

This is part of a broader trend in the alternative protein sector, where the fermentation category alone attracted $1.7 billion in investment in 2021 (Risner et al., 2023).

Traditional fungal fermented products also command significant market value. According to Future Market Insights (2025), the global miso market is expected to double to $121.07 billion by 2035 from its 2022 valuation of $67.8 billion. Similarly, tempeh and soy sauce markets are projected to reach $6.3 billion and $17.18 billion, respectively, by 2030 (Global Info Research, 2024). Mycoprotein, a key player in the modern fungal food space, had a market value of $704.19 million in 2023, with projections to reach over $1 billion by 2031, according to Future Market Insights (2024). These figures, spanning both traditional and novel products, underscore the profound and growing economic importance of fungi in the global food industry (Ahmad et al., 2022; Finnigan et al., 2024b).

7.2 The Emerging Market for Hybrid Protein Products

Beyond "pure" alternative proteins, a strategically important and rapidly emerging category is that of hybrid products. These products blend conventional animal-based proteins with alternative proteins (from plants or fungi) or combine different types of alternative proteins (e.g., fungal and plant-based) (Maseko et al., 2025; Oh & Kim, 2025; Santhapur et al., 2024). This approach is more than just a niche; it represents a critical transitional strategy that may accelerate the mainstream adoption of novel proteins more effectively than pure alternatives.

Hybrid products can function as a "Trojan Horse" for consumer acceptance by addressing head-on the primary barriers to adoption. First, they tackle the sensory challenge. Blending can significantly enhance taste and texture; for instance, adding mushrooms to a beef patty can increase juiciness and umami flavor, while incorporating cultivated animal fat into a plant-based matrix can vastly improve mouthfeel and flavor release (Maseko et al., 2025; Oh & Kim, 2025). This directly addresses the sensory gap that often deters consumers. Second, they confront the issue of cost. Replacing a portion of a high-cost ingredient like conventional meat with a lower-cost fungal protein can make the final product more affordable, tackling the second major barrier to purchase (Maseko et al., 2025). Finally, hybrids offer a pathway to better nutrition by reducing saturated fat and increasing fiber, a powerful marketing message (Neville et al., 2017). Several companies are already capitalizing on this strategy. The Better Meat Co. supplies its Rhiza mycoprotein as an ingredient for blended goods, like the "Chicken Plus" line from Perdue Farms, which is sold in thousands of supermarkets and blends chicken with plant-based ingredients (Maseko et al., 2025; Santhapur et al., 2024). Other companies are focusing on developing cultivated fat specifically for use in hybrid applications, recognizing its potential to revolutionize the taste of plant-based meat (Oh & Kim, 2025). This pragmatic, bridge-technology approach may prove essential for easing the broader market transition towards a more diverse protein landscape.

7.3 Marketing Strategies for Novel Fungal Foods

Given the sensory and psychological barriers, the marketing and communication strategies used for novel fungal foods are critical to their success. Research into consumer behavior has identified several key principles for effective marketing that move beyond simply highlighting environmental or ethical benefits (Szejda & Parry, 2020).

The most effective strategies focus on the foundational drivers of food choice: taste, price, and convenience (Szejda & Parry, 2020). Messaging should emphasize gains and sensory pleasure: messaging that emphasizes indulgent, taste-focused language on menus and packaging (e.g., "rich, savory mushroom steak") is more effective than messaging that focuses on health or restriction (e.g., "low-fat meat alternative"); choice architecture matters as well, so products are placed in or next to the conventional meat section rather than in a separate free-from or vegetarian aisle because this reduces effort for consumers choosing an alternative and frames it as competition with conventional meat (Szejda & Parry, 2020). Last, strategies should work to increase familiarity to overcome food neophobia by positioning products as new or different versions of familiar foods (e.g., a plant-based bolognese or a mycoprotein-based stir-fry) (Szejda & Parry, 2020). Clear and appealing labeling that refers to the product as "plant-based" or "plant protein" elicits more purchase intent than more exclusionary terms such as "vegan" or "meat-free," which might trigger omnivores to think that the product is not for them (Szejda & Parry, 2020). Focusing on taste, familiarity, and convenience can help marketers make novel fungal foods attractive to the wide range of consumers necessary for market growth.

8. FUTURE PROSPECTS AND IDENTIFIED RESEARCH GAPS

The journey of mycological foods from traditional staples to pillars of a future sustainable food system is well underway. However, to fully realize this potential, a concerted and strategic research and development effort is required to address key scientific bottlenecks and capitalize on emerging technological opportunities. The future of this field will be defined not by a single breakthrough, but by integrated advancements across biology, engineering, and data science that simultaneously tackle the interconnected challenges of cost, scale, and consumer acceptance.

8.1 A Roadmap for Future Research

To accelerate progress, the scientific community and industry must prioritize several critical research areas. The following represents a roadmap of identified research gaps that, if addressed, could significantly advance the field:

- **Strain Optimization and Discovery:** A critical limitation of the current industry is its heavy reliance on a very small number of fungal species (e.g., *Fusarium venenatum, Aspergillus oryzae*) (Gnaim et al., 2025). This narrow focus leaves the vast, untapped biodiversity of the fungal kingdom largely unexplored. Researchers contend that a pressing need exists to discover and domesticate new strains (Gnaim et al., 2025). Future research efforts, therefore, should prioritize high-throughput screening for fungi with naturally superior characteristics. Key targets for optimization include higher protein content, more favorable flavor profiles, enhanced tolerance to the stresses of industrial fermentation, and the ability to efficiently metabolize a wider range of low-cost, sustainable feedstocks (Ahmed et al., 2024; Pouris et al., 2024; Tamang et al., 2022).

- **Hybrid Protein Functionality:** The market for hybrid products is growing, yet the fundamental science underpinning them is still in its infancy. More research is needed to understand the complex physicochemical interactions that occur when fungal proteins are blended with plant-based or animal-based proteins (Maseko et al., 2025; Oh & Kim, 2025; Santhapur et al., 2024). Systematic studies are required to determine how factors like pH, temperature, and salt concentration influence protein gelation, water-holding capacity, emulsification, and ultimately, hybrid products' final texture and mouthfeel. This knowledge is essential for the rational design of next-generation products with superior sensory qualities (Massa et al., 2025)
- **Consumer Psychology and Sensory Science:** While taste is known to be a primary driver, a deeper understanding of the specific molecules and psychological factors that influence acceptance is needed. Research should focus on identifying the specific volatile compounds responsible for undesirable off-flavors in some alternative proteins and developing fermentation or processing strategies to mitigate them (Gnaim et al., 2025). Concurrently, more work is needed to understand the drivers of food neophobia and to develop effective communication and marketing strategies that build familiarity and trust with consumers (Chezan et al., 2022).
- **Standardization of Diagnostics and Safety Protocols:** As the industry explores novel fungal species for food production, robust and standardized diagnostic tools are essential to ensure safety. A research gap exists in developing rapid, reliable, and affordable methods for screening new strains and final products for potential mycotoxins, allergens, or other harmful secondary metabolites (Gnaim et al., 2025; Pouris et al., 2024; Scaife et al., 2024). Establishing best-practice guidelines for safety assessment will be crucial for regulatory approval and consumer confidence (Scaife et al., 2024).
- **Open-Access Techno-Economic and Life Cycle Models:** To guide R&D efforts towards the most impactful areas for improvement, the field would benefit from more open-access, flexible techno-economic analysis (TEA) and life cycle assessment (LCA) models. These tools would allow researchers and startups to simulate the effects of various process improvements—such as enhancing saccharification yield from waste substrates or optimizing energy use—on the final production cost and environmental footprint, thereby prioritizing the most promising innovation pathways (Risner et al., 2023; Upcraft et al., 2021).

8.2 Emerging Opportunities and Concluding Remarks

Alongside addressing these research gaps, the field is poised to benefit from several cross-cutting technological advancements that promise to reshape the future of food production (Pouris et al., 2024).

The integration of Artificial Intelligence (AI) and data science offers a powerful means to accelerate R&D cycles. AI models can be used to screen vast genomic databases to predict which fungal species might possess desirable traits, to optimize fermentation parameters in real-time within bioreactors, and to model complex consumer preference data to guide product development

Innovations in advanced bioprocessing and engineering are also critical (Du et al., 2022). The development of novel bioreactor designs, such as more efficient airlift fermenters, and breakthroughs in downstream processing for harvesting and purifying fungal biomass can significantly improve efficiency and reduce production costs (Cedeno et al., 2025).

Furthermore, technologies like 3D food printing present an exciting opportunity to overcome key sensory hurdles. Such technology may enable precision placement of materials derived from fungi, which have complex and customized textures more closely aligned with the fibrous and heterogeneous nature of whole-muscle meats (Barzee et al., 2021). In summary, kingdom Fungi provides a rich palette for constructing a more sustainable, resilient, and just global food system from our past practices in fermentation to the frontiers of biotechnology, but it will take an interdisciplinary effort that connects biological discovery to engineering innovation and human behavior to achieve this potential while addressing significant scientific, economic, regulatory, and social challenges.

REFERENCES

Ahlborn, J., Gmoser, R., Einhorn, P. U., & Schön, C. (2020). From stale bread and brewers spent grain to a new food source using edible filamentous fungi. *Bioengineered, 11*(1), 582–598. DOI: 10.1080/21655979.2020.1768694 PMID: 32449450

Ahmad, M. I., Farooq, S., Alhamoud, Y., Li, C., & Zhang, H. (2022). A review on mycoprotein: History, nutritional composition, production methods, and health benefits. *Trends in Food Science & Technology, 121*, 14–29. DOI: 10.1016/j.tifs.2022.01.027

Ahmed, M. G., Gouda, S. A., Donia, S., & Hassanein, N. M. (2024). Production of single cell protein by fungi from different food wastes. *Biomass Conversion and Biorefinery, 15*(4), 5447–5462. DOI: 10.1007/s13399-024-05478-5

Akata, I., Ergönül, B. Ü. L. E. N. T., & Kalyoncu, F. (2012). Chemical compositions and antioxidant activities of 16 wild edible mushroom species grown in Anatolia. *International Journal of Pharmacology, 8*(2), 134–138. DOI: 10.3923/ijp.2012.134.138

Akyüz, M., & Kirbağ, S. (2010). Nutritive value of wild edible and cultured mushrooms. *Turkish Journal of Biology = Türk Biyoloji Dergisi, 34*, 97–102. DOI: 10.3906/biy-0805-17

Allied Market Research. (2021). *Fungal Protein Market by Type (Yeast, Fusarium Venenatum, Mushroom), Nature (Organic, Conventional), and Application (Food & Beverage, Animal Feed, Pharmaceuticals, Others): Global Opportunity Analysis and Industry Forecast, 2021-2030.* https://www.alliedmarketresearch.com/fungal-protein-market-A12366

Amara, A. A., & El-Baky, N. A. (2023). Fungi as a Source of Edible Proteins and Animal Feed. *Journal of Fungi (Basel, Switzerland), 9*(1), 73. DOI: 10.3390/jof9010073 PMID: 36675894

Awasthi, M. K., Kumar, V., Hellwig, C., Wikandari, R., Harirchi, S., Sar, T., Wainaina, S., Sindhu, R., Binod, P., Zhang, Z., & Taherzadeh, M. J. (2022). Filamentous fungi for sustainable vegan food production systems within a circular economy: Present status and future prospects. *Food Research International, 164*, 112318. DOI: 10.1016/j.foodres.2022.112318 PMID: 36737911

Azizi, N. F., Kumar, M. R., Yeap, S. K., Abdullah, J. O., Khalid, M., Omar, A. R., Osman, M. A., Mortadza, S. A. S., & Alitheen, N. B. (2021). Kefir and its biological activities. *Foods, 10*(6), 1210. DOI: 10.3390/foods10061210 PMID: 34071977

Bartholomai, B. M., Ruwe, K. M., Thurston, J., Jha, P., Scaife, K., Simon, R., Abdelmoteleb, M., Goodman, R. E., & Farhi, M. (2022). Safety evaluation of Neurospora crassa mycoprotein for use as a novel meat alternative and enhancer. *Food and Chemical Toxicology, 168*, 113342. DOI: 10.1016/j.fct.2022.113342 PMID: 35963473

Barzee, T. J., Cao, L., Pan, Z., & Zhang, R. (2021). Fungi for future foods. *Journal of Future Foods, 1*(1), 25–37. DOI: 10.1016/j.jfutfo.2021.09.002

Batra, L. R., & Millner, P. D. (1974). Some Asian fermented foods and beverages, and associated fungi. *Mycologia, 66*(6), 942–950. DOI: 10.1080/00275514.1974.12019699

Bloch, J. L., Rouault, M., Langhi, C., Hignard, M., Iriantsoa, V., & Michelet, O. (2025). The novel food evaluation process delays access to food innovation in the European Union. *NPJ Science of Food, 9*(1), 117. Advance online publication. DOI: 10.1038/s41538-025-00492-x PMID: 40593922

Cedeno, F. R. P., Olubiyo, O. J., & Ferreira, S. (2025). From microbial proteins to cultivated meat for alternative meat-like products: A review on sustainable fermentation approaches. *Journal of Biological Engineering, 19*(1), 44. Advance online publication. DOI: 10.1186/s13036-025-00509-9 PMID: 40369620

Cerrone, F., & O'Connor, K. E. (2025). Cultivation of filamentous fungi in airlift bioreactors: Advantages and disadvantages. *Applied Microbiology and Biotechnology, 109*(1), 41. Advance online publication. DOI: 10.1007/s00253-025-13422-4 PMID: 39928147

Chaudhary, V., Katyal, P., Poonia, A. K., Kaur, J., Puniya, A. K., & Panwar, H. (2021). Natural pigment from Monascus: The production and therapeutic significance. *Journal of Applied Microbiology, 133*(1), 18–38. DOI: 10.1111/jam.15308 PMID: 34569683

Cheung, S. S., Chong, G., Elisia, I., Hasman, D., Lee, M., Chang, L., & Tai, J. (2023). Wild Chaga (Inonotus obliquus) modulates inflammation, neural cell survival and inhibits proliferation of cancer cells. *Pharmacological Research. Modern Chinese Medicine, 9*, 100328. DOI: 10.1016/j.prmcm.2023.100328

Chezan, D., Flannery, O., & Patel, A. (2022). Factors affecting consumer attitudes to fungi-based protein: A pilot study. *Appetite, 175*, 106043. DOI: 10.1016/j.appet.2022.106043 PMID: 35487309

Chiu, C. Y., Yen, T. E., Liu, S. H., & Chiang, M. T. (2019). Comparative effects and mechanisms of chitosan and its derivatives on hypercholesterolemia in high-fat diet-fed rats. *International Journal of Molecular Sciences*, *21*(1), 92. DOI: 10.3390/ijms21010092 PMID: 31877743

Crippa, M., Solazzo, E., Guizzardi, D., Tubiello, F. N., & Leip, A. (2021). Food systems are responsible for a third of global anthropogenic GHG emissions. *Nature Food*, *2*(3), 198–209. DOI: 10.1038/s43016-021-00225-9 PMID: 37117443

Daglio lu, O. (2000). Tarhana as a traditional Turkish fermented cereal food. Its recipe, production and composition. *Food/Nahrung, 44*(2), 85-88. https://doi.org/10.1002/(SICI)1521-3803(20000301)44:2<85::AID-FOOD85>3.0.CO;2-HDOI: 10.1002/(SICI)1521-3803(20000301)44:2<85::AID-FOOD85>3.0.CO;2-H

Dai, X., Sharma, M., & Chen, J. (2021). Fungi in sustainable food production. *Fungal Biology*. Advance online publication. DOI: 10.1007/978-3-030-64406-2

Das, A. K., Nanda, P. K., Dandapat, P., Bandyopadhyay, S., Gullón, P., Sivaraman, G. K., McClements, D. J., Gullón, B., & Lorenzo, J. M. (2021). Edible mushrooms as functional ingredients for development of healthier and more sustainable muscle foods: A flexitarian approach. *Molecules*, *26*(9), 2463. https://doi.org/10.3390/molecules26092463

De Cianni, R., Mancuso, T., Rizzo, G., & Migliore, G. (2024). Health or Environment? Understanding which informative message is more effective in replacing red meat with mushroom-based alternatives. *Appetite*, *199*, 107405. DOI: 10.1016/j.appet.2024.107405 PMID: 38723668

Deloitte. (2023). *Alternative proteins: The future of food*. https://www.deloitte.com/nl/en/Industries/consumer/perspectives/alternative-proteins.html

Deng, H., Gao, R., Liao, X., & Cai, Y. (2017). Genome editing in Shiraia bambusicola using CRISPR-Cas9 system. *Journal of Biotechnology*, *259*, 228–234. DOI: 10.1016/j.jbiotec.2017.06.1204 PMID: 28690135

Du, Y., Wang, M., Yang, L., Tong, L., Guo, D., & Ji, X. (2022). Optimization and Scale-Up of fermentation processes driven by models. *Bioengineering (Basel, Switzerland)*, *9*(9), 473. DOI: 10.3390/bioengineering9090473 PMID: 36135019

Edo, G. I., Mafe, A. N., Ali, A. B., Akpoghelie, P. O., Yousif, E., Isoje, E. F., Igbuku, U. A., Zainulabdeen, K., Owheruo, J. O., Essaghah, A. E. A., Umar, H., Ahmed, D. S., & Alamiery, A. A. (2025). A critical review on the impacts of β-glucans on gut microbiota and human health. *The Microbe*, *7*, 100394. DOI: 10.1016/j.microb.2025.100394

European Food Safety Authority. (2025, March 27). *Novel food*. https://www.efsa.europa.eu/en/topics/topic/novel-food

European Union. (2015). *Regulation (EU) 2015/2283 of the European Parliament and of the Council of 25 November 2015 on novel foods*. Official Journal of the European Union. https://eur-lex.europa.eu/eli/reg/2015/2283/oj/eng

Finnigan, T., Mach, K., & Edlin, A. (2024a). Mycoprotein: a healthy new protein with a low environmental impact. In *Sustainable protein sources* (pp. 539-566). Academic Press. DOI: 10.1016/B978-0-323-91652-3.00011-3

Finnigan, T., Theobald, H., & Bajka, B. (2024b). Mycoprotein: A healthy and sustainable source of alternative Protein-Based foods. *Annual Review of Food Science and Technology*. DOI: 10.1146/annurev-food-111523-121802 PMID: 39626232

Finnigan, T. J., Wall, B. T., Wilde, P. J., Stephens, F. B., Taylor, S. L., & Freedman, M. R. (2019). Mycoprotein: The future of nutritious nonmeat protein, a symposium review. *Current Developments in Nutrition*, *3*(6), nzz021. DOI: 10.1093/cdn/nzz021 PMID: 31187084

Food and Agriculture Organization of the United Nations. (2023). *Agrifood solutions to climate change – FAO's work on the climate crisis*. Rome. DOI: 10.4060/cc8055en

Food and Drug Administration. (2002). *GRAS Notice No. GRN 000091*. https://www.fda.gov/media/173801/download

Food and Drug Administration. (2023). *GRAS Notice No. GRN 001125 for Pea Protein Fermented by Shiitake Mycelia*. https://www.fda.gov/media/180203/download

Furey, B., Slingerland, K., Bauter, M. R., Dunn, C., Goodman, R. E., & Koo, S. (2022). Safety evaluation of Fy Protein™(Nutritional Fungi Protein), a macroingredient for human consumption. *Food and Chemical Toxicology*, *166*, 113005. DOI: 10.1016/j.fct.2022.113005 PMID: 35636642

Future Market Insights. (2024) Mycoprotein Market Trends – Alternative Protein & Industry Growth 2024-2034. https://www.futuremarketinsights.com/reports/mycoprotein-market. Accessed on 19th April, 2025.

Future Market Insights. (2025). *Miso Market Analysis by Product Type, Raw Material, End-User, Distribution Channel and Other Types Through 2035*. https://www.futuremarketinsights.com/reports/miso-market

Gamarra-Castillo, O., Echeverry-Montaña, N., Marbello-Santrich, A., Hernández-Carrión, M., & Restrepo, S. (2022). Meat Substitute Development from Fungal Protein (Aspergillus oryzae). *Foods, 11*(19), 2940. DOI: 10.3390/foods11192940 PMID: 36230014

Global Info Research. (2024). "Global Soy Sauce Supply, Demand and Key Producers, 2024-2030 - GlobalInfoResearch." *Global Info Research 2023.* https://www.globalinforesearch.com/reports/2220260/soy-sauce

Gnaim, R., Dyer, P. S., & Ledesma-Amaro, R. (2025). Fusarium-based mycoprotein: Advancements in the production of sustainable meat substitutes. *Trends in Food Science & Technology, 104981*, 104981. Advance online publication. DOI: 10.1016/j.tifs.2025.104981

Hamajima, H., Matsunaga, H., Fujikawa, A., Sato, T., Mitsutake, S., Yanagita, T., & Kitagaki, H. (2016). Japanese traditional dietary fungus koji Aspergillus oryzae functions as a prebiotic for Blautia coccoides through glycosylceramide: Japanese dietary fungus koji is a new prebiotic. *SpringerPlus, 5*(1), 1–10. DOI: 10.1186/s40064-016-2950-6 PMID: 27563516

Hamilton, C. (2006). BIODIVERSITY, BIOPIRACY AND BENEFITS: WHAT ALLEGATIONS OF BIOPIRACY TELL US ABOUT INTELLECTUAL PROPERTY. *Developing World Bioethics, 6*(3), 158–173. DOI: 10.1111/j.1471-8847.2006.00168.x PMID: 17038007

Hellwig, C., Bolton, K., Häggblom-Kronlöf, G., & Rousta, K. (2022). Aspects affecting food choice in daily life as well as drivers and barriers to engagement with fungi-based food—A qualitative perspective. *Sustainability (Basel), 15*(1), 317. DOI: 10.3390/su15010317

Hilgendorf, K., Wang, Y., Miller, M. J., & Jin, Y. (2024). Precision fermentation for improving the quality, flavor, safety, and sustainability of foods. *Current Opinion in Biotechnology, 86*, 103084. DOI: 10.1016/j.copbio.2024.103084 PMID: 38394936

Hsu, T. H., Shiao, L. H., Hsieh, C., & Chang, D. M. (2002). A comparison of the chemical composition and bioactive ingredients of the Chinese medicinal mushroom DongChongXiaCao, its counterfeit and mimic, and fermented mycelium of Cordyceps sinensis. *Food Chemistry, 78*(4), 463–469. DOI: 10.1016/S0308-8146(02)00158-9

Ichikawa, E., Hosokawa, N., Hata, Y., Abe, Y., Suginami, K., & Imayasu, S. (1991). Breeding of a sake yeast with improved ethyl caproate productivity. *Agricultural and Biological Chemistry, 55*(8), 2153–2154. DOI: 10.1271/bbb1961.55.2153

Inoue, K., Gotou, T., Kitajima, H., Mizuno, S., Nakazawa, T., & Yamamoto, N. (2009). Release of antihypertensive peptides in miso paste during its fermentation, by the addition of casein. *Journal of Bioscience and Bioengineering*, *108*(2), 111–115. DOI: 10.1016/j.jbiosc.2009.03.007 PMID: 19619856

Intergovernmental Panel on Climate Change. (2022). *Climate Change 2022: Impacts, Adaptation, and Vulnerability. Contribution of Working Group II to the Sixth Assessment Report of the Intergovernmental Panel on Climate Change*. Cambridge University Press., DOI: 10.1017/9781009325844

Ito, K., & Matsuyama, A. (2021). Koji molds for Japanese soy sauce brewing: Characteristics and key enzymes. *Journal of Fungi (Basel, Switzerland)*, *7*(8), 658. DOI: 10.3390/jof7080658 PMID: 34436196

Jiang, S., Li, H., Zhang, L., Mu, W., Zhang, Y., Chen, T., Wu, J., Tang, H., Zheng, S., Liu, Y., Wu, Y., Luo, X., Xie, Y., & Ren, J. (2025). Generic Diagramming Platform (GDP): A comprehensive database of high-quality biomedical graphics. *Nucleic Acids Research*, *53*(D1), D1670–D1676. DOI: 10.1093/nar/gkae973 PMID: 39470721

Kaur, J., Murray, P., & Collins, C. (2024). Bioprospecting of novel and biologically active compounds and enzymes from a new fungal biobank for industrial biotechnology. *Biotechnology for the Environment*, *1*(1), 14. DOI: 10.1186/s44314-024-00015-0

Kusumoto, K. I., Yamagata, Y., Tazawa, R., Kitagawa, M., Kato, T., Isobe, K., & Kashiwagi, Y. (2021). Japanese traditional Miso and Koji making. *Journal of Fungi (Basel, Switzerland)*, *7*(7), 579. DOI: 10.3390/jof7070579 PMID: 34356958

Li, C., Zhou, J., Du, G., Chen, J., Takahashi, S., & Liu, S. (2020). Developing Aspergillus niger as a cell factory for food enzyme production. *Biotechnology Advances*, *44*, 107630. DOI: 10.1016/j.biotechadv.2020.107630 PMID: 32919011

Li, X., Lu, H., Yao, J., Zhang, C., Shi, T., & Huang, H. (2025). Recent advances in the application of CRISPR/Cas-based gene editing technology in Filamentous Fungi. *Biotechnology Advances*, *81*, 108561. DOI: 10.1016/j.biotechadv.2025.108561 PMID: 40086675

Linder, T. (2024). What next for mycoprotein? *Current Opinion in Food Science*, *58*, 101199. DOI: 10.1016/j.cofs.2024.101199

Majumder, R., Miatur, S., Saha, A., & Hossain, S. (2023). Mycoprotein: production and nutritional aspects: a review. *Sustainable Food Technology*, *2*(1), 81–91. DOI: 10.1039/D3FB00169E

Marden, E., Abbott, R. J., Austerlitz, F., Ortiz-Barrientos, D., Baucom, R. S., Bongaerts, P., & Rieseberg, L. H. (2021). Sharing and reporting benefits from biodiversity research. *Molecular Ecology, 30*(5), 1103–1107. DOI: 10.1111/mec.15702 PMID: 33159357

Maseko, K. H., Regnier, T., Bartels, P., & Meiring, B. (2025). Mushroom mycelia as sustainable alternative proteins for the production of hybrid cell-cultured meat: A review. *Journal of Food Science, 90*(2), e70060. Advance online publication. DOI: 10.1111/1750-3841.70060 PMID: 39921300

Massa, A., Placeres, M. D., Axpe, E., Rothschild, L., Sanz, M., Ricatti, M., & Carrero-Carralero, C. (2025). Sensorial and chemical characterization of a new food ingredient of Rhizopus oligosporus biomass produced under solid and submerged state fermentation. *Future Foods : a Dedicated Journal for Sustainability in Food Science, 100730*. DOI: 10.1016/j.fufo.2025.100730

Mau, J. L., Lin, H. C., Ma, J. T., & Song, S. F. (2001). Non-volatile taste components of several speciality mushrooms. *Food Chemistry, 73*(4), 461–466. DOI: 10.1016/S0308-8146(00)00330-7

Michel, F., Hartmann, C., & Siegrist, M. (2021). Consumers' associations, perceptions and acceptance of meat and plant-based meat alternatives. *Food Quality and Preference, 87*, 104063. DOI: 10.1016/j.foodqual.2020.104063

Moraru, C., Mincea, M. M., Frandes, M., Timar, B., & Ostafe, V. (2018). A meta-analysis on randomised controlled clinical trials evaluating the effect of the dietary supplement chitosan on weight loss, lipid parameters and blood pressure. *Medicina, 54*(6), 109. DOI: 10.3390/medicina54060109 PMID: 30545156

Nagoya Protocol. (2011). *Nagoya protocol on access to genetic resources and the fair and equitable sharing of benefits arising from their utilization to the convention on biological diversity text and annex*. Accessed on the Cbd Website at www.cbd.int/abs

Neville, M., Tarrega, A., Hewson, L., & Foster, T. (2017). Consumer-orientated development of hybrid beef burger and sausage analogues. *Food Science & Nutrition, 5*(4), 852–864. DOI: 10.1002/fsn3.466 PMID: 28748073

Ng, Z. Y., Kee, P. E., Abdullah, R., Lan, J. C. W., Ling, T. C., Jiang, J. J., & Khoo, K. S. (2024). Conversion of lignocellulosic biomass waste into mycoprotein: Current status and future directions for sustainable protein production. *Biomass Conversion and Biorefinery, 15*(24), 31337–31363. https://doi.org/10.1007/s13399-024-06130-yDOI: 10.1007/s13399-024-06130-y

Nout, M. J. R., & Aidoo, K. E. (2010). Asian fungal fermented food. In *Springer eBooks* (pp. 29–58). https://doi.org/DOI: 10.1007/978-3-642-11458-8_2

Oh, Y., & Kim, H. (2025). Characterization of plant-based sausage quality using shiitake mushroom mycelium and soybean and wheat proteins. *Food Science of Animal Resources*. DOI: 10.5851/kosfa.2025.e37

Ohata, M., Uchida, S., Zhou, L., & Arihara, K. (2016). Antioxidant activity of fermented meat sauce and isolation of an associated antioxidant peptide. *Food Chemistry, 194*, 1034–1039. DOI: 10.1016/j.foodchem.2015.08.089 PMID: 26471650

Paul, A. K., Lim, C. L., Apu, M. A. I., Dolma, K. G., Gupta, M., de Lourdes Pereira, M., Wilairatana, P., Rahmatullah, M., Wiart, C., & Nissapatorn, V. (2023). Are fermented foods effective against inflammatory diseases? *International Journal of Environmental Research and Public Health, 20*(3), 2481. DOI: 10.3390/ijerph20032481 PMID: 36767847

Pérez-Moreno, J., Guerin-Laguette, A., Rinaldi, A. C., Yu, F., Verbeken, A., Hernández-Santiago, F., & Martínez-Reyes, M. (2021). Edible mycorrhizal fungi of the world: What is their role in forest sustainability, food security, biocultural conservation and climate change? *Plants, People, Planet, 3*(5), 471–490. DOI: 10.1002/ppp3.10199

Pouris, J., Kolyva, F., Bratakou, S., Vogiatzi, C. A., Chaniotis, D., & Beloukas, A. (2024). The role of fungi in food production and processing. *Applied Sciences (Basel, Switzerland), 14*(12), 5046. DOI: 10.3390/app14125046

Reis, F. S., Barros, L., Martins, A., & Ferreira, I. C. (2012). Chemical composition and nutritional value of the most widely appreciated cultivated mushrooms: An inter-species comparative study. *Food and Chemical Toxicology, 50*(2), 191–197. DOI: 10.1016/j.fct.2011.10.056 PMID: 22056333

Risner, D., McDonald, K. A., Jones, C., & Spang, E. S. (2023). A techno-economic model of mycoprotein production: Achieving price parity with beef protein. *Frontiers in Sustainable Food Systems, 7*, 1204307. DOI: 10.3389/fsufs.2023.1204307

Rousta, N., Aslan, M., Yesilcimen Akbas, M., Ozcan, F., Sar, T., & Taherzadeh, M. J. (2024). Effects of fungal based bioactive compounds on human health. *Critical Reviews in Food Science and Nutrition, 64*(20), 7004–7027. DOI: 10.1080/10408398.2023.2178379 PMID: 36794421

Sandhu, D. K., & Soni, S. K. (1989). Microflora associated with Indian Punjabi warri fermentation. *Journal of Food Science and Technology, 26*(1), 21–25.

Santhapur, R., Jayakumar, D., & McClements, D. J. (2024). Formation and characterization of Mycelium–Potato protein hybrid materials for application in meat analogs or substitutes. *Foods, 13*(24), 4109. DOI: 10.3390/foods13244109 PMID: 39767051

Sarkar, T., Mazumder, B., Nandi, G., & Sen, S. (2025). Reishi Mushroom (Ganoderma lucidum): Pharmacology of a potent healer in traditional Chinese medicine. *Natural Product Research*, 1–18. DOI: 10.1080/14786419.2025.2511162 PMID: 40455062

Sato, Y., Fukuda, H., Zhou, Y., & Mikami, S. (2010). Contribution of ethanol-tolerant xylanase G2 from Aspergillus oryzae on Japanese sake brewing. *Journal of Bioscience and Bioengineering, 110*(6), 679–683. DOI: 10.1016/j.jbiosc.2010.07.015 PMID: 20727822

Scaife, K., Phipps, K. R., Scalise, D., & Polgár, O. (2024). Subchronic Toxicity Evaluation of Dietary Administration of a Fungal Biomass From Rhizomucor pusillus. *Journal of Applied Toxicology*. Advance online publication. DOI: 10.1002/jat.4713 PMID: 39428601

Sekar, Santhanam & Mariappan, S. (2007). Usage of traditional fermented products by Indian rural folks and IPR. *Indian Journal of Traditional Knowledge, 6*, 111–120.

Shahid, M., Shah, P., Mach, K., Rodgers-Hunt, B., Finnigan, T., Frost, G., Neal, B., & Hadjikakou, M. (2024). The environmental impact of mycoprotein-based meat alternatives compared to plant-based meat alternatives: A systematic review. *Future Foods : a Dedicated Journal for Sustainability in Food Science, 10*, 100410. DOI: 10.1016/j.fufo.2024.100410

Shapiro, P., & Cumbelich, W. (2025). The history of Neurospora crassa in fermented foods. *Discover Food, 5*(1), 232. Advance online publication. DOI: 10.1007/s44187-025-00547-8

Sharma, S., Prasad, H. K., Roy, A., & Dey, J. (2025). Fungi as a Source of Single Cell Proteins: Current Progress and Future Prospects. *Mycological Inventions for Sustainable Agriculture and Food Production*, 163-190. https://doi.org/DOI: 10.4018/979-8-3693-8201-1.ch008

Sima, P., Vannucci, L., & Vetvicka, V. (2018). β-glucans and cholesterol. *International Journal of Molecular Medicine, 41*(4), 1799–1808. DOI: 10.3892/ijmm.2018.3411 PMID: 29393350

Soccol, C. R., Da Costa, E. S. F., Letti, L. A. J., Karp, S. G., Woiciechowski, A. L., & de Souza Vandenberghe, L. P. (2017). Recent developments and innovations in solid state fermentation. *Biotechnology Research and Innovation*, *1*(1), 52–71. DOI: 10.1016/j.biori.2017.01.002

Soni, S. K., & Sandhu, D. K. (1991). Role of yeast domination in Indian idli batter fermentation. *World Journal of Microbiology & Biotechnology*, *7*(4), 505–507. DOI: 10.1007/BF00303378 PMID: 24425138

Stamets, P. (2006). Can mushrooms help save the world? *Explore (New York, N.Y.)*, *2*(2), 152–161. DOI: 10.1016/j.explore.2005.12.011 PMID: 16781630

Surya, R., Amalia, N., Gunawan, W. B., Taslim, N. A., Ghafoor, M., Mayulu, N., Hardinsyah, H., Abdi Syahputra, R., Kartawidjajaputra, F., Rizzo, G., Tjandrawinata, R. R., Subali, D., Kurniawan, R., & Nurkolis, F. (2024). Tempe as superior functional antioxidant food: From biomechanism to future development of soybean-based functional food. *Farmatsiia (Sofia)*, *71*, 1–7. DOI: 10.3897/pharmacia.71.e116748

Szejda, K., & Parry, J. (2020). *Strategies to accelerate consumer adoption of plant-based meat: Recommendations from a comprehensive literature review*. Research Report. Washington, DC: The Good Food Institute. Available at go.gfi.org/plant-based-meat-consumer-adoption-recommendations

Tamang, Buddhiman & Sha, Shankar & Thakur, Nagendra & Tamang, Jyoti. (2012). Haria, a traditional, rice fermented alcoholic beverage of West Bengal. *International Journal of Agriculture and Food Science Technology.* 157-160.

Tamang, J. P. (Ed.). (2020). *Ethnic fermented foods and beverages of India: science history and culture*. Springer Nature., DOI: 10.1007/978-981-15-1486-9

Tamang, J. P., Anupma, A., & Shangpliang, H. N. J. (2022). Ethno-microbiology of Tempe, an Indonesian fungal-fermented soybean food and Koji, a Japanese fungal starter culture. *Current Opinion in Food Science*, *48*, 100912. DOI: 10.1016/j.cofs.2022.100912

Tamang, J. P., Thapa, N., Bhalla, T. C., & Savitri, N. (2016). Ethnic fermented foods and beverages of India. In *Springer eBooks* (pp. 17–72). https://doi.org/DOI: 10.1007/978-81-322-2800-4_2

Teng, T. S., Chin, Y. L., Chai, K. F., & Chen, W. N. (2021). Fermentation for future food systems. *EMBO Reports*, *22*(5), e52680. Advance online publication. DOI: 10.15252/embr.202152680 PMID: 33908143

Thakur, M., Kasi, I. K., Islary, P., & Bhatti, S. K. (2023). Nutritional and health-promoting effects of lichens used in food applications. *Current Nutrition Reports, 12*(4), 555–566. DOI: 10.1007/s13668-023-00489-6 PMID: 37581862

Thapa, S., & Tamang, J. P. (2006). Microbiological and physico-chemical changes during fermentation of kodo ko jaanr, a traditional alcoholic beverage of the Darjeeling hills and Sikkim. *Indian Journal of Microbiology, 46*(4), 333–341.

Tsai, S. Y., Tsai, H. L., & Mau, J. L. (2008). Non-volatile taste components of Agaricus blazei, Agrocybe cylindracea and Boletus edulis. *Food Chemistry, 107*(3), 977–983. DOI: 10.1016/j.foodchem.2007.07.080 PMID: 26065761

Tsioumani, E. (2018). Beyond access and benefit-sharing: Lessons from the law and governance of agricultural biodiversity. *The Journal of World Intellectual Property, 21*(3–4), 106–122. DOI: 10.1111/jwip.12094

United Nations Department of Economic and Social Affairs. (2023). *The sustainable development goals report 2023.* UN. https://unstats.un.org/sdgs/report/2023/

Upcraft, T., Tu, W., Johnson, R., Finnigan, T., Van Hung, N., Hallett, J., & Guo, M. (2021). Protein from renewable resources: Mycoprotein production from agricultural residues. *Green Chemistry, 23*(14), 5150–5165. DOI: 10.1039/D1GC01021B

Veiter, L., Rajamanickam, V., & Herwig, C. (2018). The filamentous fungal pellet—Relationship between morphology and productivity. *Applied Microbiology and Biotechnology, 102*(7), 2997–3006. DOI: 10.1007/s00253-018-8818-7 PMID: 29473099

Ververis, E., Ackerl, R., Azzollini, D., Colombo, P. A., De Sesmaisons, A., Dumas, C., Fernandez-Dumont, A., Da Costa, L. F., Germini, A., Goumperis, T., Kouloura, E., Matijevic, L., Precup, G., Roldan-Torres, R., Rossi, A., Svejstil, R., Turla, E., & Gelbmann, W. (2020). Novel foods in the European Union: Scientific requirements and challenges of the risk assessment process by the European Food Safety Authority. *Food Research International, 137*, 109515. DOI: 10.1016/j.foodres.2020.109515 PMID: 33233150

Vetter, J. (2019). Biological values of cultivated mushrooms—A review. *Acta Alimentaria, 48*(2), 229–240. DOI: 10.1556/066.2019.48.2.11

Violo, M. (2024). How fungi can further support the United Nation's 17 Sustainable Development Goals (SDGs). *MycoStories*. https://www.mycostories.com/post/how-fungi-can-further-support-the-united-nation-s-17-sustainable-development-goals-sdgs

Waltz, E. (2016). Gene-edited CRISPR mushroom escapes US regulation. *Nature, 532*(7599), 293. DOI: 10.1038/nature.2016.19754 PMID: 27111611

Wang, M., & Zhao, R. (2022). A review on nutritional advantages of edible mushrooms and its industrialization development situation in protein meat analogues. *Journal of Future Foods*, *3*(1), 1–7. DOI: 10.1016/j.jfutfo.2022.09.001

Yee, C. S., Asyiqin, N., Abd Rahim, M. H., Mohd Zaini, N. A., Balqis, R., Ameer, M., Ilham, Z., & Al Qadr Imad, W. A. (2025). Smart Fermentation Technologies: Microbial Process Control in Traditional Fermented Foods. *Fermentation (Basel, Switzerland)*, *11*(6), 323. DOI: 10.3390/fermentation11060323

Zhang, K., Wu, W., & Yan, Q. (2020). Research advances on sake rice, koji, and sake yeast: A review. *Food Science & Nutrition*, *8*(7), 2995–3003. DOI: 10.1002/fsn3.1625 PMID: 32724564

Zhang, R., Farouk, M. M., Realini, C. E., & Thum, C. (2025). Hybridization in Meat-Based Dual Protein Foods: Mechanisms, challenges, and consumer insights. *Comprehensive Reviews in Food Science and Food Safety*, *24*(4), e70216. DOI: 10.1111/1541-4337.70216 PMID: 40519121

KEY TERMS AND DEFINITIONS

Bioactive Compounds: These are naturally occurring components in any fermentable product that provide functional health benefits.

Fungi: These are a diverse group of heterotrophic, eukaryotic micro-organisms, such as mushrooms, molds, yeasts, and lichens, with biotechnological functions.

Koji: Koji is a Japanese food delicacy fermented using Aspergillus species that produces several other fermentable food products, including miso, sake, and soy sauce.

Life Cycle Assessment (LCA): To examine the ecological footprints rigorously, and enhance the final product valuation, LCA provides a framework guiding through all ethical aspects

Miso: Miso is a Japanese fermented paste, often derived from 'Koji' and classified into various types, such as 'white miso', 'red miso', 'dark or vintage miso', based on its koji content.

Mycoprotein (MP): MP is a protein-rich, alternative, sustainable meat analog with 45% dried protein content. *Fusarium venenatum* is commonly employed in MP production, displaying many health benefits.

Sake: *Aspergillus oryzae* and *Saccharomyces cerevisiae* are fermented on japonica rice to derive the Japanese beverage product sake.

Single Cell Protein (SCP): To ensure global food security and eradicate scarcity, SCP is a valuable alternative protein source produced by single-celled microorganisms.

Solid-State Fermentation (SSF): SSF is a fermentation technique on solid or semi-solid substrate that offers minimum water utilization, while supporting the growth of aerial hyphal networks of many fungal species, including *Aspergillus oryzae*.

Submerged Fermentation (SmF): SmF allows the growth of microbial species on liquid-rich medium, with the advantage of regulating fermentation parameters such as pH, dissolved oxygen (DO), temperature, etc.

Surface Fermentation (SF): SF offers an extensive substrate utilization with various microorganisms like bacteria, yeast, and fungi. Unlike the SSF, microbes can use liquid and solid substrates to grow.

Chapter 11
Thirty Years of Investigations on Genetic Diversity Study in Allium Genus:
Current Research and Trends Using Bibliometric Analysis

Pooja Singh
Graphic Era University, Dehradun, India

Janhvi Mishra Rawat
Graphic Era University, Dehradun, India

Shivani Negi
Graphic Era University, Dehradun, India

Vishal Tripathi
Graphic Era University, Dehradun, India

Ashish Thapliyal
Graphic Era University, Dehradun, India

Prabhakar Semwal
https://orcid.org/0000-0001-5137-5169
Graphic Era University, Dehradun, India

ABSTRACT

Genetic diversity analysis is essential for the improvement and conservation of plant species. The genus Allium is highly valued worldwide for its medicinal and culinary uses. Therefore, in this study, we present a bibliometric analysis of the literature on genetic diversity within the Allium genus. For the screening of bibliometrics analysis, a total of 272 articles were identified from the Scopus database. By employing statistical methods and visualization tools like R software and VOSviewer, the

DOI: 10.4018/979-8-3373-2003-8.ch011

analysis revealed India's prominence in the number of publications and the USA's dominance in global collaboration. Moreover, the journals "Genetic Resources and Crop Evolution", "Acta Horticulturae", and "Scientia Horticulturae" published the most. The co-occurrence map showed four different significant research themes and clusters, providing insights into active areas of study. Overall, this analysis contributes to a better understanding of genetic diversity in Allium species globally.

INTRODUCTION

The *Allium* genus, belonging to the Amaryllidaceae family, currently contains more than 1089 accepted species, making it one of the largest genera among monocotyledonous plants (Bhowmick et al., 2023; Singh et al., 2025). This genus encompasses several economically important species such as garlic (*A. sativum* L.), shallot (*A. ascalonicum* L.), chive (*A. schoenoprasum*), onion (*A. cepa* L.), and leek (*A. ampeloprasum*) (Odeny & Narina, 2011). *Allium* species are widespread across the Northern Hemisphere; however, only one species, *A. dregeanum*, has been identified in the Southern Hemisphere (Vuković et al., 2023). These species have been widely utilized since prehistoric times for their pungent flavor, along with their ornamental and medicinal properties (Gargi et al., 2024a; Taleghani et al., 2024; Yan et al., 2023). They contain many bioactive compounds that contribute to their various nutritional, biological, and pharmacological activities, such as antiviral, antimicrobial, antioxidant, antitumor, and anti-inflammatory (Gargi et al., 2024a). The major bioactive constituents include organic sulfur compounds, flavonoids, anthocyanins, polyphenols, and phenolic acids (Bastaki et al., 2021; Dai et al., 2025; Gargi et al., 2025; Hanley, 2025). Many *Allium* species have been used traditionally to treat various ailments such as high blood pressure, bronchitis, cataract, heart diseases, diabetes, and abdominal pain (Khan et al., 2017). *Allium* species, especially garlic and onion, are significant in agriculture due to their economic value, nutritional benefits, and widespread cultivation (Lanzotti, 2006). The two largest producers of *Allium* species are China (20,712,087 tons of garlic and 23,659,708 tons of onion), followed by India (2,907,000 tons of garlic and 26,738,000 tons of onion) (https://www.fao.org/faostat/en/data/QC/visualize; accessed on 30 January 2025). Given the substantial global production and economic importance of *Allium* crops, understanding their genetic diversity, phylogenetics, and breeding programs is crucial.

Genetic diversity reflects variations in gene frequencies within and among populations and is shaped by mutation, recombination, gene flow, and natural selection. These processes enable *Allium* spp. to adapt effectively across diverse environmental conditions. This diversity gives rise to variations in DNA sequence, biochemical characteristics such as the protein structure and isoenzyme properties, morphological characters, and physiological properties among individuals (Ramanatha Rao & Hodgkin, 2002; Salgotra & Chauhan, 2023). Investigations on the genetic diversity of the *Allium* genus are crucial for gaining valuable insights into its evolutionary history, population structure, and for planning effective conservation strategies for its threatened species (Wu et al., 2025). Genetic diversity has been reported in different species of *Allium*, such as onion (*A. cepa* L.) (Lyngkhoi et al., 2021), garlic (*A. sativum* L.) (Chen et al., 2013; Kumar et al., 2019), Persian shallot (*A. hirtifolium*) (Ebrahimi et al., 2009), spring onions (*A. fistulosum*) (Ijeomah et al., 2020), leek (*A. ampeloprasum*) (Guenaoui et al., 2013), snowdrop (*A. galanthum*) (Wu et al., 2025), etc. Numerous publications covering the genetic diversity of *Allium* species make it essential to conduct a bibliometric analysis to identify the current trends and research progress in this field on a global scale.

Recently, bibliometric analysis has become a popular method for analyzing large number of scientific data in form of published literature by utilizing statistical methods and graphical representations to investigate the trends in research evolution, structures and patterns of a specific research field, patterns of collaboration of researchers, their institutions and their countries (Gargi et al., 2024b; Singh et al., 2024; Singh et al., 2023). This study aimed to understand the research patterns, trends, progress, and research gaps in publications on the genetic diversity of Allium species between 1994 to 2025 (04, April). To achieve this, we formulated the following questions:

R1: What temporal trends characterize scholarly output in this specific domain?
R2: Which journals, institutions and authors are publishing the articles based on the genetic diversity of *Allium* species?
R3: Which countries lead the research on the genetic diversity of *Allium* species?
R4: What are the research trends and research themes in this field?

DATA AND METHODOLOGY

To achieve our aim, a comprehensive literature search was conducted on 04 April 2025 using the Scopus database to identify the most relevant studies on the genetic diversity of the *Allium* genus, following the standard bibliometric methodology proposed by Donthu and co-authors (2021). In this study, data were extracted through the following search terms: "genome diversity" OR "genetic diversity" AND "*Allium*", AND (LIMIT-TO (SUBSTAGE, "final")) AND (LIMIT-TO (LANGUAGE, "English"). We identified 272 studies on genetic diversity in the *Allium* genus, published from 1994 (January) to 2025 (April), for bibliometric analysis. Figure 1 represents the detailed PRISMA flow diagram for data selection and inclusion and exclusion criteria for bibliometric analysis. In this analysis, we include only studies published in English focused on genetic diversity in *Allium* species. Publications that were in press, duplicated articles, not relevant to the topic, and published in languages other than English were excluded from the study.

Figure 1. Data collection and PRISMA flow chart for data selection methodology

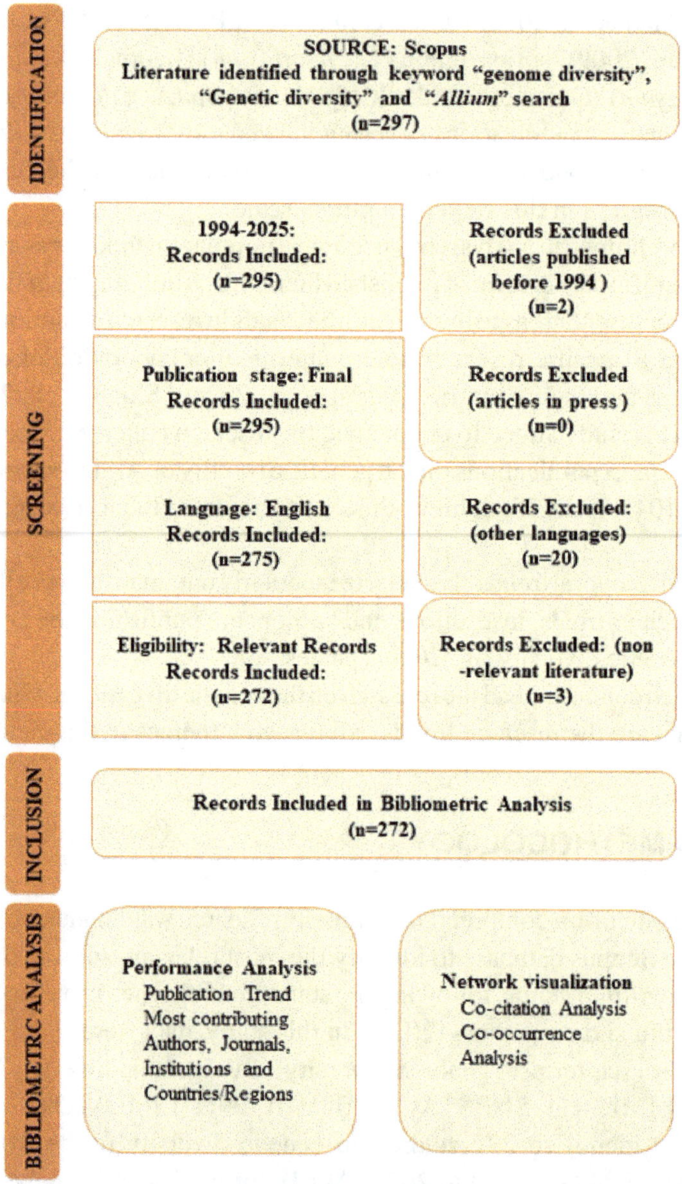

PRISMA Flow chart For data selection methodology

Literature Data Screening and Visualization

The initial search yielded 297 records; after screening for English-language publications, 275 were retained; and after excluding non-relevant literature, 272 documents were selected for analysis. In this study, to visualize the bibliometric analysis results, two different data visualization tools were used. The Bibliometrix R software (version 4.3.1) was used to generate global distribution maps, trending topics, and word clouds. At the same time, VOSviewer was employed to create and visualize bibliometric networks, co-citation, co-authorship and co-occurrence relationships.

RESULTS AND DISCUSSION

Assessment of Annual Research Output and Corresponding Citation Metrics

This bibliometric analysis examined 272 publications authored by 1,125 authors from 161 institutions across 62 countries, published in 155 journals. Figure 2a and 2b represent the number of publications and the citation distribution from 1994 to 2025, respectively. The distribution of publications in the last two decades showed an upward trend. The annual publication count was less than 10 in the first five years; since 2021, the number of publications has increased steadily through 2025. Two distinct phases are easily identifiable in this graph. The first phase, spanning from 1994 to 2017, was characterized by an initial period of steadiness. The second phase began from 2018 to 2025, demonstrating rapid development in this domain. The maximum number of articles (n=25) was published in 2023. In terms of citation trends, publications received the highest average number of citations in the year 2005. A decline in citation counts does not suggest a lack of relevance in the literature; rather, it reflects the time required for a publication to gain recognition and be widely cited (Pilkington & Meredith, 2009).

Figure 2(a). Annual publication distribution showing the trend in the number of publications per year. (b): Distribution of the mean total citations per document, illustrating citation impact over the time

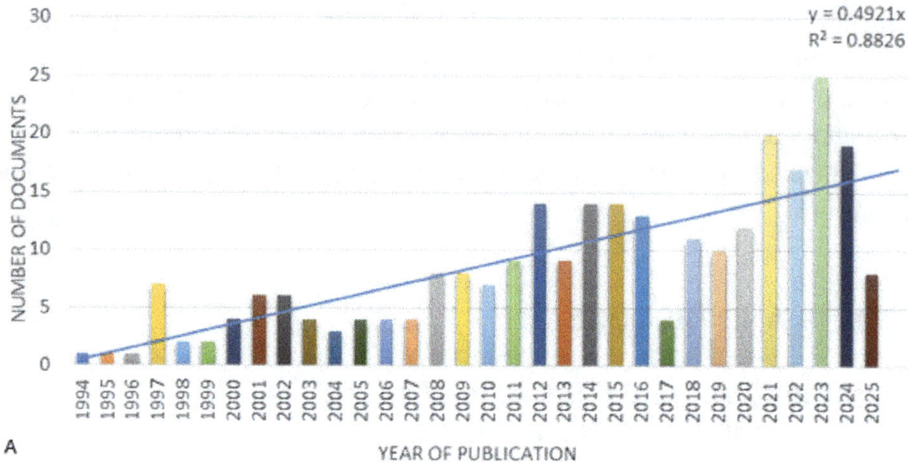

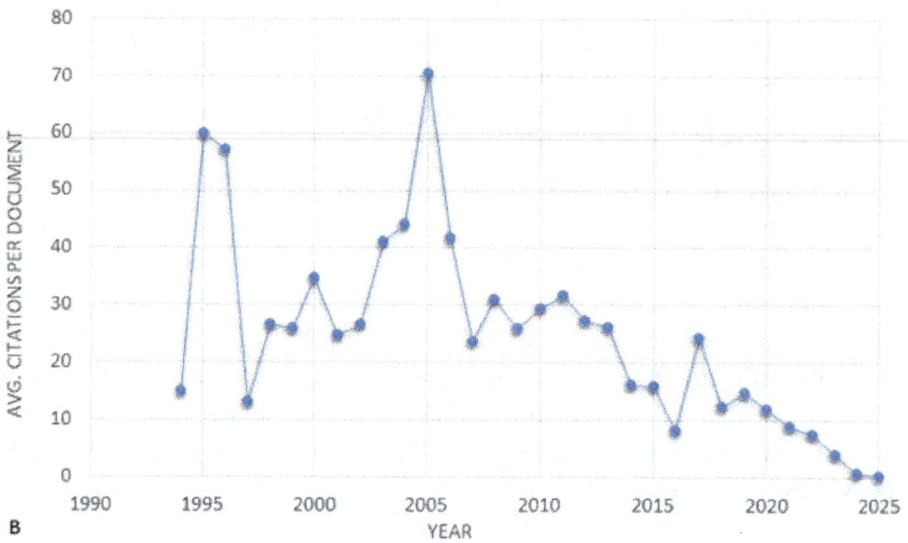

Most Productive Countries/Regions and Institutions

To evaluate the contributions of various research entities to the genetic diversity of the *Allium* genus over the past 22 years, we analyzed the most productive countries and regions as well as their institutions. Table 1 presents the top-most productive countries/regions, ranked by the number of published documents. The dataset revealed that 62 countries/regions have participated in the genetic diversity study of *Allium* species. India ranked first in the number of publications (n=57), underscoring its significant role in this domain. The United States ranked second with 41 publications, while China ranked third with 28 publications. However, in terms of citations, the top three countries follow different trends: the United States ranked first with 1064 citations, India ranked second with 615 citations, and Germany ranked third with 442 citations. Table 2 shows the most productive institutions in the field of genetic diversity of *Allium* spp. The Indian Council of Agriculture Research - Directorate of Onion and Garlic Research ranked first with 19 publications, making it the most productive institution in this domain. Following closely, the University of Wisconsin-Madison, US, ranked second with 14 publications, and the University of Tehran, Iran, ranked third with 09 publications.

Table 1. Top 10 most productive countries/regions contributing to research on genetic diversity of Allium species

Rank	Country	Documents	Total link strength	Citations
1	India	57	18	615
2	United States	41	25	1064
3	China	28	8	313
4	Iran	21	10	349
5	Turkey	16	10	299
6	Brazil	13	1	181
7	South Korea	12	5	285
8	Germany	10	4	442
8	Japan	10	5	189
10	Spain	10	6	232

Table 2. The top 10 most productive institution in research of genetic diversity of Allium spp

Rank	Affiliation	Documents	Country
1	ICAR - Directorate of Onion and Garlic Research,	19	India
2	University of Wisconsin-Madison	14	United States
3	University of Tehran	9	Iran
4	ICAR - Indian Agricultural Research Institute,	8	India
5	USDA Agricultural Research Service	7	United States
6	United States Department of Agriculture	7	United States
7	Empresa Brasileira de Pesquisa Agropecuária - Embrapa	6	Brazil
8	Chinese Academy of Agricultural Sciences	6	China
8	Rural Development Administration, Suwon	6	South Korea
10	Wageningen University & Research	5	Netherlands

Leading Journals and Influential Publications

Our dataset includes 155 sources (journals, books, etc.), with the two leading journals, "Genetic Resources and Crop Evolution" and "Acta Horticulturae", each containing 16 articles on the genetic diversity of the *Allium* genus. The second most productive journal is "Scientia Horticulturae" (13 publications with 442 citations), followed by "Journal of the American Society for Horticultural Science" (7 publications with 323 citations) and "Euphytica" (6 publications with 139 citations), respectively. The "Scientia Horticulturae", despite publishing fewer documents, achieved the highest citation count of 442. This demonstrates the high relevance and quality of its articles, highlighting their critical role in advancing knowledge on the genetic diversity of the *Allium* genus. Other contributing journals and their citations are listed in Table 3.

Table 3. Top 10 journals ranked by publication numbers and citations

Rank	Source	Documents	Citations	IF*
1	Genetic Resources and Crop Evolution	16	278	1.6
2	Acta Horticulturae	16	67	-
3	Scientia Horticulturae	13	442	3.9
4	Journal of the American Society for Horticultural Science	7	323	1.2
5	Euphytica	6	139	1.6
6	Plos One	6	97	2.9

continued on following page

Table 3. Continued

Rank	Source	Documents	Citations	IF*
7	Biochemical Systematics and Ecology	5	88	1.4
8	Theoretical and Applied Genetics	5	190	4.4
8	Frontiers in Plant Science	4	62	4.1
10	Indian Journal of Horticulture	4	29	-

(*JCR 2024)

Citation analysis is a methodology used to evaluate the articles that are most frequently cited; such citations can explain the impact of publications within a particular research domain. Table 4 presents the 10 most highly cited articles on genetic diversity in *Allium* species, published from 2003 to 2017. This decade has been marked by rapid advances in molecular marker technology and increased interest in germplasm characterization. The most cited article is "*Genetic diversity among U.S. garlic clones as detected using AFLP methods*" by Volk and co-workers (2004), published in the journal "*Journal of the American Society for Horticultural Science*" with 103 citations. This study used the AFLP method to assess genetic diversity among 211 U.S. garlic clones and found that molecular diversity corresponded with phenotypic classification. This study concludes both structured diversity in garlic clones and challenges in maintaining genetically unique germplasm resources (Volk et al., 2004). Closely following, the research article "*Comparison of AFLPs, RAPD markers, and isozymes for diversity assessment of garlic and detection of putative duplicates in germplasm collections*" published in the same journal ranked second with 100 citations. In this study, a comparative analysis of molecular markers (AFLP and RAPD) and isozymes was conducted to assess genetic diversity and detect garlic cloves in the garlic germplasm collection, highlighting methodological advances in germplasm management (Ipek et al., 2003). "*Evaluation of allicin content and botanical traits in Iranian garlic (Allium sativum) ecotypes*" with 97 citations ranked third in the list. In this study, 24 garlic ecotypes in Iran were mainly focused on botanical traits and allicin compound. All ecotypes exhibited high allicin levels, with significant variation. There is no correlation between allicin content and geographic origin, indicating that genetic factors mainly determine allicin levels. Six ecotype clusters were identified, with one group showing superior botanical and pharmaceutical potential and recommended for future breeding. The most relevant articles in *Allium* genetic study research have shaped the field by introducing robust molecular and phenotypic assessment methods, developing key genetic resources, and linking genetic diversity to agronomic and nutritional traits (Baghalian et al., 2005).

Table 4. Top 10 cited articles and their detailed information on the genetic diversity of Allium species

S. No.	First Author	Title	Year of publication	Source title	Total citations	Document Type	References
1.	Volk G.M.	Genetic diversity among U.S. garlic clones as detected using AFLP methods	2004	Journal of the American Society for Horticultural Science	103	Article	(Volk et al., 2004)
2.	Ipek M.	Comparison of AFLPs, RAPD markers, and isozymes for diversity assessment of garlic and detection of putative duplicates in germplasm collections	2003	Journal of the American Society for Horticultural Science	100	Article	(Ipek et al., 2003)
3.	Baghalian K.	Evaluation of allicin content and botanical traits in Iranian garlic (Allium sativum) ecotypes.	2005	Scientia Horticulturae	97	Article	(Baghalian et al., 2005)
4.	Panthee D.R.	Diversity analysis of garlic (Allium sativum) germplasms available in Nepal based on morphological characters	2006	Genetic Resources and Crop Evolution	91	Article	(Panthee et al., 2006)
5.	Bernaert N.	Antioxidant capacity, total phenolic and ascorbate content as a function of the genetic diversity of leek (Allium ampeloprasum var. porrum)	2012	Food Chemistry	89	Article	(Bernaert et al., 2012)
6.	Egea L.A.	Assessment of genetic diversity and structure of large garlic (Allium sativum) germplasm bank, by diversity arrays technology "genotyping-by-sequencing" platform (DArTseq)	2017	Frontiers in Genetics	84	Article	(Egea et al., 2017)
7.	Baldwin S.	Development of robust genomic simple sequence repeat markers for estimation of genetic diversity within and among bulb onion (Allium cepa L.) populations	2012	Molecular Breeding	75	Article	(Baldwin et al., 2012)
8.	Havey M.J.	Diversity among male-sterility-inducing and male-fertile cytoplasms of onion	2000	Theoretical and Applied Genetics	71	Article	(Havey, 2000)
9.	Kunkalikar S.R.	Importance and genetic diversity of vegetable-infecting tospoviruses in India	2011	Phytopathology	71	Article	(Kunkalikar et al., 2011)
10.	Ma K.-H	Isolation and characteristics of eight novel polymorphic microsatellite loci from the genome of garlic (Allium sativum)	2009	Scientia Horticulturae	67	Article	(Ma et al., 2009)

Global Co-Authorship Networks Analysis

We created a collaboration map among participating countries based on co-authorship. We used VOSviewer software to make the co-authorship network (Figure 3a). The node size reflects the number of publications, while the connecting lines represent collaborations among countries and regions; more connecting lines indicate a high level of collaboration.

In this co-authorship network of countries and regions, publication counts above 4 were used as the threshold for analysis and in total, 29 countries out of 62 met the threshold. The analysis of the collaboration map indicates highly active collaboration among the countries in this field. The United States has the strongest international collaboration, with a total link strength (TLS) of 25, followed by India (TLS 18) and Iran and Turkey (TLS 18). The research on genetic diversity in *Allium* spp. first emerged in the United States and India, followed by Spain and Japan. These two countries demonstrate high centrality and strong collaboration among the other countries, highlighting their critical role in this research area.

Figure 3a. Collaboration map between the countries/ regions. 3b: Co-citation map of authors, demonstrating the connections among researchers in the field of genetic diversity in Allium species.

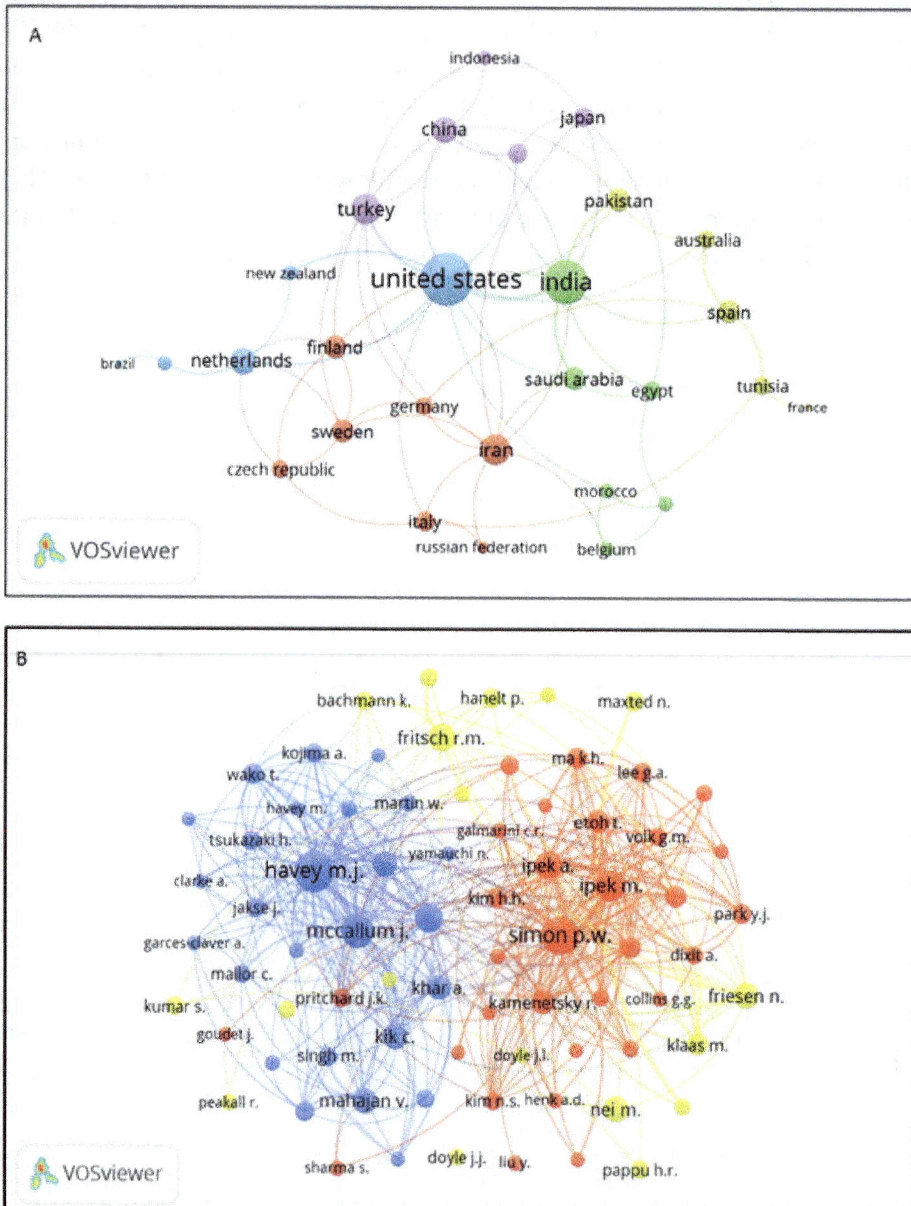

Analysis of the Most Productive Authors and Co-Cited Authors

A total of 1,125 authors have contributed to 272 documents related to Genetic Diversity Research in *Allium* species. The analysis of author productivity highlights Mahajan, V., as the most productive author with 15 articles published and Singh, M., with 12 publications from the same institution. Havey, M.J., affiliated with the University of Wisconsin-Madison, ranks third with 9 articles and has the highest overall h-index (40) among the top ten authors, with insight into both productivity and impact by h-index values. The importance of authors from ICAR institutions also highlights India's leadership in this area of research. At the same time, the h-index of Havey, M.J., reflects the significant contributions from the United States. Table 4 presents the top-most productive authors.

A co-citation map of authors, demonstrating the connections among researchers in the field of genetic diversity in *Allium* species, was prepared using VOSviewer software, as presented in Fig. 3b. Authors with a minimum of 30 citations were selected for the co-citation analysis. In total, 76 out of 18,313 authors meet the minimum threshold for the analysis. In terms of citations, Havey, M.J., has the highest citation count of 196 and TLS of 6183; this value demonstrates the most significant collaborative relationship among 75 authors. Closely followed by Simon, P.W., ranked second with 176 citations, 74 collaborations linked with 74 authors, and TLS of 5843. McCallum, J., ranked third with 137 citations and TLS of 5683. In terms of collaborations among the authors, Meena, V.S., demonstrates the most significant collaborative relationship among 253 authors, with TLS of 21398, followed by Kumar A. with 262 collaborations and 16957 TLS, and Glick B.R. with 255 collaborations and 13860 TLS.

Table 5. Top 10 most prolific authors in genetic diversity research of Allium species

Rank	Authors	Articles	Affiliation	h-index
1	Mahajan, V.	15	ICAR - Directorate of Onion and Garlic Research, Pune	11
2	Singh, M.	12	ICAR - Directorate of Onion and Garlic Research, Pune	28
3	Havey, M.J.	9	University of Wisconsin-Madison	40
4	Khar, A.	8	ICAR - Indian Agricultural Research Institute	13
5	Gupta, A.J.	7	ICAR - Directorate of Onion and Garlic Research	7
6	Manjunathagowda, D.C.	6	ICAR - Indian Institute of Horticultural Research,	9
7	Benke, A.P.	5	ICAR - Directorate of Onion and Garlic Research, Pune	6
8	Ipek, A.	5	University of Bursa Uludag, Bursa, Turkey	17

continued on following page

Table 5. Continued

Rank	Authors	Articles	Affiliation	h-index
8	Ipek, M.	5	Bursa Uludağ ÜniversitesiThe institution will open in a new tab, Bursa, Turkey	15
10	Khade, Y.P.	5	ICAR - Directorate of Onion and Garlic Research	5

Table 6. Top 10 authors ranked by citation count and collaborative network strength

Rank	Authors	Citations	Total Link Strength
1	Havey M.J.	196	6183
2	Simon P.W.	176	5843
3	Mccallum J.	137	5683
4	Ipek M.	111	3898
5	Shigyo M.	102	5043
6	Fritsch R.M.	96	2081
7	Ipek A.	95	3467
8	Friesen N.	86	2238
9	Pither-Joyce M.	84	3856
10	Kik C.	83	3034

Analysis of Keyword Clusters and Trending Hotspots

Co-occurrence analysis of keywords is a method for identifying semantic relationships and underlying themes in a research area. The size corresponds to the keyword nodes in the co-occurrence map, which indicate frequency of occurrence. The relatedness between the topics is visualized by proximity and thickness of connecting lines between two nodes. At the same time, colored clusters suggest conceptual similarity among the keywords, highlighting different themes under the research field.

Cluster 1 in red represents "*Assessment of genetic diversity and population structures in Allium species by molecular markers*", this theme collectively represents the use of various molecular markers such as (RAPD, SSR, ISSR, AFLP and microsatellites) and analytical methods (PCA, UPGMA, and cluster analysis) for the evaluation of genetic diversity, polymorphism, genetic divergence, yield associated variability among the genotypes of *Allium* species and transcriptomic approaches and the identification of core collection for the breeding and conservation purposes (Chalbi et al., 2023; Egea et al., 2017; Kumar et al., 2019; Papaioannou et al., 2023).

Figure 4a. Co-occurrence network of keywords cluster; (b): Topic trend on the Allium genetic diversity

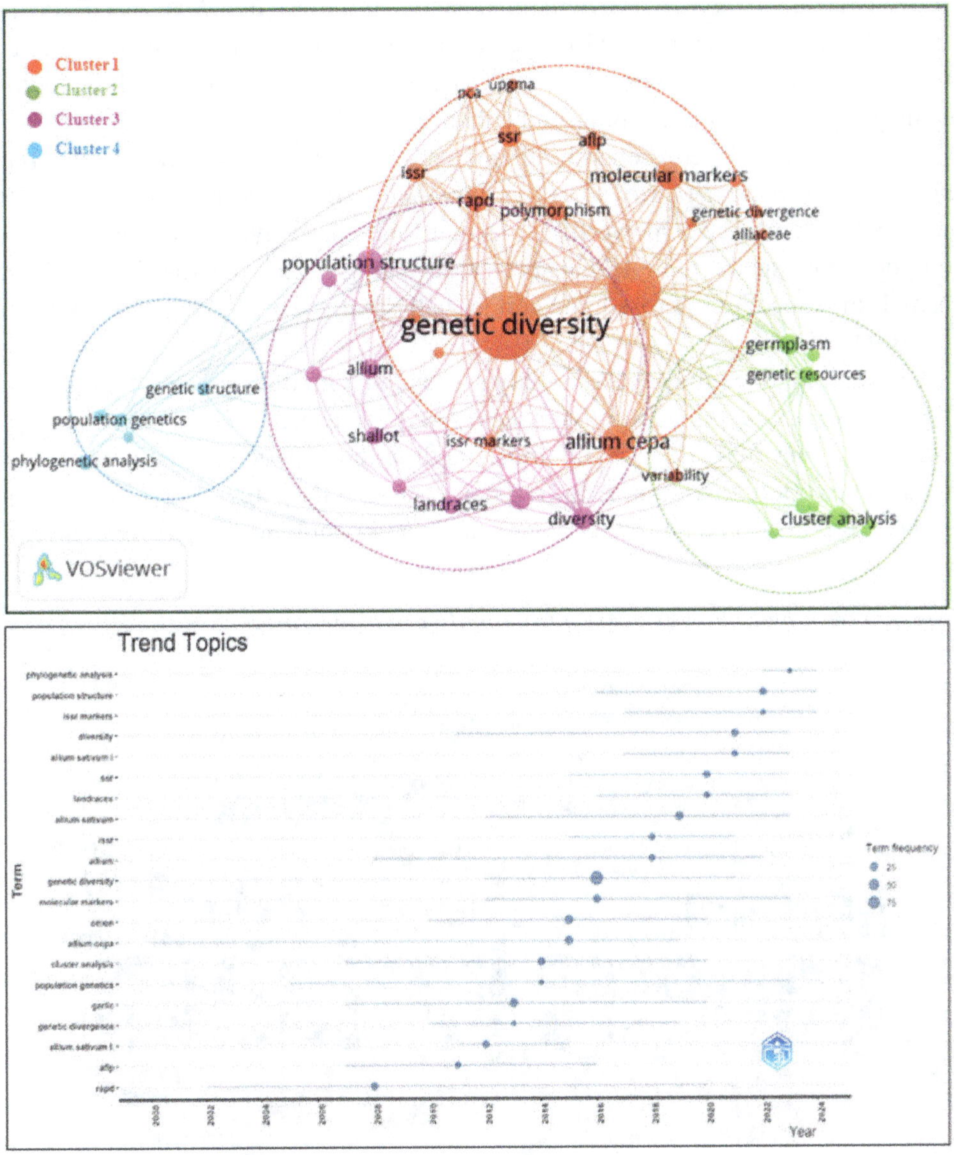

Cluster 2 in green represents the theme "*Morphological and biochemical characterization of Allium using multivariate analysis for genetic resource enhancement*". This theme focuses on evaluating genetic variability in *Allium* species using morphological and biochemical traits, employing statistical tools such as cluster analysis, principal component analysis (PCA), and other multivariate methodologies to analyze germplasm collections and enhance genetic resources.

Cluster 3 in pink represents the theme "*Molecular characterization and conservation of genetic diversity in Allium landraces using SSR markers*". The SSR markers have emerged as powerful molecular tools for analyzing genetic diversity and population structure in *Allium* species, revealing significant variability within germplasm collections across India and globally (Kumar et al., 2019; Lyngkhoi et al., 2021).

Cluster 4 in blue represents the study associated with the theme of "*Assessment of genetic structure and population genetics in Allium species using molecular approaches*". This theme focuses on understanding the genetic structure, gene flow and genetic differentiation in Allium species through the integration of molecular markers, RT-PCR, and advanced phylogenetic and population genetic analysis.

Figure 5. Word cloud map of most frequently used keywords

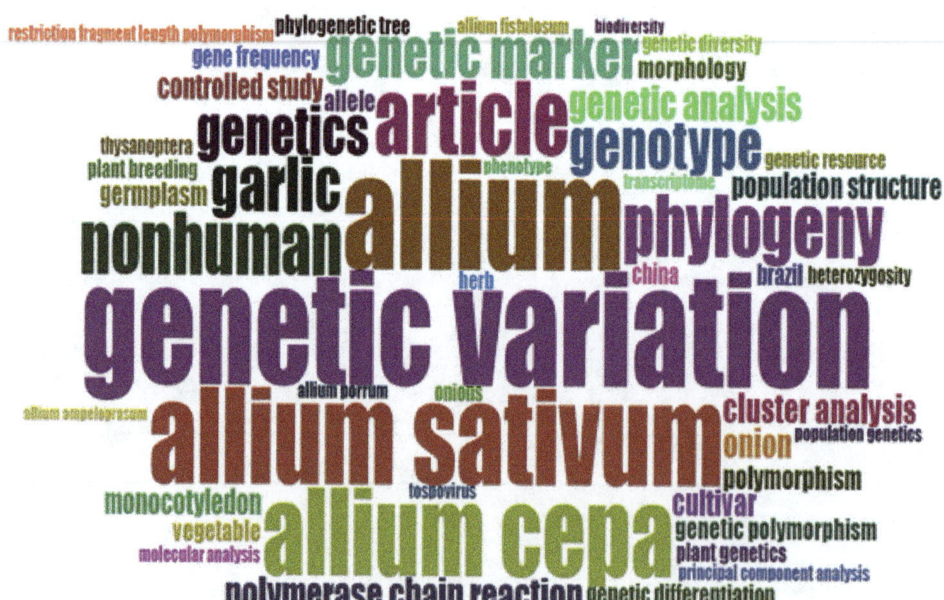

A word cloud is generated using Bibliometrix to visually represent the keywords with high frequency in the dataset. The size of each word corresponds to its frequency; the most frequently occurring words are displayed in the data, and the larger the word, the higher the frequency. The smaller-sized words are less common. According to the word cloud, the most frequently used keywords by authors in this field are genetic variation, *Allium sativum*, and *Allium cepa*. The prominence of these keywords also reflects that researchers are actively exploring the genetic resources of *Allium sativum* and *Allium cepa* to support breeding and crop improvement efforts. According to trends in this field, phylogenetic analysis, population structure, and ISSR markers are the leading topics (Fig. 5). Currently, genetic data from chloroplast genomes or highly variable DNA regions are used to construct evolutionary relationships among *Allium* species. Recent advances have enabled the resolution of complex lineages and the discovery of novel relationships, including those among Daghestanica and other subgroups (Xie et al., 2019). The ISSR markers are highly effective for detecting polymorphism and classifying *Allium spp*., making them a valuable tool for phylogenetic and population genetic studies and for guiding breeding programs (Papaioannou et al., 2023).

This study aimed to identify emerging research hotspots and evolving trends in the genetic diversity of *Allium* species. The bibliometric analysis of literature on *Allium's* genetic diversity from 1994-2025 shows an upward trend in publication output, indicating increasing scientific interest and research efforts in this field. The low number of publications in this area is attributed to challenges such as their high heterozygosity, inbreeding depression, and large genome size, which lag behind research development in this field (Hancı & Paşazade, 2025; Khandagale et al., 2020). We analyzed the most productive countries, journals, institutions, and authors in this bibliometric analysis. "*Genetic Resources and Crop Evolution*", "*Acta Horticulturae*", and "*Scientia Horticulturae*" are the most productive journals in this field. Mahajan V. from the ICAR - Directorate of Onion and Garlic Research is the most relevant author, with the most publications. Seven out of the ten most prolific authors are affiliated with the same institution, representing their important role in the field. According to our dataset, the United States and India have a strong presence, with considerable research output in this domain. The global contributions of institutions from countries such as China, Brazil, and South Korea reflect interest in research on the conservation and sustainable utilization of *Allium* species, which can ultimately advance research in this field. The co-occurrence analysis of keywords provides clusters representing four main research themes in this field.

Cluster 1 emphasizes the use and development of diverse molecular markers, such as RAPD, AFLP, ISSR, and SSR, for evaluating genetic diversity among *Allium* species, which is crucial for conservation. Cluster 2 mainly focused on multivariate analysis to assess genetic variability and enhance germplasm collection, as demonstrated in studies of garlic ecotypes using allcin biochemical markers (Baghalian et al., 2005; Khar et al., 2011). Cluster 3 highlights the power of SSR markers in molecular characterization and conservation of *Allium* landraces, revealing significant genetic diversity in India and globally. Among all molecular markers, SSRs have emerged as the preferred tools for genetic diversity assessment due to their reproducibility, co-dominant inheritance, cross-species transferability, and high polymorphism (Khade et al., 2025; Son et al., 2012). Cluster 4 highlights the integration of molecular approaches, including RT-PCR, with phylogenetic databases like *Allium*DB. The development of resources such as *Allium*DB, a unified genomic, transcriptomic, and phenotypic database, now accelerates trait discovery, molecular marker identification, and comparative genomics, thereby supporting both fundamental research and breeding innovations (Yang et al., 2024). The trend analysis reveals current research hotspots in this field. According to Figure 4b, the current topic includes phylogenetic analysis, population structure, and the use of ISSR markers, which are effective tools for detecting polymorphism and can be further applied in breeding programs and the development of improved crops (Anwar et al., 2017; Hancı & Paşazade, 2025). An evaluation of the most-cited articles in *Allium* genetic diversity research is conducted to understand the foundational knowledge in this domain. According to our analysis, early studies such as Volk et al. (2004) and Ipek et al. (2003) laid the foundation for diversity assessment using AFLP and RAPD markers to detect genetic variation and duplicates in garlic germplasm.

CONCLUSION AND FUTURE PROSPECTS

This study represents the first comprehensive bibliometric analysis of the research field discussing the genetic diversity in *Allium*. Over recent years, publications on the genetic diversity of *Allium spp*. have increased steadily, reflecting growing scientific interest and advances in genomic technology. Leading journals in this field include "*Genetic Resources and Crop Evolution*", "*Acta Horticulturae*", and "*Scientia Horticulturae*". Countries leading research in this field include India, the U.S.A., Iran, and China, driven by both economic importance and the native diversity of *Allium* species in these regions. Mainly, four research themes were identified in the keyword cluster analysis, and currently, phylogenetic and population structure analyses, along with omics approaches, are focusing on genome sequencing and the conservation of genetic resources. According to our analysis, genetic studies in *Allium* species are limited to the economically important species such as *A. cepa* and *A. sativum*, leaving the majority of the genus unexplored. A comprehensive sequencing across the genus is required. This bibliometric study could be used in future analyses of data obtained from different databases.

ACKNOWLEDGEMENT

The authors acknowledge Graphic Era (Deemed to be University), Dehradun, for providing all necessary facilities and support during this study.

DECLARATION

The authors used Grammarly (www.grammarly.com), an AI-based English language editing software, to check grammar, punctuation, and writing clarity in the manuscript.

REFERENCES

Anwar, G. M., Helmey, R. K., & Moustafa, Y. M. M. (2017). Genetic diversity among some species of the genus *Allium* L. using SSR and ISSR markers. *Egyptian Journal of Genetics and Cytology*, *46*(1), 221–233. DOI: 10.21608/ejgc.2018.9516

Baghalian, K., Ziai, S. A., Naghavi, M. R., Badi, H. N., & Khalighi, A. (2005). Evaluation of allicin content and botanical traits in Iranian garlic (*Allium sativum* L.) ecotypes. *Scientia Horticulturae*, *103*(2), 155–166. DOI: 10.1016/j.scienta.2004.07.001

Baldwin, S., Pither-Joyce, M., Wright, K., Chen, L., & McCallum, J. (2012). Development of robust genomic simple sequence repeat markers for estimation of genetic diversity within and among bulb onion (*Allium cepa* L.) populations. *Molecular Breeding*, *30*(3), 1401–1411. DOI: 10.1007/s11032-012-9727-6

Bastaki, S. M., Ojha, S., Kalasz, H., & Adeghate, E. (2021). Chemical constituents and medicinal properties of *Allium* species. *Molecular and Cellular Biochemistry*, *476*(12), 4301–4321. DOI: 10.1007/s11010-021-04213-2 PMID: 34420186

Bernaert, N., De Paepe, D., Bouten, C., De Clercq, H., Stewart, D., Van Bockstaele, E., De Loose, M., & Van Droogenbroeck, B. (2012). Antioxidant capacity, total phenolic and ascorbate content as a function of the genetic diversity of leek (*Allium ampeloprasum* var. porrum). *Food Chemistry*, *134*(2), 669–677. DOI: 10.1016/j.foodchem.2012.02.159 PMID: 23107677

Bhowmick, B. K., Sarkar, S., Roychowdhury, D., Patil, S. D., Lekhak, M. M., Ohri, D., Rao, S. R., Yadav, S., Verma, R. C., Dhar, M. K., Raina, S. N., & Jha, S. (2023). *Allium* cytogenetics: A critical review on the Indian taxa. *Comparative Cytogenetics*, *17*, 129–156. DOI: 10.3897/CompCytogen.17.98903 PMID: 37304149

Chalbi, A., Chikh-Rouhou, H., Mezghani, N., Slim, A., Fayos, O., Bel-Kadhi, M. S., & Garcés-Claver, A. (2023). Genetic Diversity Analysis of Onion (*Allium cepa* L.) from the Arid Region of Tunisia Using Phenotypic Traits and SSR Markers. *Horticulturae*, *9*(10), 1098. https://www.mdpi.com/2311-7524/9/10/1098. DOI: 10.3390/horticulturae9101098

Chen, S., Zhou, J., Chen, Q., Chang, Y., Du, J., & Meng, H. (2013). Analysis of the genetic diversity of garlic (*Allium sativum* L.) germplasm by SRAP. *Biochemical Systematics and Ecology*, *50*, 139–146. DOI: 10.1016/j.bse.2013.03.004

Dai, X., Azi, F., Alnadari, F., & Yu, Z. (2025). Developing organosulfur compounds in Allium as the next-generation flavor and bioactive ingredients for food and medicine. *Critical Reviews in Food Science and Nutrition*, *65*(33), 1–19. DOI: 10.1080/10408398.2025.2500674 PMID: 40366718

Donthu, N., Kumar, S., Mukherjee, D., Pandey, N., & Lim, W. M. (2021). How to conduct a bibliometric analysis: An overview and guidelines. *Journal of Business Research*, *133*, 285–296. DOI: 10.1016/j.jbusres.2021.04.070

Ebrahimi, R., Zamani, Z., & Kashi, A. (2009). Genetic diversity evaluation of wild Persian shallot (*Allium hirtifolium* Boiss.) using morphological and RAPD markers. *Scientia Horticulturae*, *119*(4), 345–351. DOI: 10.1016/j.scienta.2008.08.032

Egea, L. A., Mérida-García, R., Kilian, A., Hernandez, P., & Dorado, G. (2017). Assessment of genetic diversity and structure of large garlic (*Allium sativum*) germplasm bank, by diversity arrays technology "genotyping-by-sequencing" platform (DArTseq). *Frontiers in Genetics*, *8*, 98. DOI: 10.3389/fgene.2017.00098 PMID: 28775737

Gargi, B., Painuli, S., Semwal, P., Singh, D. P., Sharma, R., Rauf, A., Khalil, A. A., Khalid, A., Hemeg, H. A., & Wilairatana, P. (2024). Global status, recent trends, and knowledge mapping of olive oil research and cardiovascular disease: 50 years of investigations through bibliometric analysis. *Food Science & Nutrition*, *12*(3), 1465–1478. DOI: 10.1002/fsn3.3885 PMID: 38455210

Gargi, B., Semwal, P., Jaiswal, D. K., Painuli, S., Trivedi, V. L., Tripathi, V., & Rai, N. (2025). Impact of circadian rhythm and seasonal variability on the essential oil of *Allium stracheyi* baker from Uttarakhand, Himalaya. *Food Chemistry*, *X*, 102720. DOI: 10.1016/j.fochx.2025.102720 PMID: 40686884

Gargi, B., Singh, P., Painuli, S., Rai, N., Semwal, P., Cruz-Martins, N., & Sharma, R. (2024a). Literature-based screening and bibliometric analysis of the chemical composition, antioxidant and antimicrobial potential of essential oils isolated from Allium genus: 23 years of investigation. *Pharmacological Research. Modern Chinese Medicine, 10*, 100354. DOI: 10.1016/j.prmcm.2023.100354

Guenaoui, C., Mang, S., Figliuolo, G., & Neffati, M. (2013). Diversity in *Allium ampeloprasum*: From small and wild to large and cultivated. *Genetic Resources and Crop Evolution, 60*(1), 97–114. DOI: 10.1007/s10722-012-9819-5

Hancı, F., & Paşazade, E. (2025). A comparison of efficiency parameters of SRAP and ISSR markers in revealing variation in *Allium* Germplasm. *Horticulturae, 11*(3), 294. DOI: 10.3390/horticulturae11030294

Hanley, B. (2025). Naturally-Occurring Sulphur Compounds. In *Natural and Unnatural Product Chemistry: From Molecules to Systems* (pp. 77–103). Springer., DOI: 10.1007/978-3-031-81698-7_3

Havey, M. J. (2000). Diversity among male-sterility-inducing and male-fertile cytoplasms of onion. *Theoretical and Applied Genetics, 101*(5-6), 778–782. DOI: 10.1007/s001220051543

Ijeomah, C., Amuda, O., Babatunde, B., & Abutu, P. (2020). Evaluation of genetic diversity of spring onions (*Allium fistulosum*) based on DNA markers. *Journal of Experimental Agriculture International, 42*, 23–33. DOI: 10.9734/jeai/2020/v42i330481

Ipek, M., Ipek, A., & Simon, P. W. (2003). Comparison of AFLPs, RAPD markers, and isozymes for diversity assessment of garlic and detection of putative duplicates in germplasm collections. *Journal of the American Society for Horticultural Science, 128*(2), 246–252. DOI: 10.21273/JASHS.128.2.246

Khade, Y. P., Sinhasane, S. R., Mainkar, P., Rai, K. M., Salunkhe, S., Singh, P. R., Jadhav, M. S., Bhandari, H. R., More, S. J., Gupta, A. J., Mahajan, V., & Singh, M. (2025). Exploring sequence-related amplified polymorphism (SRAP) markers for assessing genetic diversity in onion (*Allium cepa* L.) genotypes and their wild relatives. *Genetic Resources and Crop Evolution, 72*(1), 399–415. DOI: 10.1007/s10722-024-01978-5

Khan, S. A., Jameel, M., Kanwal, S., & Shahid, S. (2017). Medicinal importance of *Allium* species: A current review. *International Journal of Pharmaceutical Sciences and Research, 2*(1), 29–39.

Khandagale, K., Krishna, R., Roylawar, P., Ade, A., Benke, A., Shinde, B., Singh, M., Gawande, S., & Rai, A. (2020). Omics approaches in *Allium* research: Progress and way ahead. *PeerJ*, *8*, e9824. DOI: 10.7717/peerj.9824 PMID: 32974094

Khar, A., Lawande, K. E., & Negi, K. S. (2011). Microsatellite marker based analysis of genetic diversity in short day tropical Indian onion and cross amplification in related *Allium* spp. *Genetic Resources and Crop Evolution*, *58*(5), 741–752. DOI: 10.1007/s10722-010-9616-y

Kumar, M., Rakesh Sharma, V., Kumar, V., Sirohi, U., Chaudhary, V., Sharma, S., Saripalli, G., Naresh, R., Yadav, H. K., & Sharma, S. (2019). Genetic diversity and population structure analysis of Indian garlic (*Allium sativum* L.) collection using SSR markers. *Physiology and Molecular Biology of Plants : an International Journal of Functional Plant Biology*, *25*(2), 377–386. DOI: 10.1007/s12298-018-0628-y PMID: 30956421

Kunkalikar, S. R., Poojari, S., Arun, B. M., Rajagopalan, P. A., Chen, T. C., Yeh, S. D., Naidu, R. A., Zehr, U. B., & Ravi, K. S. (2011). Importance and genetic diversity of vegetable-infecting tospoviruses in India. *Phytopathology*, *101*(3), 367–376. DOI: 10.1094/PHYTO-02-10-0046 PMID: 21299415

Lanzotti, V. (2006). The analysis of onion and garlic. *Journal of Chromatography. A*, *1112*(1-2), 3–22. DOI: 10.1016/j.chroma.2005.12.016 PMID: 16388813

Lyngkhoi, F., Saini, N., Gaikwad, A. B., Thirunavukkarasu, N., Verma, P., Silvar, C., Yadav, S., & Khar, A. (2021). Genetic diversity and population structure in onion (Allium cepa L.) accessions based on morphological and molecular approaches. *Physiology and Molecular Biology of Plants : an International Journal of Functional Plant Biology*, *27*(11), 2517–2532. DOI: 10.1007/s12298-021-01101-3 PMID: 34924708

Ma, K. H., Kwag, J. G., Zhao, W., Dixit, A., Lee, G. A., Kim, H. H., Chung, I. M., Kim, N. S., Lee, J. S., Ji, J. J., Kim, T. S., & Park, Y. J. (2009). Isolation and characteristics of eight novel polymorphic microsatellite loci from the genome of garlic (*Allium sativum* L.). *Scientia Horticulturae*, *122*(3), 355–361. DOI: 10.1016/j.scienta.2009.06.010

Odeny, D. A., & Narina, S. S. (2011). *Allium*. In *Wild crop relatives: genomic and breeding resources: vegetables* (pp. 1-10). Springer.

Panthee, D. R., Kc, R. B., Regmi, H. N., Subedi, P. P., Bhattarai, S., & Dhakal, J. (2006). Diversity analysis of garlic (*Allium sativum* L.) germplasms available in Nepal based on morphological characters [Article]. *Genetic Resources and Crop Evolution*, *53*(1), 205–212. DOI: 10.1007/s10722-004-6690-z

Papaioannou, C., Fassou, G., Petropoulos, S. A., Lamari, F. N., Bebeli, P. J., & Papasotiropoulos, V. (2023). Evaluation of the genetic diversity of Greek garlic (*Allium sativum* L.) accessions using DNA markers and association with phenotypic and chemical variation. *Agriculture, 13*(7), 1408. DOI: 10.3390/agriculture13071408

Pilkington, A., & Meredith, J. (2009). The evolution of the intellectual structure of operations management—1980–2006: A citation/co-citation analysis. *Journal of Operations Management, 27*(3), 185–202. DOI: 10.1016/j.jom.2008.08.001

Ramanatha Rao, V., & Hodgkin, T. (2002). Genetic diversity and conservation and utilization of plant genetic resources. *Plant Cell, Tissue and Organ Culture, 68*(1), 1–19. DOI: 10.1023/A:1013359015812

Salgotra, R. K., & Chauhan, B. S. (2023). Genetic diversity, conservation, and utilization of plant genetic resources. *Genes, 14*(1), 174. DOI: 10.3390/genes14010174 PMID: 36672915

Singh, N., Gusain, A., Nigam, M., & Mishra, A. P. (2025). The pharmacological and therapeutic versatility of *Allium* species: A comprehensive exploration of bioactive constituents and biological activities. *Discover Applied Sciences, 7*(4), 349. DOI: 10.1007/s42452-025-06800-0

Singh, P., Gargi, B., Semwal, P., & Tripathi, V. (2023). Global trends, research progress and knowledge mapping of plant–pollinator interactions through bibliometric analysis (1984 to 2023). *Biodiversity (Nepean), 24*(4), 200–211. DOI: 10.1080/14888386.2023.2258836

Singh, P., Gargi, B., Semwal, P., & Verma, S. (2024). Global research and research progress on climate change and their impact on plant phenology: 30 years of investigations through bibliometric analysis. *Theoretical and Applied Climatology, 155*(6), 4909–4923. DOI: 10.1007/s00704-024-04919-5

Son, J. H., Park, K. C., Lee, S. I., Kim, J. H., & Kim, N. S. (2012). Species relationships among *Allium* species by ISSR analysis. *Horticulture, Environment and Biotechnology, 53*(3), 256–262. DOI: 10.1007/s13580-012-0130-3

Taleghani, A., Ayati, Z., Eghbali, S., Emami, S. A., & Tayarani-Najaran, Z. (2024). Health benefits of *Allium* spp. in metabolic syndrome: A review. *South African Journal of Botany, 167*, 217–255. DOI: 10.1016/j.sajb.2024.01.040

Volk, G. M., Henk, A. D., & Richards, C. M. (2004). Genetic diversity among U.S. garlic clones as detected using AFLP methods. *Journal of the American Society for Horticultural Science, 129*(4), 559–569. DOI: 10.21273/JASHS.129.4.559

Vuković, S., Popović-Djordjević, J. B., Kostić, A. Ž., Pantelić, N. D., Srećković, N., Akram, M., Laila, U., & Katanić Stanković, J. S. (2023). *Allium* species in the Balkan region—Major metabolites, antioxidant and antimicrobial properties. *Horticulturae*, *9*(3), 408. DOI: 10.3390/horticulturae9030408

Wu, J., Liu, D., Wariss, H. M., Zhang, H., Su, M., Li, W., & Han, Z. (2025). Genetic diversity and construction of core collection provides new insight for the conservation of edible *Allium galanthum* in Xinjiang. *Scientia Horticulturae*, *341*, 113961. DOI: 10.1016/j.scienta.2025.113961

Xie, D.-F., Yu, H.-X., Price, M., Xie, C., Deng, Y.-Q., Chen, J.-P., Yu, Y., Zhou, S.-D., & He, X.-J. (2019). Phylogeny of Chinese *Allium* species in section Daghestanica and adaptive evolution of *Allium* (Amaryllidaceae, Allioideae) species revealed by the chloroplast complete genome. *Frontiers in Plant Science*, *10*, 460. DOI: 10.3389/fpls.2019.00460 PMID: 31114591

Yan, J.-K., Zhu, J., Liu, Y., Chen, X., Wang, W., Zhang, H., & Li, L. (2023). Recent advances in research on *Allium* plants: Functional ingredients, physiological activities, and applications in agricultural and food sciences. *Critical Reviews in Food Science and Nutrition*, *63*(26), 8107–8135. DOI: 10.1080/10408398.2022.2056132 PMID: 35343832

Yang, P., Yuan, Y., Yan, C., Jia, Y., You, Q., Da, L., Lou, A., Lv, B., Zhang, Z., & Liu, Y. (2024). AlliumDB: A central portal for comparative and functional genomics in *Allium*. *Horticulture Research*, *11*(2), uhad285. Advance online publication. DOI: 10.1093/hr/uhad285 PMID: 38371639

Compilation of References

Aarif, K. O. M., Alam, A., & Hotak, Y. (2025). Smart Sensor Technologies Shaping the Future of Precision Agriculture: Recent Advances and Future Outlooks. *Journal of Sensors*, *2025*(1), 2460098. DOI: 10.1155/js/2460098

Abed, N., Murgun, R., Deldari, A., Sankarannair, S., & Ramesh, M. V. (2025). IoT and AI-driven solutions for human-wildlife conflict: Advancing sustainable agriculture and biodiversity conservation. *Smart Agricultural Technology*, *10*, 100829. DOI: 10.1016/j.atech.2025.100829

Agbotui, D. K., Ingold, M., Wiehle, M., & Buerkert, A. (2023). Can carbon payments improve profitability of traditional conventional and organic cocoa agroforests? A case study in the Eastern Region of Ghana. *Agroforestry Systems*, *97*(5), 813–831. DOI: 10.1007/s10457-023-00828-0

Aguilar-Paredes, A., Valdés, G., Araneda, N., Valdebenito, E., Hansen, F., & Nuti, M. (2023). Microbial community in the composting process and its positive impact on the soil biota in sustainable agriculture. *Agronomy (Basel)*, *13*(2), 542. DOI: 10.3390/agronomy13020542

Ahlborn, J., Gmoser, R., Einhorn, P. U., & Schön, C. (2020). From stale bread and brewers spent grain to a new food source using edible filamentous fungi. *Bioengineered*, *11*(1), 582–598. DOI: 10.1080/21655979.2020.1768694 PMID: 32449450

Ahmad, F., Talukdar, N. R., Goparaju, L., Biradar, C., Dhyani, S. K., & Rizvi, J. (2022). GIS-based assessment of land-agroforestry potentiality of Jharkhand State, India. *Regional Sustainability*, *3*(3), 254–268. https://www.scopus.com/inward/record.uri?eid=2-s2.0-85141875839&doi=10.1016%2Fj.regsus.2022.10.004&partnerID=40&md5=70195ae0f51802cf59dceb27996ef053

Ahmad, M. I., Farooq, S., Alhamoud, Y., Li, C., & Zhang, H. (2022). A review on mycoprotein: History, nutritional composition, production methods, and health benefits. *Trends in Food Science & Technology, 121*, 14–29. DOI: 10.1016/j.tifs.2022.01.027

Ahmed, A. (2022). Marketing 4.0: The Unseen Potential of AI in Consumer Relations. *International Journal of New Media Studies: International Peer Reviewed Scholarly Indexed Journal, 9*(1), 5–12.

Ahmed, M. G., Gouda, S. A., Donia, S., & Hassanein, N. M. (2024). Production of single cell protein by fungi from different food wastes. *Biomass Conversion and Biorefinery, 15*(4), 5447–5462. DOI: 10.1007/s13399-024-05478-5

Ahmed, S. K., Mohammed, R. A., Nashwan, A. J., Ibrahim, R. H., Abdalla, A. Q., Ameen, B. M. M., & Khdhir, R. M. (2025). Using thematic analysis in qualitative research. *Journal of Medicine, Surgery, and Public Health, 6*, 100198. DOI: 10.1016/j.glmedi.2025.100198

Ajayi, O. O., Toromade, A. S., & Ayeni, O. (2024). Circular agro-economies (CAE): reducing waste and increasing profitability in agriculture. *International Journal of Advanced Economics*.

Akata, I., Ergönül, B. Ü. L. E. N. T., & Kalyoncu, F. (2012). Chemical compositions and antioxidant activities of 16 wild edible mushroom species grown in Anatolia. *International Journal of Pharmacology, 8*(2), 134–138. DOI: 10.3923/ijp.2012.134.138

Akhuli, S. (2025). Organic farming: A pathway to sustainable development. *International Journal of Research and Review, 12*(1), 208–211. DOI: 10.52403/ijrr.20250126

Akyüz, M., & Kirbağ, S. (2010). Nutritive value of wild edible and cultured mushrooms. *Turkish Journal of Biology = Türk Biyoloji Dergisi, 34*, 97–102. DOI: 10.3906/biy-0805-17

Alfano, M. R., Cantabene, C., Lepore, A., & Palermo, S. (2023). The green to circular bioeconomy transition: Innovation and resilience among Italian enterprises. *Business Strategy and the Environment, 32*(8), 6094–6105. DOI: 10.1002/bse.3474

Ali, H., Rafiq, M., & Shang, Z. (2024). Ecological Conservation and Restoration in the Indian Himalaya Region. In *Sustainable Ecological Restoration and Conservation in the Hindu Kush Himalayan Region: A Comprehensive Review* (pp. 117-135). GB: CABI. DOI: 10.1079/9781800622579.0009

Alifa, A. N., Isnawati, H., Syamsu, R. M. &, & Imam, M. (2024). Socioeconomic and ecological sustainability of agroforestry. *Biodiversity and Bioeconomy: Status Quo, Challenges, and Opportunities*, 281–294. DOI: 10.1016/B978-0-323-95482-2.00013-4

Allied Market Research. (2021). *Fungal Protein Market by Type (Yeast, Fusarium Venenatum, Mushroom), Nature (Organic, Conventional), and Application (Food & Beverage, Animal Feed, Pharmaceuticals, Others): Global Opportunity Analysis and Industry Forecast, 2021-2030.* https://www.alliedmarketresearch.com/fungal-protein-market-A12366

Almond, R. E. A., Grooten, M., & Petersen, T. (2020). *Living planet report 2020 - Bending the curve of biodiversity loss.* World Wildlife Fund. https://files.worldwildlife.org/wwfcmsprod/files/Publication/file/279c656a32_ENGLISH_FULL.pdf

Al-Saeedi, N. K., & Massad, M. Q. (2025, April). Exploring the Efficacy of Integrated Multi-Trophic Aquaculture (IMTA) for Nutrient Recycling and Productivity Enhancement in Marshlands. In *IOP Conference Series: Earth and Environmental Science* (Vol. 1487).

Altieri, M. A. (2018). *Agroecology: The science of sustainable agriculture* (2nd ed.). CRC Press., DOI: 10.1201/9780429495465

Altieri, M. A., Nicholls, C. I., de Molina, M. G., & Rojas, A. S. (2024). Landscape agroecology: Methodologies and applications for the design of sustainable agroecosystems. *Land (Basel), 13*(11), 1746. DOI: 10.3390/land13111746

Amara, A. A., & El-Baky, N. A. (2023). Fungi as a Source of Edible Proteins and Animal Feed. *Journal of Fungi (Basel, Switzerland), 9*(1), 73. DOI: 10.3390/jof9010073 PMID: 36675894

Amare, D., & Darr, D. (2023). Profitability of Smallholder Agroforestry Woodlot Innovations. *Human Ecology: an Interdisciplinary Journal, 51*(5), 1009–1019. DOI: 10.1007/s10745-023-00446-5

Ancillai, C., Terho, H., Cardinali, S., & Pascucci, F. (2019). Advancing social media driven sales research: Establishing conceptual foundations for B-to-B social selling. *Industrial Marketing Management, 82*, 293–308. DOI: 10.1016/j.indmarman.2019.01.002

Andriatsitohaina, R. N. N., Laby, P., Llopis, J. C., & Martin, D. A. (2024). Agroforestry in Madagascar: Past, present, and future. *Agroforestry Systems, 98*(6), 1659–1680. DOI: 10.1007/s10457-024-00975-y PMID: 39086741

Anikwe, M., & Ife, K. (2023). The role of soil ecosystem services in the circular bioeconomy. *Frontiers in Soil Science*, *3*, 1209100. DOI: 10.3389/fsoil.2023.1209100

Ansar, A., Du, J., Javed, Q., Adnan, M., & Javaid, I. (2025). Biodegradable Waste in Compost Production: A Review of Its Economic Potential. *Nitrogen*, *6*(2), 24. DOI: 10.3390/nitrogen6020024

Anthelme, G., Junior, P. F. P., Boué, V. B. B. N., Larissa, G., Marie-Solange, T., & Yves, A. Y. C. (2024). Evaluating carbon sequestration and biodiversity in Cocoa agroforestry Systems: A case study from the Man Region, Côte d'Ivoire. *International Journal of Environment and Climate Change*, *14*(10), 691–700. DOI: 10.9734/ijecc/2024/v14i104517

Anwar, G. M., Helmey, R. K., & Moustafa, Y. M. M. (2017). Genetic diversity among some species of the genus *Allium* L. using SSR and ISSR markers. *Egyptian Journal of Genetics and Cytology*, *46*(1), 221–233. DOI: 10.21608/ejgc.2018.9516

Arifanti, V. B., Kauffman, J. B., Subarno, , Ilman, M., Tosiani, A., & Novita, N. (2022). Contributions of mangrove conservation and restoration to climate change mitigation in Indonesia. *Global Change Biology*, *28*(15), 4523–4538. DOI: 10.1111/gcb.16216 PMID: 35470521

Arshad, M. A., Rouf, S., Abbas, R. N., Shahbaz, Z., Aleem, K., Shahbaz, H., Pervaiz, R., Sarwar, A., & Rehman, H. (2024). Navigating synergies: A comprehensive review of agroforestry system and agronomy crops. *Haya: The Saudi Journal of Life Sciences*, *9*(04), 97–113. DOI: 10.36348/sjls.2024.v09i04.003

Artola, A., Font, X., Moral-Vico, J., & Sánchez, A. (2024). The role of solid-state fermentation to transform existing waste treatment plants based on composting and anaerobic digestion into modern organic waste-based biorefineries, in the framework of circular bioeconomy. *Frontiers in Chemical Engineering*, *6*, 1463785. DOI: 10.3389/fceng.2024.1463785

Asare, R., Afari-Sefa, V., Osei-Owusu, Y., & Pabi, O. (2014). Cocoa agroforestry for increasing forest connectivity in a fragmented landscape in Ghana. *Agroforestry Systems*, *88*(6), 1143–1156. DOI: 10.1007/s10457-014-9688-3

Ascoli, D., Plana, E., Oggioni, S. D., Tomao, A., Colonico, M., Corona, P., Giannino, F., Moreno, M., Xanthopoulos, G., Kaoukis, K., Athanasiou, M., Colaço, M. C., Rego, F., Sequeira, A. C., Acácio, V., Serra, M., & Barbati, A. (2023). Fire-smart solutions for sustainable wildfire risk prevention: Bottom-up initiatives meet top-down policies under EU green deal. *International Journal of Disaster Risk Reduction*, *92*, 103715. DOI: 10.1016/j.ijdrr.2023.103715

Auttri, B., Chaitanya, K., Daida, S., & Jain, S. K. (2023). Digital Transformation in Customer Relationship Management: Enhancing Engagement and Loyalty. [EEL]. *European Economic Letters, 13*(3), 1140–1149.

Awasthi, M. K., Kumar, V., Hellwig, C., Wikandari, R., Harirchi, S., Sar, T., Wainaina, S., Sindhu, R., Binod, P., Zhang, Z., & Taherzadeh, M. J. (2022). Filamentous fungi for sustainable vegan food production systems within a circular economy: Present status and future prospects. *Food Research International, 164*, 112318. DOI: 10.1016/j.foodres.2022.112318 PMID: 36737911

Awazi, N. P., Tientcheu-Avana, M.-L., Temgoua, L. F., Tsufac, A. R., Forje, G. W., Cedric, C. D., Nfornkah, B. N., & Tchamba, M. N. (2023). Agroforestry-based bioeconomy transition in Cameroon: realities and the way forward. In *Biodiversity and Bioeconomy*. Status Quo, Challenges, and Opportunities., DOI: 10.1016/B978-0-323-95482-2.00014-6

Awewomom, J., Dzeble, F., Takyi, Y. D., Ashie, W. B., Ettey, E. N. Y. O., Afua, P. E., Sackey, L. N. A., Opoku, F., & Akoto, O. (2024). Addressing global environmental pollution using environmental control techniques: A focus on environmental policy and preventive environmental management. *Discover Environment, 2*(1), 8. DOI: 10.1007/s44274-024-00033-5

Azizi, N. F., Kumar, M. R., Yeap, S. K., Abdullah, J. O., Khalid, M., Omar, A. R., Osman, M. A., Mortadza, S. A. S., & Alitheen, N. B. (2021). Kefir and its biological activities. *Foods, 10*(6), 1210. DOI: 10.3390/foods10061210 PMID: 34071977

Bagada, P. J., Damor, P. A., Patel, R. J., Patel, K. C., Patel, D. J., Parmar, H. V., Rank, P. H., & Vekariya, P. B. (2025). Enhancing soil moisture sensors for optimized agricultural water management: A comprehensive review. *Plant Archives*, 25 (Special Issue, ICTPAIRS-JAU), 879–888.

Baghalian, K., Ziai, S. A., Naghavi, M. R., Badi, H. N., & Khalighi, A. (2005). Evaluation of allicin content and botanical traits in Iranian garlic (*Allium sativum* L.) ecotypes. *Scientia Horticulturae, 103*(2), 155–166. DOI: 10.1016/j.scienta.2004.07.001

Baharin, A. T., Ishak, N. A., Ismail, N., Osman, M. F., Isa, E. V. M., & Ariffin, W. N. M. (2024). Bioeconomy Sustainability: Integrating Circular Economy Principles with Big Data and IoT for Sustainable Farming in Agriculture 4.0. *International Journal of Biomass Utilization and Sustainable Energy (IJBUSE), 2.*

Balasubramanian, S. (2017). *Traditional knowledge and patent issues: An overview of Turmeric, Basmati, Neem cases*. Intellectual Property and Technology Law updates, S&A IP-Tech. https://www.manupatrafast.in/NewsletterArchives/listing/SAIPTech%20Singh%20Associates/2017/Mar/IPTech%20March17.pdf

Baldwin, S., Pither-Joyce, M., Wright, K., Chen, L., & McCallum, J. (2012). Development of robust genomic simple sequence repeat markers for estimation of genetic diversity within and among bulb onion (*Allium cepa* L.) populations. *Molecular Breeding, 30*(3), 1401–1411. DOI: 10.1007/s11032-012-9727-6

Barrios, E., Sileshi, G. W., Shepherd, K., & Sinclair, F. (2012). Agroforestry and soil health: linking trees, soil biota, and ecosystem services. In *Soil Ecology and Ecosystem Services* (pp. 315–330). Oxford University Press., DOI: 10.1093/acprof:oso/9780199575923.003.0028

Bartholomai, B. M., Ruwe, K. M., Thurston, J., Jha, P., Scaife, K., Simon, R., Abdelmoteleb, M., Goodman, R. E., & Farhi, M. (2022). Safety evaluation of Neurospora crassa mycoprotein for use as a novel meat alternative and enhancer. *Food and Chemical Toxicology, 168,* 113342. DOI: 10.1016/j.fct.2022.113342 PMID: 35963473

Baruah, N., Bora, A., & Gogoi, N. (2024). Agricultural Waste in Circular Economy: An Indian Scenario. *Emerging Trends and Techniques in Biofuel Production from Agricultural Waste*, 15–42.

Barzee, T. J., Cao, L., Pan, Z., & Zhang, R. (2021). Fungi for future foods. *Journal of Future Foods, 1*(1), 25–37. DOI: 10.1016/j.jfutfo.2021.09.002

Bastaki, S. M., Ojha, S., Kalasz, H., & Adeghate, E. (2021). Chemical constituents and medicinal properties of *Allium* species. *Molecular and Cellular Biochemistry, 476*(12), 4301–4321. DOI: 10.1007/s11010-021-04213-2 PMID: 34420186

Batra, L. R., & Millner, P. D. (1974). Some Asian fermented foods and beverages, and associated fungi. *Mycologia, 66*(6), 942–950. DOI: 10.1080/00275514.1974.12019699

Batsioula, M., Malamakis, A., Patsios, S. I., Geroliolios, D., Skoutida, S., Melas, L., & Banias, G. F. (2023). *Innovative Bio-based Circular Economy Schemes: The Case of Biomass and Food Waste Utilization as an Enabler of Regional Resilience* (pp. 1795–1806). DOI: 10.1007/978-3-031-23721-8_138

Belfrid, D. M., Arcadius, D., N'tcha Yekanbessoun, M., Mariano, S., Fidèle, S., Jean, A., & Pepin, A. M. (2022). Inventory of agroecosystem services and perceptions potential implications due to climate change: A case study from Benin in West Africa. *Environmental Impact Assessment Review, 95,* •••. https://www.scopus.com/inward/record.uri?eid=2-s2.0-85128162100&doi=10.1016%2Fj.eiar.2022.106792&partnerID=40&md5=ec9437a14903ffcb1192bc7c9822efb4

Bency Baby, T., & Suriyaprakash, T. N. K. (2021). Intellectual property rights: Bioprospecting, biopiracy and protection of traditional knowledge - An Indian perspective. In Prabu, S. L., & Suriyaprakash, T. N. K. (Eds.), *Intellectual Property* (p. 6). IntechOpen.

Bernaert, N., De Paepe, D., Bouten, C., De Clercq, H., Stewart, D., Van Bockstaele, E., De Loose, M., & Van Droogenbroeck, B. (2012). Antioxidant capacity, total phenolic and ascorbate content as a function of the genetic diversity of leek (*Allium ampeloprasum* var. porrum). *Food Chemistry, 134*(2), 669–677. DOI: 10.1016/j.foodchem.2012.02.159 PMID: 23107677

Bhagwat, S. A., Willis, K. J., Birks, H. J. B., & Whittaker, R. J. (2008). Agroforestry: A refuge for tropical biodiversity? *Trends in Ecology & Evolution, 23*(5), 261–267. DOI: 10.1016/j.tree.2008.01.005 PMID: 18359125

Bhardwaj, S., Venkatesan, S., Rawat, S., & Nath, P. (2024). Transforming Agriculture with IoT for Precision Agriculture and Sustainable Crop Management. In *The Future of Agriculture* (pp. 164–200). IoT, AI and Blockchain Technology for Sustainable Farming. DOI: 10.2174/9789815274349124010012

Bhattacharya, P. (2023). *Indigenous knowledge and resilience among Indian tribes: Cultural practices in the face of modernity.*

Bhowmick, B. K., Sarkar, S., Roychowdhury, D., Patil, S. D., Lekhak, M. M., Ohri, D., Rao, S. R., Yadav, S., Verma, R. C., Dhar, M. K., Raina, S. N., & Jha, S. (2023). *Allium* cytogenetics: A critical review on the Indian taxa. *Comparative Cytogenetics, 17*, 129–156. DOI: 10.3897/CompCytogen.17.98903 PMID: 37304149

Bhujel, R. R., & Joshi, H. G. (2021). *Organic Farming for sustainability: A case study of farmers in Sikkim*. ResearchGate.

Bhutia, Dr. D. (2019). DZUMSA SYSTEM IN SIKKIM AS AN ALTERNATIVE MODE TOWARDS DISPUTE RESOLUTION. *International Journal of Research and Analytical Reviews, 6*(2), 797–798.

Bisheko, M. J., & G, R. (2025). Challenges and interventions in water resource management for horticultural crop production in rural and tribal communities in India. *Cogent Food & Agriculture, 11*(1), 2452335. DOI: 10.1080/23311932.2025.2452335

Bisht, D., Joshi, V., Yadava, A. K., & Pant, H. (2021). COMMUNITY FOREST MANAGEMENT IN UTTARAKHAND: a COMPARATIVE CASE STUDY OF WESTERN HIMALAYAN VAN PANCHAYATS. *ResearchGate*.

Bisht, I. S., Rana, J. C., Yadav, R., & Ahlawat, S. P. (2020). Mainstreaming agricultural biodiversity in traditional production Landscapes for Sustainable Development: The Indian Scenario. *Sustainability (Basel), 12*(24), 10690. DOI: 10.3390/su122410690

Bloch, J. L., Rouault, M., Langhi, C., Hignard, M., Iriantsoa, V., & Michelet, O. (2025). The novel food evaluation process delays access to food innovation in the European Union. *NPJ Science of Food, 9*(1), 117. Advance online publication. DOI: 10.1038/s41538-025-00492-x PMID: 40593922

Boadi, S. A., Bosselmann, A. S., Owusu, K., Asare, R., & Olwig, M. F. (2023). Household economics of cocoa agroforestry: costs and benefits. In *Agroforestry as Climate Change Adaptation* (pp. 121–145). Springer International Publishing., DOI: 10.1007/978-3-031-45635-0_5

Boeri, P., Piñuel, L., Dalzotto, D., & Sharry, S. (2020). Native biodiversity: A strategic resource to accelerate bioeconomy development in Latin America and the Caribbe-an. In *Agricultural, Forestry and Bioindustry Biotechnology and Biodiscovery* (pp. 163–174). Springer International Publishing. DOI: 10.1007/978-3-030-51358-0_10

Boiardo, M. (2025). Assessing the Role of Organic Farming in Enhancing Soil Health and Biodiversity in Europe. *Europian Journal of Sustainable Agriculture P-ISSN 3051-0112 E-ISSN 3051-0120, 1*(01), 9-19.

Bonoli, A. (2023). *Circular Economy and Circular Bioeconomy as pillars of the ecological transition for a worldwide enhanced sustainability* (Doctoral dissertation, Universidad de Granada).

Botero-Valencia, J., García-Pineda, V., Valencia-Arias, A., Valencia, J., Reyes-Vera, E., Mejia-Herrera, M., & Hernández-García, R. (2025). Machine Learning in Sustainable Agriculture: Systematic Review and Research Perspectives. *Agriculture, 15*(4), 377. DOI: 10.3390/agriculture15040377

Braun, V., & Clarke, V. (2006). Using thematic analysis in psychology. *Qualitative Research in Psychology, 3*(2), 77–101. DOI: 10.1191/1478088706qp063oa

Bucheli, V. J. P., & Bokelmann, W. (2017). Agroforestry systems for biodiversity and ecosystem services: The case of the sibundoy valley in the colombian province of putumayo. *The International Journal of Biodiversity Science, Ecosystem Services & Management, 13*(1), 380–397. DOI: 10.1080/21513732.2017.1391879

Cambria, E., Schuller, B., Xia, Y., & Havasi, C. (2013). New avenues in opinion mining and sentiment analysis. [Technical University of Munich]. *IEEE Intelligent Systems, 28*(2), 15–21. DOI: 10.1109/MIS.2013.30

Campos, P., Álvarez, A., Oviedo, J. L., Mesa, B., Caparrós, A., & Ovando, P. (2020). Environmental incomes: Refined standard and extended accounts applied to cork oak open woodlands in Andalusia, Spain. *Ecological Indicators, 117*, 106551. https://www.scopus.com/inward/record.uri?eid=2-s2.0-85085523786&doi=10.1016%2Fj.ecolind.2020.106551&partnerID=40&md5=aac1770e738494d1ad2cf29c7e232ab4. DOI: 10.1016/j.ecolind.2020.106551

Campoy-Muñoz, P., Cardenete, M. A., Pérez-Hernández, P. P., Martín-Lozano, J. M., & Martínez-Rueda, J. (2022). A SAM Approach for the Analysis of Small-Scale Forestry-Based Activities: A Case Study of a Landowner Cooperative for Non-wooded Products from Pinus pinea L. in Southern Spain. *Small-scale Forestry, 21*(2), 249–273. DOI: 10.1007/s11842-021-09497-2

Cardinale, B. J., Duffy, J. E., Gonzalez, A., Hooper, D. U., Perrings, C., Venail, P., Narwani, A., Mace, G. M., Tilman, D., Wardle, D. A., Kinzig, A. P., Daily, G. C., Loreau, M., Grace, J. B., Larigauderie, A., Srivastava, D. S., & Naeem, S. (2012). Biodiversity loss and its impact on humanity. *Nature, 486*(7401), 59–67. DOI: 10.1038/nature11148 PMID: 22678280

Cedeno, F. R. P., Olubiyo, O. J., & Ferreira, S. (2025). From microbial proteins to cultivated meat for alternative meat-like products: A review on sustainable fermentation approaches. *Journal of Biological Engineering, 19*(1), 44. Advance online publication. DOI: 10.1186/s13036-025-00509-9 PMID: 40369620

Cerrone, F., & O'Connor, K. E. (2025). Cultivation of filamentous fungi in airlift bioreactors: Advantages and disadvantages. *Applied Microbiology and Biotechnology, 109*(1), 41. Advance online publication. DOI: 10.1007/s00253-025-13422-4 PMID: 39928147

Chalbi, A., Chikh-Rouhou, H., Mezghani, N., Slim, A., Fayos, O., Bel-Kadhi, M. S., & Garcés-Claver, A. (2023). Genetic Diversity Analysis of Onion (*Allium cepa* L.) from the Arid Region of Tunisia Using Phenotypic Traits and SSR Markers. *Horticulturae, 9*(10), 1098. https://www.mdpi.com/2311-7524/9/10/1098. DOI: 10.3390/horticulturae9101098

Chan, C. K. Y. (2023). A comprehensive AI policy education framework for university teaching and learning. *International Journal of Educational Technology in Higher Education, 20*(1), 38. DOI: 10.1186/s41239-023-00408-3

Chaudhary, V., Katyal, P., Poonia, A. K., Kaur, J., Puniya, A. K., & Panwar, H. (2021). Natural pigment from Monascus: The production and therapeutic significance. *Journal of Applied Microbiology, 133*(1), 18–38. DOI: 10.1111/jam.15308 PMID: 34569683

Chavan, V. (2021). *State Biodiversity Boards: Towards Better Governance*. ResearchGate.

Chen, S., Zhou, J., Chen, Q., Chang, Y., Du, J., & Meng, H. (2013). Analysis of the genetic diversity of garlic (*Allium sativum* L.) germplasm by SRAP. *Biochemical Systematics and Ecology*, *50*, 139–146. DOI: 10.1016/j.bse.2013.03.004

Cheung, S. S., Chong, G., Elisia, I., Hasman, D., Lee, M., Chang, L., & Tai, J. (2023). Wild Chaga (Inonotus obliquus) modulates inflammation, neural cell survival and inhibits proliferation of cancer cells. *Pharmacological Research. Modern Chinese Medicine*, *9*, 100328. DOI: 10.1016/j.prmcm.2023.100328

Chezan, D., Flannery, O., & Patel, A. (2022). Factors affecting consumer attitudes to fungi-based protein: A pilot study. *Appetite*, *175*, 106043. DOI: 10.1016/j.appet.2022.106043 PMID: 35487309

Chiu, C. Y., Yen, T. E., Liu, S. H., & Chiang, M. T. (2019). Comparative effects and mechanisms of chitosan and its derivatives on hypercholesterolemia in high-fat diet-fed rats. *International Journal of Molecular Sciences*, *21*(1), 92. DOI: 10.3390/ijms21010092 PMID: 31877743

Coletta, M., Monticelli, M., D'Alessandro, A., Gentili, C., Torresi, A., Waris, N., & La Terza, A. (2025). Managing soil to support soil biodiversity in protected areas agroecosystems: A comparison between arable lands, olive groves, and vineyards in the Conero Park (Italy). *Environmental Monitoring and Assessment*, *197*(2), 200. DOI: 10.1007/s10661-025-13658-7 PMID: 39869200

Convention on Biological Diversity. (2021). *Global biodiversity outlook 5*. Retrieved from https://www.cbd.int/gbo5

Convention on Biological Diversity. (2022). *Kunming-Montreal Global Biodiversity Framework*.

Cordeiro, C. M., & Sindhøj, E. (2024). Situating the discourse of recycled nutrient fertilizers in circular economy principles for sustainable agriculture. *Frontiers in Sustainability*, *5*, 1465752. DOI: 10.3389/frsus.2024.1465752

Cortina-Segarra, J., García-Sánchez, I., Grace, M., Andrés, P., Baker, S., Bullock, C., Decleer, K., Dicks, L. V., Fisher, J. L., Frouz, J., Klimkowska, A., Kyriazopoulos, A. P., Moreno-Mateos, D., Rodríguez-González, P. M., Sarkki, S., & Ventocilla, J. L. (2021). Barriers to ecological restoration in Europe: Expert perspectives. *Restoration Ecology*, *29*(4), e13346. Advance online publication. DOI: 10.1111/rec.13346

Crippa, M., Solazzo, E., Guizzardi, D., Tubiello, F. N., & Leip, A. (2021). Food systems are responsible for a third of global anthropogenic GHG emissions. *Nature Food*, *2*(3), 198–209. DOI: 10.1038/s43016-021-00225-9 PMID: 37117443

Dagar, J. C. (2017). Agroforestry : Four decades of research development. *Indian J. Agrofor, January 1999*, 0–32. https://www.researchgate.net/profile/J-Dagar-2/publication/353259684_Agroforestry_Four_decades_of_research_development/links/60efdb639541032c6d3ec0aa/Agroforestry-Four-decades-of-research-development.pdf

Daglio lu, O. (2000). Tarhana as a traditional Turkish fermented cereal food. Its recipe, production and composition. *Food/Nahrung, 44*(2), 85-88. https://doi.org/10.1002/(SICI)1521-3803(20000301)44:2<85::AID-FOOD85>3.0.CO;2-HDOI: 10.1002/(SICI)1521-3803(20000301)44:2<85::AID-FOOD85>3.0.CO;2-H

Dai, X., Azi, F., Alnadari, F., & Yu, Z. (2025). Developing organosulfur compounds in Allium as the next-generation flavor and bioactive ingredients for food and medicine. *Critical Reviews in Food Science and Nutrition*, *65*(33), 1–19. DOI: 10.1080/10408398.2025.2500674 PMID: 40366718

Dai, X., Sharma, M., & Chen, J. (2021). Fungi in sustainable food production. *Fungal Biology*. Advance online publication. DOI: 10.1007/978-3-030-64406-2

Danielsen, F., Eicken, H., Funder, M., Johnson, N., Lee, O., Theilade, I., Argyriou, D., & Burgess, N. D. (2022). Community monitoring of natural resource systems and the environment. *Annual Review of Environment and Resources*, *47*(1), 637–670. DOI: 10.1146/annurev-environ-012220-022325

Das, A. K., Nanda, P. K., Dandapat, P., Bandyopadhyay, S., Gullón, P., Sivaraman, G. K., McClements, D. J., Gullón, B., & Lorenzo, J. M. (2021). Edible mushrooms as functional ingredients for development of healthier and more sustainable muscle foods: A flexitarian approach. *Molecules*, *26*(9), 2463. https://doi.org/10.3390/molecules26092463

Das, A., Gujre, N., Devi, R. J., & Mitra, S. (2021). A review on traditional ecological knowledge and its role in natural resources management: North East India, a cultural paradise. *Environmental Management*, •••, 1–22. PMID: 34791537

Das, B. R., Bhuyan, M. J., Deka, N., & Bhattacharya, P. (2025). Assessing homestead agroforestry-based livelihoods in Sivasagar District of Assam, India, using livelihood significance index (LSI). *Forest Policy and Economics*, *170*, 103388. Advance online publication. DOI: 10.1016/j.forpol.2024.103388

Das, B., Bhardwaj, P. K., Chaudhary, S. K., Pathaw, N., Singh, H. K., Tampha, S., Singh, K. K., Sharma, N., & Mukherjee, P. K. (2024). Bioeconomy and ethnopharmacology – Translational perspective and sustainability of the bioresources of northeast region of India. *Journal of Ethnopharmacology*, *330*(April), 118203. DOI: 10.1016/j.jep.2024.118203 PMID: 38641075

Dasgupta, S., Badola, R., Ali, S. Z., & Tariyal, P. (2022). Bioresource, energy and forest sustainability: A case study from Indian Himalayan region. *Journal of Cleaner Production*, *337*, 130497. DOI: 10.1016/j.jclepro.2022.130497

De Cianni, R., Mancuso, T., Rizzo, G., & Migliore, G. (2024). Health or Environment? Understanding which informative message is more effective in replacing red meat with mushroom-based alternatives. *Appetite*, *199*, 107405. DOI: 10.1016/j.appet.2024.107405 PMID: 38723668

de Ciencias Agrarias y Forestales, L. (2025). 9 Bioeconomy and Biotechnology: Innovative Strategies in the Utilization of Resources in the Argentine North Patagonia. *Plant Biotechnology and Sustainable Agriculture*, 113.

De Penning & De Penning. (2025). The National Biodiversity Authority: New Rules and Their Implications. https://depenning.com/blog/the-national-biodiversity-authority-new-rules-and-their-implications/

de Queiroz-Stein, G., & Siegel, K. M. (2023). Possibilities for mainstreaming biodiversity? Two perspectives on the concept of bioeconomy. *Earth System Governance*, *17*(May), 100181. DOI: 10.1016/j.esg.2023.100181

Deloitte. (2023). *Alternative proteins: The future of food*. https://www.deloitte.com/nl/en/Industries/consumer/perspectives/alternative-proteins.html

Deng, H., Gao, R., Liao, X., & Cai, Y. (2017). Genome editing in Shiraia bambusicola using CRISPR-Cas9 system. *Journal of Biotechnology*, *259*, 228–234. DOI: 10.1016/j.jbiotec.2017.06.1204 PMID: 28690135

Department of Economic and Social Affairs. United Nations. (n.d.). The 17 Goals. https://sdgs.un.org/goals

Dhanya, B., Purushothaman, S., & Viswanath, S. (2016). Economic rationale of traditional agroforestry systems: A case-study of Ficus trees in semiarid agroecosystems of Karnataka, southern India. *Forests, Trees and Livelihoods*, *25*(4), 267–281. DOI: 10.1080/14728028.2016.1218800

Dominicis, L. F., Lima, M. F. B., Souza, Á. N., Joaquim, M. S., Araújo, J. B. C. N., Coelho, L. M.Junior, Ribeiro, J. F., & do Santos, P. P. (2023). Payment for Environmental Services and the Financial Viability of Agroforestry Systems: An Integrated Analysis of Socio-Environmental Projects in the Descoberto Basin—Federal District. *Forests, 14*(10), 2110. Advance online publication. DOI: 10.3390/f14102110

Dondina, O., Saura, S., Bani, L., & Mateo-Sánchez, M. C. (2018). Enhancing connectivity in agroecosystems: Focus on the best existing corridors or on new pathways? *Landscape Ecology, 33*(10), 1741–1756. DOI: 10.1007/s10980-018-0698-9

Dong, S., Shang, Z., Gao, J., & Boone, R. (2022). Enhancing the ecological services of the Qinghai-Tibetan Plateau's grasslands through sustainable restoration and management in era of global change. *Agriculture, Ecosystems & Environment, 326*, 107756. DOI: 10.1016/j.agee.2021.107756

Donthu, N., Kumar, S., Mukherjee, D., Pandey, N., & Lim, W. M. (2021). How to conduct a bibliometric analysis: An overview and guidelines. *Journal of Business Research, 133*, 285–296. DOI: 10.1016/j.jbusres.2021.04.070

Duan, W., Khurshid, A., Khan, K., & Calin, A. C. (2024). Transforming industry: Investigating 4.0 technologies for sustainable product evolution in china through a novel fuzzy three-way decision-making process. *Technological Forecasting and Social Change, 200*, 123125. DOI: 10.1016/j.techfore.2023.123125

Dubey, P. K., Singh, A., Raghubanshi, A., & Abhilash, P. C. (2021). Steering the restoration of degraded agroecosystems during the United Nations Decade on Ecosystem Restoration. *Journal of Environmental Management, 280*, 111798. DOI: 10.1016/j.jenvman.2020.111798 PMID: 33309393

Du, Y., Wang, M., Yang, L., Tong, L., Guo, D., & Ji, X. (2022). Optimization and Scale-Up of fermentation processes driven by models. *Bioengineering (Basel, Switzerland), 9*(9), 473. DOI: 10.3390/bioengineering9090473 PMID: 36135019

Ebrahimi, R., Zamani, Z., & Kashi, A. (2009). Genetic diversity evaluation of wild Persian shallot (*Allium hirtifolium* Boiss.) using morphological and RAPD markers. *Scientia Horticulturae, 119*(4), 345–351. DOI: 10.1016/j.scienta.2008.08.032

Edgar, G. J., Stuart-Smith, R. D., Willis, T. J., Kininmonth, S., Baker, S. C., Banks, S., Barrett, N. S., Becerro, M. A., Bernard, A. T. F., Berkhout, J., Buxton, C. D., Campbell, S. J., Cooper, A. T., Davey, M., Edgar, S. C., Försterra, G., Galván, D. E., Irigoyen, A. J., Kushner, D. J., & Thomson, R. J. (2014). Global conservation outcomes depend on marine protected areas with five key features. *Nature, 506*(7487), 216–220. DOI: 10.1038/nature13022 PMID: 24499817

Edlinger, A., Herzog, C., Garland, G., Walder, F., Banerjee, S., Keel, S. G., Mayer, J., Philippot, L., Romdhane, S., Schiedung, M., Schmidt, M. W. I., Seitz, B., Wüst-Galley, C., & Van Der Heijden, M. G. A. (2025). Compost application Enhances soil health and maintains crop yield: Insights from 56 Farmer-Managed Arable Fields. *Journal of Sustainable Agriculture and Environment*, *4*(1), e70041. DOI: 10.1002/sae2.70041

Edo, G. I., Mafe, A. N., Ali, A. B., Akpoghelie, P. O., Yousif, E., Isoje, E. F., Igbuku, U. A., Zainulabdeen, K., Owheruo, J. O., Essaghah, A. E. A., Umar, H., Ahmed, D. S., & Alamiery, A. A. (2025). A critical review on the impacts of β-glucans on gut microbiota and human health. *The Microbe*, *7*, 100394. DOI: 10.1016/j.microb.2025.100394

Edo, M., Entling, M. H., & Rösch, V. (2023). Agroforestry supports high bird diversity in European farmland. *Agronomy for Sustainable Development*, *44*(1), 1. Advance online publication. DOI: 10.1007/s13593-023-00936-2

Egea, L. A., Mérida-García, R., Kilian, A., Hernandez, P., & Dorado, G. (2017). Assessment of genetic diversity and structure of large garlic (*Allium sativum*) germplasm bank, by diversity arrays technology “genotyping-by-sequencing” platform (DArTseq). *Frontiers in Genetics*, *8*, 98. DOI: 10.3389/fgene.2017.00098 PMID: 28775737

Egwuatu, D. C., Ogunye, R. O., Anene, P. C., Alhassan, Y., Nwafor, I. R., Udoh, J. I., Asenuga, O. O., & Nwokafor, C. V. (2024). A review on conversion of agricultural waste to bioenergy: Processes and environmental impacts. *Biotechnology Journal International*, *28*(6), 18–34. DOI: 10.9734/bji/2024/v28i6746

Ermgassen, S. O. Z., & Löfqvist, S. (2024). Financing ecosystem restoration. *Current Biology*, *34*(9), R412–R417. DOI: 10.1016/j.cub.2024.02.031 PMID: 38714174

Estreguil, C., Caudullo, G., Rega, C., & Paracchini, M. L. (2016). *Enhancing connectivity improving green enhancing connectivity improving green infrastructure. Cost-benefit solutions for forest and agri-environment A pilot study in Lombardy.* Issue October., DOI: 10.2788/170924

European Commission. (2020). *The European Green Deal.* Retrieved from https://ec.europa.eu/info/publications/communication-european-green-deal_en

European Food Safety Authority. (2025, March 27). *Novel food.* https://www.efsa.europa.eu/en/topics/topic/novel-food

European Union. (2015). *Regulation (EU) 2015/2283 of the European Parliament and of the Council of 25 November 2015 on novel foods*. Official Journal of the European Union. https://eur-lex.europa.eu/eli/reg/2015/2283/oj/eng

Eyasu, G., & Gebrewahid, Y. (2024). Potential of agroforestry practices on woody species diversity and composition: Implication for conservation of indigenous trees in selected districts of Tigray, Northern Ethiopia. *Asian Journal of Research in Agriculture and Forestry, 10*(4), 88–100. DOI: 10.9734/ajraf/2024/v10i4319

Fahad, S., Chavan, S. B., Chichaghare, A. R., Uthappa, A. R., Kumar, M., Kakade, V., Pradhan, A., Jinger, D., Rawale, G., Yadav, D. K., Kumar, V., Farooq, T. H., Ali, B., Sawant, A. V., Saud, S., Chen, S., & Poczai, P. (2022). Agroforestry systems for soil health improvement and maintenance. *Sustainability (Basel), 14*(22), 14877. Advance online publication. DOI: 10.3390/su142214877

Farsari, I. (2023). Exploring the nexus between sustainable tourism governance, resilience and complexity research. *Tourism Recreation Research, 48*(3), 352–367. DOI: 10.1080/02508281.2021.1922828

Faucon, M.-P., Aussenac, T., Debref, R., Firmin, S., Houben, D., Marraccini, E., Sauvée, L., Trinsoutrot-Gattin, I., & Gloaguen, R. (2023). Combining agroecology and bioeconomy to meet the societal challenges of agriculture. *Plant and Soil, 492*(1-2), 61–78. DOI: 10.1007/s11104-023-06294-y

Feliziani, G., Bordoni, L., & Gabbianelli, R. (2025). Regenerative Organic Agriculture and Human Health: The Interconnection Between Soil, Food Quality, and Nutrition. *Antioxidants, 14*(5), 530. DOI: 10.3390/antiox14050530 PMID: 40427412

Fikry, M. Y., & Sarjan, M. (2024). Peran agroforestri dalam mendukung pengelolaan sumberdaya alam berkelanjutan. *LAMBDA : Jurnal Ilmiah Pendidikan MIPA Dan Aplikasinya, 4*(1), 16–22. DOI: 10.58218/lambda.v4i1.846

Filho, J. J. D. S., Paço, A. D., & Gaspar, P. D. (2025). Artificial Intelligence and MCDA in Circular Economy: Governance Strategies and Optimization for Reverse Supply Chains of Solid Waste. *Applied Sciences (Basel, Switzerland), 15*(9), 4758. DOI: 10.3390/app15094758

Financial Express. (2024). World Water Day 2024: As Sonam Wangchuk continues his 'climate fast', know about the Ice Stupas of Ladakh.

Finnigan, T., Mach, K., & Edlin, A. (2024a). Mycoprotein: a healthy new protein with a low environmental impact. In *Sustainable protein sources* (pp. 539-566). Academic Press. DOI: 10.1016/B978-0-323-91652-3.00011-3

Finnigan, T. J., Wall, B. T., Wilde, P. J., Stephens, F. B., Taylor, S. L., & Freedman, M. R. (2019). Mycoprotein: The future of nutritious nonmeat protein, a symposium review. *Current Developments in Nutrition*, *3*(6), nzz021. DOI: 10.1093/cdn/nzz021 PMID: 31187084

Finnigan, T., Theobald, H., & Bajka, B. (2024b). Mycoprotein: A healthy and sustainable source of alternative Protein-Based foods. *Annual Review of Food Science and Technology*. DOI: 10.1146/annurev-food-111523-121802 PMID: 39626232

Fiore, M., Demichelis, F., Deorsola, F. A., Fino, D., Saracco, G., Pugliese, M., & Tommasi, T. (2025). Optimizing biomethane production and plants growth with biochar-enhanced anaerobic digestion. *Results in Engineering*, *26*, 104883. DOI: 10.1016/j.rineng.2025.104883

Food and Agriculture Organization of the United Nations. (2023). *Agrifood solutions to climate change – FAO's work on the climate crisis*. Rome. DOI: 10.4060/cc8055en

Food and Agriculture Organization. (2020). *The state of the world's forests*. Retrieved from https://www.fao.org/state-of-forests

Food and Agriculture Organization. (n.d.). Reports on sustainable and circular bioeconomies.

Food and Drug Administration. (2002). *GRAS Notice No. GRN 000091*. https://www.fda.gov/media/173801/download

Food and Drug Administration. (2023). *GRAS Notice No. GRN 001125 for Pea Protein Fermented by Shiitake Mycelia*. https://www.fda.gov/media/180203/download

Francis, C. A., Lieblein, G., Gliessman, S. R., Breland, T. A., Creamer, N., Harwood, R., Salomonsson, L., Helenius, J., Rickerl, D., Salvador, R., Wiedenhoeft, M., Simmons, S., Allen, P., Altieri, M., Flora, C., & Poincelot, R. (2003). Agroecology: The ecology of food systems. *Journal of Sustainable Agriculture*, *22*(3), 99–118. DOI: 10.1300/J064v22n03_10

Fremout, T., Thomas, E., Taedoumg, H., Briers, S., Gutiérrez-Miranda, C. E., Alcázar-Caicedo, C., Lindau, A., Mounmemi Kpoumie, H., Vinceti, B., Kettle, C., Ekué, M., Atkinson, R., Jalonen, R., Gaisberger, H., Elliott, S., Brechbühler, E., Ceccarelli, V., Krishnan, S., Vacik, H., & Muys, B. (2022). Diversity for Restoration (D4R): Guiding the selection of tree species and seed sources for climate-resilient restoration of tropical forest landscapes. *Journal of Applied Ecology*, *59*(3), 664–679. DOI: 10.1111/1365-2664.14079 PMID: 35524517

Fujihara, M., & Inoue-Murayama, M. (2024). The wildlife biobanking of germ cells for in situ and ex situ conservation in Japan. *Theriogenology Wild*, *4*, 100086. DOI: 10.1016/j.therwi.2024.100086

Furey, B., Slingerland, K., Bauter, M. R., Dunn, C., Goodman, R. E., & Koo, S. (2022). Safety evaluation of Fy Protein™(Nutritional Fungi Protein), a macroingredient for human consumption. *Food and Chemical Toxicology*, *166*, 113005. DOI: 10.1016/j.fct.2022.113005 PMID: 35636642

Future Market Insights. (2024) Mycoprotein Market Trends – Alternative Protein & Industry Growth 2024-2034. https://www.futuremarketinsights.com/reports/mycoprotein-market. Accessed on 19th April, 2025.

Future Market Insights. (2025). *Miso Market Analysis by Product Type, Raw Material, End-User, Distribution Channel and Other Types Through 2035*. https://www.futuremarketinsights.com/reports/miso-market

Gain, A. K., Giupponi, C., Renaud, F. G., & Vafeidis, A. T. (2020). Sustainability of complex social-ecological systems: Methods, tools, and approaches. *Regional Environmental Change*, *20*(3), 102. DOI: 10.1007/s10113-020-01692-9

Galappaththi, E. K., Susarla, V. B., Loutet, S. J. T., Ichien, S. T., Hyman, A. A., & Ford, J. D. (2022). Climate change adaptation in fisheries. *Fish and Fisheries*, *23*(1), 4–21. DOI: 10.1111/faf.12595

Gamaralalage, D., Rodgers, S., Gill, A., Meredith, W., Bott, T., West, H., Alce, J., Snape, C., & McKechnie, J. (2025). Biowaste to biochar: A techno-economic and life cycle assessment of biochar production from food-waste digestate and its agricultural applications. *Biochar*, *7*(1), 50. DOI: 10.1007/s42773-025-00456-0 PMID: 40078517

Gamarra-Castillo, O., Echeverry-Montaña, N., Marbello-Santrich, A., Hernández-Carrión, M., & Restrepo, S. (2022). Meat Substitute Development from Fungal Protein (Aspergillus oryzae). *Foods*, *11*(19), 2940. DOI: 10.3390/foods11192940 PMID: 36230014

García-Cornejo, B., Pérez-Méndez, J. A., Wall, A., & Castrillo-Cachón, D. (2025). The effect of management accounting practices and ICT on the efficiency of organic farms. *Journal of Rural Studies*, *114*, 103554. DOI: 10.1016/j.jrurstud.2024.103554

Gargi, B., Painuli, S., Semwal, P., Singh, D. P., Sharma, R., Rauf, A., Khalil, A. A., Khalid, A., Hemeg, H. A., & Wilairatana, P. (2024). Global status, recent trends, and knowledge mapping of olive oil research and cardiovascular disease: 50 years of investigations through bibliometric analysis. *Food Science & Nutrition*, *12*(3), 1465–1478. DOI: 10.1002/fsn3.3885 PMID: 38455210

Gargi, B., Semwal, P., Jaiswal, D. K., Painuli, S., Trivedi, V. L., Tripathi, V., & Rai, N. (2025). Impact of circadian rhythm and seasonal variability on the essential oil of *Allium stracheyi* baker from Uttarakhand, Himalaya. *Food Chemistry*, *X*, 102720. DOI: 10.1016/j.fochx.2025.102720 PMID: 40686884

Gargi, B., Singh, P., Painuli, S., Rai, N., Semwal, P., Cruz-Martins, N., & Sharma, R. (2024a). Literature-based screening and bibliometric analysis of the chemical composition, antioxidant and antimicrobial potential of essential oils isolated from Allium genus: 23 years of investigation. *Pharmacological Research. Modern Chinese Medicine*, *10*, 100354. DOI: 10.1016/j.prmcm.2023.100354

George, A. S. (2023). Future Economic Implications of Artificial Intelligence. *Partners Universal International Research Journal*, *2*(3), 20–39.

Ghosh, A., Kumar, R. V., Manna, M. C., Singh, A. K., Parihar, C. M., Kumar, S., Roy, A. K., & Koli, P. (2021). Eco-restoration of degraded lands through trees and grasses improves soil carbon sequestration and biological activity in tropical climates. *Ecological Engineering*, *162*, 106176. DOI: 10.1016/j.ecoleng.2021.106176

Giri, K., Mishra, G., Rawat, M., Pandey, S., Bhattacharyya, R., Bora, N., & Rai, J. P. N. (2020). Traditional farming systems and agro-biodiversity in eastern Himalayan region of India. In *Rhizosphere biology* (pp. 71–89). DOI: 10.1007/978-981-15-1902-4_5

Global Bioeconomy Summit. (2020). *Expanding the Sustainable Bioeconomy-Vision and Way Forward* (Communiqu'e of the Global Bioeconomy Summit (ed.)). Communiqu'e of the Global Bioeconomy Summit.

Global Info Research. (2024). "Global Soy Sauce Supply, Demand and Key Producers, 2024-2030 - GlobalInfoResearch." *Global Info Research 2023*. https://www.globalinforesearch.com/reports/2220260/soy-sauce

Glockow, T., Kaster, A., Rabe, K. S., & Niemeyer, C. M. (2024). Sustainable agriculture: Leveraging microorganisms for a circular economy. *Applied Microbiology and Biotechnology*, *108*(1), 452. DOI: 10.1007/s00253-024-13294-0 PMID: 39212740

Gnaim, R., Dyer, P. S., & Ledesma-Amaro, R. (2025). Fusarium-based mycoprotein: Advancements in the production of sustainable meat substitutes. *Trends in Food Science & Technology, 104981*, 104981. Advance online publication. DOI: 10.1016/j.tifs.2025.104981

Goncalves, N., Andrade, D., Batista, A., Cullen, L.Jr, Souza, A., Gomes, H., & Uezu, A. (2021). Potential economic impact of carbon sequestration in coffee agroforestry systems. *Agroforestry Systems*, *95*(2), 419–430. DOI: 10.1007/s10457-020-00569-4

Government of India. (2024). *The Constitution of India.* Government of India. Art. 21. https://cdnbbsr.s3waas.gov.in/s380537a945c7aaa788ccfcdf1b99b5d8f/uploads/2024/07/20240716890312078.pdf

Greiber, T., Moreno, S. P., Ahren, M., Carrasco, J. N., Kamau, E. C., Medaglia, J. C., Oliva, M. J., & Perron-Welch, F. (2012). *An explanatory guide to the Nagoya Protocol on access and benefit-sharing*. International Union for Conservation of Nature.

Guenaoui, C., Mang, S., Figliuolo, G., & Neffati, M. (2013). Diversity in *Allium ampeloprasum*: From small and wild to large and cultivated. *Genetic Resources and Crop Evolution*, *60*(1), 97–114. DOI: 10.1007/s10722-012-9819-5

Guo, C., Li, M., & Chen, H. (2025). Study on the Influencing Factors of Green Agricultural Subsidies on Straw Resource Utilization Technology Adopted by Farmers in Heilongjiang Province, China. *Agriculture*, *15*(1), 93. DOI: 10.3390/agriculture15010093

Gupta, R., Gabrielsen, B., & Ferguson, S. M. (2005). Nature's medicines: Traditional knowledge and intellectual property management. Case studies from the National Institutes of Health, USA. *Current Drug Discovery Technologies*, *2*(4), 16–17. DOI: 10.2174/1570163057775202937 PMID: 16475917

Gupta, V. (2024). An Empirical Evaluation of a Generative Artificial Intelligence Technology Adoption Model from Entrepreneurs' Perspectives. *Systems*, *12*(3), 103. DOI: 10.3390/systems12030103

Gurung, J., Chettri, N., Sharma, E., Ning, W., Chaudhary, R. P., Badola, H. K., Wangchuk, S., Uprety, Y., Gaira, K. S., Bidha, N., Phuntsho, K., Uddin, K., & Shah, G. M. (2019). Evolution of a transboundary landscape approach in the Hindu Kush Himalaya: Key learnings from the Kangchenjunga Landscape. *Global Ecology and Conservation*, *17*, e00599. DOI: 10.1016/j.gecco.2019.e00599

Halim-Lim, S. A., Jamaludin, A. A., Islam, A. T., Weerabahu, S., & Priyono, A. (2025). Unlocking potential for a circular bioeconomy transition through digital innovation, lean manufacturing and green practices: A review. *Management of Environmental Quality, 36*(1), 130–154. DOI: 10.1108/MEQ-11-2023-0386

Hamajima, H., Matsunaga, H., Fujikawa, A., Sato, T., Mitsutake, S., Yanagita, T., & Kitagaki, H. (2016). Japanese traditional dietary fungus koji Aspergillus oryzae functions as a prebiotic for Blautia coccoides through glycosylceramide: Japanese dietary fungus koji is a new prebiotic. *SpringerPlus, 5*(1), 1–10. DOI: 10.1186/s40064-016-2950-6 PMID: 27563516

Hamilton, C. (2006). BIODIVERSITY, BIOPIRACY AND BENEFITS: WHAT ALLEGATIONS OF BIOPIRACY TELL US ABOUT INTELLECTUAL PROPERTY. *Developing World Bioethics, 6*(3), 158–173. DOI: 10.1111/j.1471-8847.2006.00168.x PMID: 17038007

Hanafiah, M. H., Asyraff, M. A., Ismail, M. N. I., & Sjukriana, J. (2024). Understanding the key drivers in using mobile payment (M-Payment) among Generation Z travellers. *Young Consumers, 25*(5), 645–664. DOI: 10.1108/YC-08-2023-1835

Hancı, F., & Paşazade, E. (2025). A comparison of efficiency parameters of SRAP and ISSR markers in revealing variation in *Allium* Germplasm. *Horticulturae, 11*(3), 294. DOI: 10.3390/horticulturae11030294

Hanisch, A. L. (2024). Improving an agroforestry system with livestock in southern Brazil. *Tropical Forest Issues, 62*(62), 2–5. DOI: 10.55515/JCWC8581

Hanley, B. (2025). Naturally-Occurring Sulphur Compounds. In *Natural and Unnatural Product Chemistry: From Molecules to Systems* (pp. 77–103). Springer., DOI: 10.1007/978-3-031-81698-7_3

Hartono, R., Yoeseph, N. M., Purnomo, F. A., Safi'ie, M. A., & Bawono, S. A. T. (2024). Portable internet of things-based soil nutrients monitoring for precision and efficient smart farming. *Bulletin of Electrical Engineering and Informatics, 13*(5), 3326–3333. DOI: 10.11591/eei.v13i5.7928

Havey, M. J. (2000). Diversity among male-sterility-inducing and male-fertile cytoplasms of onion. *Theoretical and Applied Genetics, 101*(5-6), 778–782. DOI: 10.1007/s001220051543

Hawkins, J. W., Gallagher, E. J., van der Haar, S., Sevor, M. K. E., Weng, X., Rufino, M. C., & Schoneveld, G. C. (2024). Low-emissions and profitable cocoa through moderate-shade agroforestry: Insights from Ghana. *Agriculture, Ecosystems & Environment, 367*(January), 108961. DOI: 10.1016/j.agee.2024.108961

Hazarika, A., & Rahmati, M. (2023). Towards an evolved immersive experience: Exploring 5G-and beyond-enabled ultra-low-latency communications for augmented and virtual reality. *Sensors (Basel), 23*(7), 3682. DOI: 10.3390/s23073682 PMID: 37050742

Hebishima, H., & Inage, S. I. (2025). Development of a new plant factory concept with sustainable water and energy supply. *Energy Nexus, 17*, 100341. DOI: 10.1016/j.nexus.2024.100341

Hellwig, C., Bolton, K., Häggblom-Kronlöf, G., & Rousta, K. (2022). Aspects affecting food choice in daily life as well as drivers and barriers to engagement with fungi-based food—A qualitative perspective. *Sustainability (Basel), 15*(1), 317. DOI: 10.3390/su15010317

Helmayuni, & Mardianto. (2021). Economic valuation of "Parak" traditional agroforestry west sumatera (case study in Nagari Paninggahan, Solok regency. *INFLUENCE International Journal of Science Review, 3*(3), 34–42.

Hermoso, V., Carvalho, S. B., Giakoumi, S., Goldsborough, D., Katsanevakis, S., Leontiou, S., Markantonatou, V., Rumes, B., Vogiatzakis, I. N., & Yates, K. L. (2022). The EU Biodiversity Strategy for 2030: Opportunities and challenges on the path towards biodiversity recovery. *Environmental Science & Policy, 127*, 263–271. DOI: 10.1016/j.envsci.2021.10.028

Hidalgo, P. J., Hernández, H., Sánchez-Almendro, A. J., López-Tirado, J., Vessella, F., & Porras, R. (2021). Fragmentation and connectivity of island forests in agricultural mediterranean environments: A comparative study between the guadalquivir valley (Spain) and the apulia region (Italy). *Forests, 12*(9), 1201. Advance online publication. DOI: 10.3390/f12091201

Hilgendorf, K., Wang, Y., Miller, M. J., & Jin, Y. (2024). Precision fermentation for improving the quality, flavor, safety, and sustainability of foods. *Current Opinion in Biotechnology, 86*, 103084. DOI: 10.1016/j.copbio.2024.103084 PMID: 38394936

Hilmi, Y. S., Tóth, J., Gabnai, Z., Király, G., & Temesi, Á. (2024). Farmers' resilience to climate change through the circular economy and sustainable agriculture: A review from developed and developing countries. *Renewable Agriculture and Food Systems, 39*, e15. DOI: 10.1017/S1742170524000097

Hoban, S., Archer, F. I., Bertola, L. D., Bragg, J. G., Breed, M. F., Bruford, M. W., Coleman, M. A., Ekblom, R., Funk, W. C., Grueber, C. E., Hand, B. K., Jaffé, R., Jensen, E., Johnson, J. S., Kershaw, F., Liggins, L., MacDonald, A. J., Mergeay, J., Miller, J. M., & Hunter, M. E. (2022). Global genetic diversity status and trends: Towards a suite of Essential Biodiversity Variables (<scp>EBVs</scp>) for genetic composition. *Biological Reviews of the Cambridge Philosophical Society, 97*(4), 1511–1538. DOI: 10.1111/brv.12852 PMID: 35415952

Hoeben, A. D., Stern, T., & Lloret, F. (2023). A Review of Potential Innovation Pathways to Enhance Resilience in Wood-Based Value Chains. *Current Forestry Reports, 9*(5), 301–318. DOI: 10.1007/s40725-023-00191-4

Holyoak, M. (2008). Connectance and connectivity. *Encyclopedia of Ecology, Five-Volume Set, 1–5*, 737–743. DOI: 10.1016/B978-008045405-4.00095-1

Hsu, T. H., Shiao, L. H., Hsieh, C., & Chang, D. M. (2002). A comparison of the chemical composition and bioactive ingredients of the Chinese medicinal mushroom DongChongXiaCao, its counterfeit and mimic, and fermented mycelium of Cordyceps sinensis. *Food Chemistry, 78*(4), 463–469. DOI: 10.1016/S0308-8146(02)00158-9

Hussain, S., Sharma, S., Bhatti, R. C., & Singh, A. N. (2024). Sustainability in the Indian Himalayan Region: Understandings the Ecosystem Services, Climate Change Impacts, Land Use Shifts and Their Threats. In *The Himalayas in the Anthropocene: Environment and Development* (pp. 33–57).

Hutto, C. J., & Gilbert, E. (2014). *VADER: A parsimonious rule-based model for sentiment analysis of social media text*. In *Proceedings of the International AAAI Conference on Weblogs and Social Media*. DOI: 10.1609/icwsm.v8i1.14550

Ichikawa, E., Hosokawa, N., Hata, Y., Abe, Y., Suginami, K., & Imayasu, S. (1991). Breeding of a sake yeast with improved ethyl caproate productivity. *Agricultural and Biological Chemistry, 55*(8), 2153–2154. DOI: 10.1271/bbb1961.55.2153

Ijeomah, C., Amuda, O., Babatunde, B., & Abutu, P. (2020). Evaluation of genetic diversity of spring onions (*Allium fistulosum*) based on DNA markers. *Journal of Experimental Agriculture International, 42*, 23–33. DOI: 10.9734/jeai/2020/v42i330481

India code: Biological diversity act, 2002. (2002). https://www.indiacode.nic.in/handle/123456789/2046?sam_handle=123456789/1362

Indratna, A. D., Putri, A. P. I. K. S., Azzahra, E. O., Oktavia, M. S., Widodo, T., Agustina, A., & Pertiwi, Y. A. B. (2023). Potential economic value of non-wood forest products in agroforestry system at Gempolan Village, Kerjo District, Karanganyar Regency: An economic perspective. *IOP Conference Series. Earth and Environmental Science, 1220*(1), 012014. Advance online publication. DOI: 10.1088/1755-1315/1220/1/012014

Ingle, A. P., Philippini, R. R., Martiniano, S., Marcelino, P. R. F., Gupta, I., Prasad, S., & Silva, S. S. (2020). Bioresources and their significance: Prospects and obstacles. In Kataki, R., Pandey, A., Khanal, S. K., & Pant, D. (Eds.), *Current Developments in Biotechnology and Bioengineering* (pp. 3–4). Elsevier., DOI: 10.1016/B978-0-444-64309-4.00001-5

Inoue, K., Gotou, T., Kitajima, H., Mizuno, S., Nakazawa, T., & Yamamoto, N. (2009). Release of antihypertensive peptides in miso paste during its fermentation, by the addition of casein. *Journal of Bioscience and Bioengineering, 108*(2), 111–115. DOI: 10.1016/j.jbiosc.2009.03.007 PMID: 19619856

Intergovernmental Panel on Climate Change. (2022). *Climate Change 2022: Impacts, Adaptation, and Vulnerability. Contribution of Working Group II to the Sixth Assessment Report of the Intergovernmental Panel on Climate Change.* Cambridge University Press., DOI: 10.1017/9781009325844

Intergovernmental Science-Policy Platform on Biodiversity and Ecosystem Services. (2019). *Global assessment report on biodiversity and ecosystem services.* Retrieved from https://ipbes.net/global-assessment

Ipek, M., Ipek, A., & Simon, P. W. (2003). Comparison of AFLPs, RAPD markers, and isozymes for diversity assessment of garlic and detection of putative duplicates in germplasm collections. *Journal of the American Society for Horticultural Science, 128*(2), 246–252. DOI: 10.21273/JASHS.128.2.246

Islam, K., Sabau, G., Dawson, J., Cheema, M., Daraio, J., & Galagedara, L. (2025). Evaluating agricultural sustainability in Newfoundland, Canada: Insights from a data envelopment analysis (DEA) approach. SSRN Electronic Journal. DOI: 10.2139/ssrn.5127875

Islam, M. (2024). Climate Change Adaptation through a Social Innovation Lens: A Framework Analysis of Ladakh, India's Community-based Approach for Sustainable Agriculture Practice. In *Conference Paper.*

Islam, S., & Manaloor, V. (2021). Millets for food and nutrition security in India: Determinants and policy implications. *Journal of Nutrition and Food Security, 6*(2), 185–198. DOI: 10.18502/jnfs.v6i2.6074

Ito, K., & Matsuyama, A. (2021). Koji molds for Japanese soy sauce brewing: Characteristics and key enzymes. *Journal of Fungi (Basel, Switzerland)*, *7*(8), 658. DOI: 10.3390/jof7080658 PMID: 34436196

Jambut, A., Hassan Zaki, P., & Hasmadi Ismail, M. (2022). Profitability of agroforestry in tumpangsari-gunung forest reserve arjuna-lalijiwo, East Java, Indonesia. *Вестник Атырауского Университета Имени Халела Досмухамедова*, *64*(1), 63–73. DOI: 10.47649/vau.2022.v64.i1.07

Jaramillo, R., Marmol-Guijarro, A., Doskocil, L. G., Chimbolema, S., & Suárez, E. (2023). Using 3D models to quantify aboveground biomass in the cushion-forming species Plantago rigida in an Andean páramo peatland. *Mires and Peat*, *29*, 29. DOI: 10.19189/MAP.2023.RSC.SC.1829791

Jiang, S., Li, H., Zhang, L., Mu, W., Zhang, Y., Chen, T., Wu, J., Tang, H., Zheng, S., Liu, Y., Wu, Y., Luo, X., Xie, Y., & Ren, J. (2025). Generic Diagramming Platform (GDP): A comprehensive database of high-quality biomedical graphics. *Nucleic Acids Research*, *53*(D1), D1670–D1676. DOI: 10.1093/nar/gkae973 PMID: 39470721

Jishnu, L. (2012, October 15). Curious case of seaweed. *Down to Earth*. https://www.downtoearth.org.in/coverage/curious-case-of-seaweed-39207

Jitendra, A. (2024). Linking ecological restoration and biodiversity conservation with bioeconomy. *Biodiversity and Bioeconomy: Status Quo, Challenges, and Opportunities*, 61–79. DOI: 10.1016/B978-0-323-95482-2.00004-3

Jose, S. (2009). Agroforestry for ecosystem services and environmental benefits: An overview. *Agroforestry Systems*, *76*(1), 1–10. https://www.scopus.com/inward/record.uri?eid=2-s2.0-63649140240&doi=10.1007%2Fs10457-009-9229-7&partnerID=40&md5=c297b224f129b06436d775ac90490152. DOI: 10.1007/s10457-009-9229-7

Jose, S., & Udawatta, R. P. (2021). Agroforestry for ecosystem services: An introduction. In *Agroforestry and Ecosystem Services* (pp. 1–17). Springer International Publishing., DOI: 10.1007/978-3-030-80060-4_1

Joshi, P. (2023). Silvopasture: A sustainable agro-system practice for rural farmers. *Agriculture Extension in Developing Countries*, *1*(May), 27–30. DOI: 10.26480/aedc.02.2023.46.48

Joshi, R., Sharma, B., Singh, H., Dhakal, N., Ayer, S., & Maraseni, T. (2024). Poplar Plantation as an Agroforestry Approach: Economic Benefits and Its Role in Carbon Sequestration in North India. *Journal of Resources and Ecology*, *15*(4), 880–888. DOI: 10.5814/j.issn.1674-764x.2024.04.009

Kabato, W., Getnet, G. T., Sinore, T., Nemeth, A., & Molnár, Z. (2025). Towards climate-smart agriculture: Strategies for sustainable agricultural production, food security, and greenhouse gas reduction. *Agronomy (Basel), 15*(3), 565. DOI: 10.3390/agronomy15030565

Kacprzak, A., & Hensel, P. (2023). Exploring online customer experience: A systematic literature review and research agenda. *International Journal of Consumer Studies, 47*(6), 2583–2608. DOI: 10.1111/ijcs.12966

Kadhila, T., Amesho, K. T. T., Shangdiar, S., Shihomeka, S. P., Shopati, A. K., & Edoun, E. I. (2024). Sustainable Bioeconomy for Next-Generation Industrial Ecology. In *Bioeconomy for Sustainability* (pp. 293–317). Springer Nature Singapore., DOI: 10.1007/978-981-97-1837-5_13

Kamal, M., & Himel, A. S. (2023). Redefining Modern Marketing: An Analysis of AI and NLP's Influence on Consumer Engagement, Strategy, and Beyond. *Eigenpub Review of Science and Technology, 7*(1), 203–223.

Karmwar, M., & Smita. (2022). Tribal development in India: A case of the Siddi community. *Mekal Insight, 6*(1–2), 45–56. https://www.igntu.ac.in/MekalInsight/Vol6-Issue1-2-year-2022/7.pdf

Karolyi, G. A., & Tobin-de la Puente, J. (2023). Biodiversity finance: A call for research into financing nature. *Financial Management, 52*(2), 231–251. DOI: 10.1111/fima.12417

Kattel, G. R. (2022). Climate warming in the Himalayas threatens biodiversity, ecosystem functioning and ecosystem services in the 21st century: Is there a better solution? *Biodiversity and Conservation, 31*(8), 2017–2044. DOI: 10.1007/s10531-022-02417-6

Kaur, A., & Monga, R. (2024). Recent trends in agroforestry research and development. In *Futuristic Trends in Agriculture Engineering & Food Sciences: Vol. 3 Book 14* (pp. 183–188). Iterative International Publisher, Selfypage Developers Pvt Ltd. DOI: 10.58532/V3BCAG14P5CH2

Kaur, G., & Sharma, S. (2025). Sustainable Methods for Solid Waste Management in High-Altitude Tourist Regions. In *Solid Waste Management and Disposal Practices in Rural Tourism* (pp. 395–416). IGI Global.

Kaur, J., Murray, P., & Collins, C. (2024). Bioprospecting of novel and biologically active compounds and enzymes from a new fungal biobank for industrial biotechnology. *Biotechnology for the Environment, 1*(1), 14. DOI: 10.1186/s44314-024-00015-0

Kaur, S., Kaur, D., Tiwana, A. S., & Gupta, S. (2022). Climate change-related governance and policies in Indian Himalayas. In *Climate change: Impacts, responses and sustainability in the Indian Himalaya* (pp. 309–330). Springer International Publishing. DOI: 10.1007/978-3-030-92782-0_16

Kaushal, V., & Yadav, R. (2023). Learning successful implementation of Chatbots in businesses from B2B customer experience perspective. *Concurrency and Computation, 35*(1), e7450. DOI: 10.1002/cpe.7450

Kay, S., Graves, A., Palma, J. H. N., Moreno, G., Roces-Díaz, J. V., Aviron, S., Chouvardas, D., Crous-Duran, J., Ferreiro-Domínguez, N., García de Jalón, S., Măcicăşan, V., Mosquera-Losada, M. R., Pantera, A., Santiago-Freijanes, J. J., Szerencsits, E., Torralba, M., Burgess, P. J., & Herzog, F. (2019). Agroforestry is paying off – Economic evaluation of ecosystem services in European landscapes with and without agroforestry systems. *Ecosystem Services, 36*, 100896. https://www.scopus.com/inward/record.uri?eid=2-s2.0-85060894709&doi=10.1016%2Fj.ecoser.2019.100896&partnerID=40&md5=d17aa01b3e132884047dedd94fd84244. DOI: 10.1016/j.ecoser.2019.100896

Kelleher, J. D., Mac Namee, B., & D'Arcy, A. (2020). *Fundamentals of machine learning for predictive data analytics: Algorithms, worked examples, and case studies*. MIT Press.

Kennedy, S., Fuchs, M., van Ingen, W., & Schoenmaker, D. (2023). A resilience approach to corporate biodiversity impact measurement. *Business Strategy and the Environment, 32*(5), 2567–2582. DOI: 10.1002/bse.3140

Khade, Y. P., Sinhasane, S. R., Mainkar, P., Rai, K. M., Salunkhe, S., Singh, P. R., Jadhav, M. S., Bhandari, H. R., More, S. J., Gupta, A. J., Mahajan, V., & Singh, M. (2025). Exploring sequence-related amplified polymorphism (SRAP) markers for assessing genetic diversity in onion (*Allium cepa* L.) genotypes and their wild relatives. *Genetic Resources and Crop Evolution, 72*(1), 399–415. DOI: 10.1007/s10722-024-01978-5

Khan, R. W. A., & Shaheen, H. (2022). Biomass carbon stock estimation in lesser Himalayan subtropical broadleaf forests of Kashmir. *DOAJ (DOAJ: Directory of Open Access Journals)*.

Khandagale, K., Krishna, R., Roylawar, P., Ade, A., Benke, A., Shinde, B., Singh, M., Gawande, S., & Rai, A. (2020). Omics approaches in *Allium* research: Progress and way ahead. *PeerJ, 8*, e9824. DOI: 10.7717/peerj.9824 PMID: 32974094

Khan, N., Jhariya, M. K., Raj, A., Banerjee, A., Meena, R. S., Bargali, S. S., Yadav, S. K., & Kumawat, A. (2021). Agroforestry and its services for soil management and sustainability. In *Sustainable Intensification for Agroecosystem Services and Management* (pp. 353–377). Springer Singapore., DOI: 10.1007/978-981-16-3207-5_11

Khanna, M., Zilberman, D., Hochman, G., & Basso, B. (2024). An economic perspective of the circular bioeconomy in the food and agricultural sector. *Communications Earth & Environment, 5*(1), 507. DOI: 10.1038/s43247-024-01663-6

Khan, S. A., Jameel, M., Kanwal, S., & Shahid, S. (2017). Medicinal importance of *Allium* species: A current review. *International Journal of Pharmaceutical Sciences and Research, 2*(1), 29–39.

Khar, A., Lawande, K. E., & Negi, K. S. (2011). Microsatellite marker based analysis of genetic diversity in short day tropical Indian onion and cross amplification in related *Allium* spp. *Genetic Resources and Crop Evolution, 58*(5), 741–752. DOI: 10.1007/s10722-010-9616-y

Khatami, F., Cagno, E., & Khatami, R. (2024). Circular Economy in the Agri-Food System at the Country Level—Evidence from European Countries. *Sustainability (Basel), 16*(21), 9497. DOI: 10.3390/su16219497

King, K. (2022). *AI Strategy for Sales and Marketing: Connecting Marketing, Sales and Customer Experience*. Kogan Page Publishers.

Kiprotich, K., Muema, E., Wekesa, C., Ndombi, T., Muoma, J., Omayio, D., Ochieno, D., Motsi, H., Mncedi, S., & Tarus, J. (2025). Unveiling the roles, mechanisms and prospects of soil microbial communities in sustainable agriculture. *Discover Soil, 2*(1).

Klaus, V. H., & Kiehl, K. (2021). A conceptual framework for urban ecological restoration and rehabilitation. *Basic and Applied Ecology, 52*, 82–94. DOI: 10.1016/j.baae.2021.02.010

Knez, S., Štrbac, S., & Podbregar, I. (2022). Climate change in the Western Balkans and EU Green Deal: Status, mitigation and challenges. *Energy, Sustainability and Society, 12*(1), 1. DOI: 10.1186/s13705-021-00328-y

Kohl, K., Hopkins, C., Barth, M., Michelsen, G., Dlouhá, J., Razak, D. A., Abidin Bin Sanusi, Z., & Toman, I. (2022). A whole-institution approach towards sustainability: A crucial aspect of higher education's individual and collective engagement with the SDGs and beyond. *International Journal of Sustainability in Higher Education, 23*(2), 218–236. DOI: 10.1108/IJSHE-10-2020-0398

Komyshev, D., Kapral, A., Drebot, O., Sakharnatska, L., & Karpyk, J. (2024). Advancing waste reduction and resource conservation through circular economy practices: A rational review. *Grassroots Journal of Natural Resources*, *7*(3), 1–18. DOI: 10.33002/nr2581.6853.0703ukr09

Krause, H. M., Mäder, P., Fliessbach, A., Jarosch, K. A., Oberson, A., & Mayer, J. (2024). Organic cropping systems balance environmental impacts and agricultural production. *Scientific Reports*, *14*(1), 1–15. DOI: 10.1038/s41598-024-76776-1 PMID: 39462042

Kremmydas, D., Beber, C., Baldoni, E., Ciaian, P., Fellmann, T., Gocht, A., Hristov, J., Pignotti, D., Vicario, D. R., Stepanyan, D., & Tillie, P. (2025). The EU target for organic farming: Potential economic and environmental impacts of two alternative pathways. *Applied Economic Perspectives and Policy*, *47*(2), 602–623. DOI: 10.1002/aepp.13470

Kriesch, L., & Losacker, S. (2024). Bioeconomy firms and where to find them. *Region*, *11*(1), 55–78. DOI: 10.18335/region.v11i1.523

Kulkarni, S. D. (2025). Sustainable precision agriculture: Integrating artificial intelligence and IoT for optimized green farming practices. *Journal of Informatics Education and Research*, *5*(1).

Kumar, A., & Kumar, S. (2025). Impact of Climate Change on Himalayan Ecosystem and Traditional Crops. *Climate Change Impact on Himalayan Biodiversity*, 393-413.

Kumar, J., Pradhan, M., & Singh, N. (2017). Sustainable Organic Farming in Sikkim: An Inclusive perspective. In *Lecture notes in electrical engineering* (pp. 367–378).

Kumaresan, M., Kovvasu, K., Kathari, L., & Vijai, A. A. (2025). Vertical farming technologies for flower production: Current trends, challenges, and future prospects. [Laurin Publishers.]. *Current Trends in Agronomy & Agricultural Research*, *2*(1), 1–10.

Kumar, M. (2023). Forestry policies and practices to promote climate change adaptation in the Indian Western Himalayan states. In *Climate Change Adaptation, Risk Management and Sustainable Practices in the Himalaya* (pp. 65–87). Springer International Publishing. DOI: 10.1007/978-3-031-24659-3_4

Kumar, M., Rakesh Sharma, V., Kumar, V., Sirohi, U., Chaudhary, V., Sharma, S., Saripalli, G., Naresh, R., Yadav, H. K., & Sharma, S. (2019). Genetic diversity and population structure analysis of Indian garlic (*Allium sativum* L.) collection using SSR markers. *Physiology and Molecular Biology of Plants : an International Journal of Functional Plant Biology*, *25*(2), 377–386. DOI: 10.1007/s12298-018-0628-y PMID: 30956421

Kumar, P., Sharma, P. K., Kumar, P., Sharma, M., & Butail, N. P. (2021). Agricultural sustainability in Indian Himalayan region: Constraints and potentials. *Indian Journal of Ecology*, *48*(3), 649–661.

Kumar, R., Singh, A., Datta, A., Yadav, R. P., Dinesh, D., & Verma, K. (2022). Carbon sequestration in degraded lands: current prospects, practices, and future strategies. In *Plans and Policies for Soil Organic Carbon Management in Agriculture* (pp. 221–255). Springer Nature Singapore. DOI: 10.1007/978-981-19-6179-3_9

Kunkalikar, S. R., Poojari, S., Arun, B. M., Rajagopalan, P. A., Chen, T. C., Yeh, S. D., Naidu, R. A., Zehr, U. B., & Ravi, K. S. (2011). Importance and genetic diversity of vegetable-infecting tospoviruses in India. *Phytopathology*, *101*(3), 367–376. DOI: 10.1094/PHYTO-02-10-0046 PMID: 21299415

Kusumoto, K. I., Yamagata, Y., Tazawa, R., Kitagawa, M., Kato, T., Isobe, K., & Kashiwagi, Y. (2021). Japanese traditional Miso and Koji making. *Journal of Fungi (Basel, Switzerland)*, *7*(7), 579. DOI: 10.3390/jof7070579 PMID: 34356958

Kuyah, S., Öborn, I., & Jonsson, M. (2017). Regulating ecosystem services delivered in agroforestry systems. In *Agroforestry: Anecdotal to Modern Science* (pp. 797–815). Springer Singapore. DOI: 10.1007/978-981-10-7650-3_33

Lakshmanan, P. K. (2015). An inquiry into the Biological Diversity Act, 2002 in the light of the Nagoya Protocol on access and benefit Sharing. *SSRN*, *6*, 35–55. DOI: 10.2139/ssrn.3048709

Lanzotti, V. (2006). The analysis of onion and garlic. *Journal of Chromatography. A*, *1112*(1-2), 3–22. DOI: 10.1016/j.chroma.2005.12.016 PMID: 16388813

Lasmar, D. J., Ferreira, I. S., & Silva, M. L. A. (2024). Bioeconomia: perspectiva de desenvolvimento de cadeias produtivas da biodiversidade do amazonas. In *A multidisciplinaridade em foco: ensino, pesquisa e extensão* (Vol. 3, pp. 42–66). DOI: 10.37885/241017841

Laudari, H. K., Sapkota, L. M., Maraseni, T., Subedi, P., Pariyar, S., Kaini, T. R., Lopchan, S. B., Weston, C., & Volkova, L. (2023). Community forestry in a changing context: A perspective from Nepal's mid-hill. *Land Use Policy*, *138*, 107018. DOI: 10.1016/j.landusepol.2023.107018

Lavuri, R., & Akram, U. (2023). Role of virtual reality authentic experience on affective responses: Moderating role virtual reality attachment. *Journal of Ecotourism*, •••, 1–19.

Lee, B., Rhee, H., Kim, S., Lee, J.-W., Koo, S., Lee, S.-J., Alounsavath, P., & Kim, Y.-S. (2021). Assessing sustainable bamboo-based income generation using a value chain approach: Case study of nongboua village in Lao PDR. *Forests, 12*(2), 153. Advance online publication. DOI: 10.3390/f12020153

Lehmann, L. M., Smith, J., Westaway, S., Pisanelli, A., Russo, G., Borek, R., Sandor, M., Gliga, A., Smith, L., & Ghaley, B. B. (2020). Productivity and economic evaluation of agroforestry systems for sustainable production of food and non-food products. *Sustainability (Basel), 12*(13), 5429. Advance online publication. DOI: 10.3390/su12135429

Lenze, S. (2022). Bioeconomy as a Normative Concept of Resilience: Challenges and Opportunities. In *Bioeconomy and Sustainability* (pp. 307–321). Springer International Publishing., DOI: 10.1007/978-3-030-87402-5_18

Levis, C., Flores, B. M., Campos-Silva, J. V., Peroni, N., Staal, A., Padgurschi, M. C., & Clement, C. R. (2024). *Contributions of human cultures to biodiversity and ecosystem conservation.* Nature Eco. DOI: 10.1038/s41559-024-02356-1

Li, C., Zhou, J., Du, G., Chen, J., Takahashi, S., & Liu, S. (2020). Developing Aspergillus niger as a cell factory for food enzyme production. *Biotechnology Advances, 44*, 107630. DOI: 10.1016/j.biotechadv.2020.107630 PMID: 32919011

Li, H., Mao, D., Wang, Z., Huang, X., Li, L., & Jia, M. (2022). Invasion of Spartina alterniflora in the coastal zone of mainland China: Control achievements from 2015 to 2020 towards the Sustainable Development Goals. *Journal of Environmental Management, 323*, 116242. DOI: 10.1016/j.jenvman.2022.116242 PMID: 36261984

Lin, B. B. (2007). Agroforestry management as an adaptive strategy against potential climate change impacts. *Mitigation and Adaptation Strategies for Global Change, 12*(5), 835–853. DOI: 10.1007/s11027-007-9101-x

Linder, T. (2024). What next for mycoprotein? *Current Opinion in Food Science, 58*, 101199. DOI: 10.1016/j.cofs.2024.101199

Liu, B. (2012). *Sentiment analysis and opinion mining* (Vol. 5, No. 1). Synthesis Lectures on Human Language Technologies.

Li, X., Lu, H., Yao, J., Zhang, C., Shi, T., & Huang, H. (2025). Recent advances in the application of CRISPR/Cas-based gene editing technology in Filamentous Fungi. *Biotechnology Advances, 81*, 108561. DOI: 10.1016/j.biotechadv.2025.108561 PMID: 40086675

Li, Y., Chen, Z., Wagg, C., Castellano, M. J., Zhang, N., & Ding, W. (2024). Soil organic carbon loss decreases biodiversity but stimulates multitrophic interactions that promote belowground metabolism. *Global Change Biology*, *30*(1), e17101. DOI: 10.1111/gcb.17101 PMID: 38273560

Lori, M., Kundel, D., Mäder, P., Singh, A., Patel, D., Sisodia, B. S., Riar, A., & Krause, H. M. (2024). Organic farming systems improve soil quality and shape microbial communities across a cotton-based crop rotation in an Indian Vertisol. *FEMS Microbiology Ecology*, *100*(11), fiae127. DOI: 10.1093/femsec/fiae127 PMID: 39289000

Luo, D., Nan, H., Zhang, Y., Sher, F., & Wang, C. (2025). Phosphorus recovery from wastewater by Ca-Al layered double hydroxide/biochar as potential agricultural phosphorus for closed-loop phosphorus recycling. *Process Safety and Environmental Protection*, *194*, 1538–1548. DOI: 10.1016/j.psep.2024.12.110

Lyngkhoi, F., Saini, N., Gaikwad, A. B., Thirunavukkarasu, N., Verma, P., Silvar, C., Yadav, S., & Khar, A. (2021). Genetic diversity and population structure in onion (Allium cepa L.) accessions based on morphological and molecular approaches. *Physiology and Molecular Biology of Plants : an International Journal of Functional Plant Biology*, *27*(11), 2517–2532. DOI: 10.1007/s12298-021-01101-3 PMID: 34924708

Maciejczak, M. (2023). Nature based innovations in the development of bioeconomy. *Annals of the Polish Association of Agricultural and Agribusiness Economists*, *25*(2), 97–108. DOI: 10.5604/01.3001.0053.6818

Magry, M. A., Cahill, D., Rookes, J., & Narula, S. A. (2024). Non-timber forest products: Evolution, development and research. *Biodiversity (Nepean)*, *25*(2), 120–141. DOI: 10.1080/14888386.2024.2336216

Majumder, R., Miatur, S., Saha, A., & Hossain, S. (2023). Mycoprotein: production and nutritional aspects: a review. *Sustainable Food Technology*, *2*(1), 81–91. DOI: 10.1039/D3FB00169E

Ma, K. H., Kwag, J. G., Zhao, W., Dixit, A., Lee, G. A., Kim, H. H., Chung, I. M., Kim, N. S., Lee, J. S., Ji, J. J., Kim, T. S., & Park, Y. J. (2009). Isolation and characteristics of eight novel polymorphic microsatellite loci from the genome of garlic (*Allium sativum* L.). *Scientia Horticulturae*, *122*(3), 355–361. DOI: 10.1016/j.scienta.2009.06.010

Malhotra, A., Nandigama, S., & Bhattacharya, K. S. (2021). Food, fields and forage: A socio-ecological account of cultural transitions among the Gaddis of Himachal Pradesh in India. *Heliyon*, *7*(7), e07569. DOI: 10.1016/j.heliyon.2021.e07569 PMID: 34355085

Malik, A., Sharma, P., Pereira, V., & Temouri, Y. (2020). From regional innovation systems to global innovation hubs: Evidence of a Quadruple Helix from an emerging economy. *Journal of Business Research*, *128*, 587–598. DOI: 10.1016/j.jbusres.2020.12.009

Manning, C. D., Raghavan, P., & Schütze, H. (2008). *Introduction to information retrieval.* Cambridge University Press. DOI: 10.1017/CBO9780511809071

Marden, E., Abbott, R. J., Austerlitz, F., Ortiz-Barrientos, D., Baucom, R. S., Bongaerts, P., & Rieseberg, L. H. (2021). Sharing and reporting benefits from biodiversity research. *Molecular Ecology*, *30*(5), 1103–1107. DOI: 10.1111/mec.15702 PMID: 33159357

Marttila, M., Havukainen, J., Uusitalo, V., Linnanen, L., & Mikkilä, M. (2024). Multicriteria analysis of sustainable agri-food waste management for an agroecosystem in Finland. *Frontiers in Sustainable Food Systems*, *8*, 1426890. DOI: 10.3389/fsufs.2024.1426890

Maseko, K. H., Regnier, T., Bartels, P., & Meiring, B. (2025). Mushroom mycelia as sustainable alternative proteins for the production of hybrid cell-cultured meat: A review. *Journal of Food Science*, *90*(2), e70060. Advance online publication. DOI: 10.1111/1750-3841.70060 PMID: 39921300

Mashood, K., Kayani, H. U. R., Malik, A. A., & Tahir, A. (2023). ARTIFICIAL INTELLIGENCE RECENT TRENDS AND APPLICATIONS IN INDUSTRIES. *Pakistan Journal of Science*, *75*(02). Advance online publication. DOI: 10.57041/pjs.v75i02.855

Mason, J. G., Eurich, J. G., Lau, J. D., Battista, W., Free, C. M., Mills, K. E., Tokunaga, K., Zhao, L. Z., Dickey-Collas, M., Valle, M., Pecl, G. T., Cinner, J. E., McClanahan, T. R., Allison, E. H., Friedman, W. R., Silva, C., Yáñez, E., Barbieri, M. Á., & Kleisner, K. M. (2022). Attributes of climate resilience in fisheries: From theory to practice. *Fish and Fisheries*, *23*(3), 522–544. DOI: 10.1111/faf.12630

Massa, A., Placeres, M. D., Axpe, E., Rothschild, L., Sanz, M., Ricatti, M., & Carrero-Carralero, C. (2025). Sensorial and chemical characterization of a new food ingredient of Rhizopus oligosporus biomass produced under solid and submerged state fermentation. *Future Foods : a Dedicated Journal for Sustainability in Food Science*, *100730*. DOI: 10.1016/j.fufo.2025.100730

Mathur, I., & Bhattacharya, P. (2024). Perception of agroforestry practices and factors influencing adoption among shifting cultivators in Tripura, India. *Forests, Trees and Livelihoods, 33*(1), 23–41. DOI: 10.1080/14728028.2023.2286022

Mau, J. L., Lin, H. C., Ma, J. T., & Song, S. F. (2001). Non-volatile taste components of several speciality mushrooms. *Food Chemistry, 73*(4), 461–466. DOI: 10.1016/S0308-8146(00)00330-7

Mehrabi, S., & Giagnocavo, C. (2024). Business models and strategies for the internalization of externalities in agri-food value chains. Agricultural and Food Economics, 12(1), 46.

Mekonnen, M., Worku, T., Yitaferu, B., Cerdà, A., & Keesstra, S. (2023). Economics of agroforestry land use system, Upper Blue Nile Basin, northwest Ethiopia. *Agroforestry Systems, 97*(3), 305–317. DOI: 10.1007/s10457-021-00612-y

Michel, F., Hartmann, C., & Siegrist, M. (2021). Consumers' associations, perceptions and acceptance of meat and plant-based meat alternatives. *Food Quality and Preference, 87*, 104063. DOI: 10.1016/j.foodqual.2020.104063

Ministry of Environment. Forest and Climate Change. (2018). *Implementation of Nagoya Protocol on access and benefit sharing India's experience.* Ministry of Environment, Forest and Climate Change. http://nbaindia.org/uploaded/pdf/Implementation%20of%20Nagoya%20Protocol%20in%20India.pdf

Minz, N. K., & Khattri, N. (2023). *Impact of Microfinancing and Self-Help Groups (SHGs) on women's empowerment.* ResearchGate.

Mishra, P. K., Sena, D. R., Pande, V. C., & Singh, R. K. (2024). Framing land degradation and restoration policy in India following the IPBES pathway. *Indian Journal of Soil Conservation, 52*, 14–25.

Mishra, A. K., Tiwari, A., Maurya, P. K., & Sharma, S. (2024). Enhancing ecosystem dynamics: Organic and regenerative practices in rice-wheat systems and their impact on soil arthropod biodiversity. *Frontiers in Environmental Science, 12*, 1407954. DOI: 10.3389/fenvs.2024.1407954

Mitra, A., & Paul, S. (2025). Empowering Indigenous Communities Through Regenerative Tourism: Cultural Preservation and Socio-Economic Resilience in Northeast India. In Sousa, B., Arantes, L., & Bhartiya, S. (Eds.), *Regenerative Tourism for Social Development* (pp. 281–310). IGI Global Scientific Publishing. DOI: 10.4018/979-8-3373-0235-5.ch012

Mlambo, D., Sebata, A., Chichinye, A., & Mabidi, A. (2024). Agroforestry and biodiversity conservation. *Agroforestry for Carbon and Ecosystem Management*, 63–78. DOI: 10.1016/B978-0-323-95393-1.00008-7

Mohan, D. (2023). Enhancing capacity building initiatives at sub-national level for supporting climate change adaptation. *Climate and Development, 15*(9), 808–815. DOI: 10.1080/17565529.2022.2163845

Molua, E. L. (2005). The economics of tropical agroforestry systems: The case of agroforestry farms in Cameroon. *Forest Policy and Economics*, *7*(2), 199–211. DOI: 10.1016/S1389-9341(03)00032-7

Mondal, S., Angon, P. B., & Roy, A. R. (2023). Ecological Advancements and Developments of Agroforestry. *Turkish Journal of Agriculture - Food Science and Technology*, *11*(12), 2476–2480. DOI: 10.24925/turjaf.v11i12.2476-2480.6172

Mony, C., Landais, Q., Georges, R., Butet, A., Burel, F., Jambon, O., Gouesbet, V., & Ernoult, A. (2022). Effects of connectivity on seed dispersal patterns in hedgerows. *Journal of Vegetation Science*, *33*(1), e13113. Advance online publication. DOI: 10.1111/jvs.13113

Moraru, C., Mincea, M. M., Frandes, M., Timar, B., & Ostafe, V. (2018). A meta-analysis on randomised controlled clinical trials evaluating the effect of the dietary supplement chitosan on weight loss, lipid parameters and blood pressure. *Medicina*, *54*(6), 109. DOI: 10.3390/medicina54060109 PMID: 30545156

Mukherjee, P. K., Das, B., Bhardwaj, P. K., Tampha, S., Singh, H. K., Chanu, L. D., Sharma, N., & Devi, S. I. (2023). Socio-economic sustainability with circular economy—An alternative approach. *The Science of the Total Environment*, *904*, 166630. DOI: 10.1016/j.scitotenv.2023.166630 PMID: 37643712

Muscat, A., De Olde, E. M., Ripoll-Bosch, R., Van Zanten, H. H. E., Metze, T. P., Termeer, C. J. M., Van Ittersum, M. K., & De Boer, I. J. M. (2021). Principles, drivers and opportunities of a circular bioeconomy. *Nature Food*, *2*(8), 561–566. DOI: 10.1038/s43016-021-00340-7 PMID: 37118163

Nagahama, K. (2016). Van (Forest) Panchayat as a system of sustainable common forest resource management in India. *SDRP Journal of Earth Sciences & Environmental Studies, 1*(3).

Nagoya Protocol. (2011). *Nagoya protocol on access to genetic resources and the fair and equitable sharing of benefits arising from their utilization to the convention on biological diversity text and annex*. Accessed on the Cbd Website at www.cbd.int/abs

Nahon, S. M. R., Trindade, F. C., Yoshiura, C. A., Martins, G. C., Costa, I. R. C. D., Costa, P. H. D. O., Herrera, H., Balestrin, D., Godinho, T. D. O., Marchiori, B. M., & Valadares, R. B. D. S. (2024). Impact of Agroforestry Practices on Soil Microbial Diversity and Nutrient Cycling in Atlantic Rainforest Cocoa Systems. *International Journal of Molecular Sciences, 25*(21), 11345. DOI: 10.3390/ijms252111345 PMID: 39518901

Nair, P. K. R. (1998). Directions in tropical agroforestry research: past, present, and future. *Directions in Tropical Agroforestry Research: Adapted from Selected Papers Presented to a Symposium on Tropical Agroforestry Organized in Connection with the Annual Meetings of the American Society of Agronomy,* 5 November 1996, Indianapolis, Indiana, USA, 223–245.

Nair, P. K. R., Mohan Kumar, B., & Nair, V. D. (2009). Agroforestry as a strategy for carbon sequestration. *Journal of Plant Nutrition and Soil Science, 172*(1), 10–23. DOI: 10.1002/jpln.200800030

Nair, P. R. (2011). Agroforestry systems and environmental quality: Introduction. *Journal of Environmental Quality, 40*(3), 784–790. DOI: 10.2134/jeq2011.0076 PMID: 21546663

Naji, K. K., Gunduz, M., Alhenzab, F., Al-Hababi, H., & Al-Qahtani, A. (2024). A Systematic Review of the Digital Transformation of the Building Construction Industry. *IEEE Access : Practical Innovations, Open Solutions, 12*, 31461–31487. DOI: 10.1109/ACCESS.2024.3365934

Nakamoto, T., Nomura, A., & Ikeda, Y. (2021). Making Policy Recommendations in International Organizations. In *Big Data Analysis on Global Community Formation and Isolation* (pp. 467–499). Springer Singapore., DOI: 10.1007/978-981-15-4944-1_14

National Biodiversity Authority. (2013). Revised People's Biodiversity Register Guidelines. http://nbaindia.org/uploaded/pdf/PBR%20Format%202013.pdf

National Biodiversity Authority. (2021). *Annual Report 2020-21.* National Biodiversity Authority. http://nbaindia.org/uploaded/pdf/AR_English_2020-21.pdf

National Biodiversity Authority. (2025, January 23). People's Biodiversity Register. http://nbaindia.org/content/105/30/1/pbr.html

National Biodiversity Authority. (n.d.) *Access and benefit sharing in India a handbook for researchers.* National Biodiversity Authority. http://nbaindia.org/uploaded/pdf/IDB_ABS.pdf

National Biodiversity Authority. (n.d.). *Access and benefits sharing experiences from India.* http://nbaindia.org/uploaded/pdf/ABS_Factsheets_1.pdf

National Law School of India University. (2019). *Handbook on biodiversity laws, access and benefit sharing.* National Law School of India University. https://ceerapub.nls.ac.in/wp-content/uploads/2019/05/Handbook-on-Biodiversity-law-Access-and-Benifit-Sharing.pdf

Nduche, M. U., Magos Brehm, J., Parra-Quijano, M., & Maxted, N. (2023). In situ and ex situ conservation gap analyses of West African priority crop wild relatives. *Genetic Resources and Crop Evolution, 70*(2), 333–351. DOI: 10.1007/s10722-022-01507-2

Neuwirth, R. J. (2023). Prohibited artificial intelligence practices in the proposed EU artificial intelligence act (AIA). *Computer Law & Security Review, 48*, 105798. DOI: 10.1016/j.clsr.2023.105798

Neville, M., Tarrega, A., Hewson, L., & Foster, T. (2017). Consumer-orientated development of hybrid beef burger and sausage analogues. *Food Science & Nutrition, 5*(4), 852–864. DOI: 10.1002/fsn3.466 PMID: 28748073

Ng, Z. Y., Kee, P. E., Abdullah, R., Lan, J. C. W., Ling, T. C., Jiang, J. J., & Khoo, K. S. (2024). Conversion of lignocellulosic biomass waste into mycoprotein: Current status and future directions for sustainable protein production. *Biomass Conversion and Biorefinery, 15*(24), 31337–31363. https://doi.org/10.1007/s13399-024-06130-yDOI: 10.1007/s13399-024-06130-y

Nihal, A. (2022). *Deciphering the role of environmental law in bio-pharmaceutical research: In the context of IPR* [Paper presentation]. Proceedings of International Conference on Recent Advances in Biotechnology, Abstract Book, Dr. B.R. Ambedkar National Institute of Technology, Jalandhar, 152.

Nitu, & Singh, R. B. (2021). Diversification of Agriculture Through Medicinal Plants in Kangra District, Himachal Pradesh. *Resource Management, Sustainable Development and Governance: Indian and International Perspectives*, 377-395.

Nout, M. J. R., & Aidoo, K. E. (2010). Asian fungal fermented food. In *Springer eBooks* (pp. 29–58). https://doi.org/DOI: 10.1007/978-3-642-11458-8_2

Nurrochmat, N. A., Purnomo, H., & Erbaugh, J. T. (2023). Developing sustainable models of agroforest-landscape in Bogor, Indonesia. *Biodiversitas (Surakarta), 24*(8), 4324–4333. DOI: 10.13057/biodiv/d240814

Obeng, E. A., Obiri, B. D., Oduro, K. A., Pentsil, S., Anglaaere, L. C., Foli, E. G., & Ofori, D. A. (2020). Economic value of non-market ecosystem services derived from trees on cocoa farms. *Current Research in Environmental Sustainability, 2*, 100019. DOI: 10.1016/j.crsust.2020.100019

Odeny, D. A., & Narina, S. S. (2011). *Allium*. In *Wild crop relatives: genomic and breeding resources: vegetables* (pp. 1-10). Springer.

Ogwu, M. C., Kosoe, E. A., Iyiola, A. O., & Fawole, W. O. (2025). Circular Economy Approaches in Bioeconomy in the Global South. In *Sustainable Bioeconomy Development in the Global South: Volume II Bioeconomy Techniques* (pp. 131-157). Singapore: Springer Nature Singapore. DOI: 10.1007/978-981-96-0305-3_5

Ohata, M., Uchida, S., Zhou, L., & Arihara, K. (2016). Antioxidant activity of fermented meat sauce and isolation of an associated antioxidant peptide. *Food Chemistry, 194*, 1034–1039. DOI: 10.1016/j.foodchem.2015.08.089 PMID: 26471650

Oh, Y., & Kim, H. (2025). Characterization of plant-based sausage quality using shiitake mushroom mycelium and soybean and wheat proteins. *Food Science of Animal Resources*. DOI: 10.5851/kosfa.2025.e37

Oktoyoki, H., Pratama, B., Putra, Y. P., Novrian, W., Douni, J. B., & Ansiska, P. (2024). Financial feasibility study and development of technical specifications for coffee agroforestry for community forest farmers with revolving fund support for productivity optimization. *Jendela Pengetahuan, 17*(2), 239–251. DOI: 10.30598/jp17iss2pp239-251

Olga Tellis v. Bombay Municipal Corporation (1985) 3 SCC 545.

Oliveira, G. M. T., Santana, A. C., Oliveira, E. S., Silva, R. J., Santos, W. A. S., Santana, Á. L., & Costa, V. C. N. (2020). The value of agroforestry ecosystem services provided in rural communities in the Eastern Amazon (Tomé-Açu – PA, Brazil). *Journal of Agricultural Studies, 8*(4), 202. DOI: 10.5296/jas.v8i4.17338

Organisation for Economic Co-operation and Development (OECD). (n.d.). Reports on the bioeconomy (e.g., *The Bioeconomy to 2030: Designing a Policy Agenda*).

Otter, V., & Deutsch, M. (2023). Did policy lose sight of the wood for the trees? An UTAUT-based partial least squares estimation of farmers acceptance of innovative sustainable land use systems. *Land Use Policy, 126*, 106467. Advance online publication. DOI: 10.1016/j.landusepol.2022.106467

Pa, C. M., Salimath, S. K., Hegde, R., v, M., Nl, D. D., & Gooli, M. (2024). Agro-biodiversity vis-à-vis agroforestry: Cultivating a sustainable future. *International Journal of Research in Agronomy, 7*(3S), 176–186. DOI: 10.33545/2618060X.2024.v7.i3Sc.428

Paliwal, S. (2023). *A Study on Value Chain of Medicinal and Aromatic Crops in Garhwal Division of Uttarakhand* (Doctoral dissertation, GB Pant University of Agriculture & Technology, Pantnagar-263145).

Pancholi, R., Yadav, R., Gupta, H., Vasure, N., Choudhary, S., Singh, M. N., & Rastogi, M. (2023). The Role of Agroforestry Systems in Enhancing Climate Resilience and Sustainability- A Review. *International Journal of Environment and Climate Change, 13*(11), 4342–4353. DOI: 10.9734/ijecc/2023/v13i113615

Pandey, J. N. (2021). *Constitutional Law of India* (58th ed.). Central Law Agency.

Pang, B., & Lee, L. (2008). Opinion mining and sentiment analysis. *Foundations and Trends in Information Retrieval, 2*(1–2), 1–135. DOI: 10.1561/1500000011

Pantera, M.-L., R., M., Herzog, F., & Herder, M. (2021). Agroforestry and the environment. *Agroforestry Systems, 95*(ue 5), 767–774. DOI: 10.1007/s10457-021-00640-8

Panthee, D. R., Kc, R. B., Regmi, H. N., Subedi, P. P., Bhattarai, S., & Dhakal, J. (2006). Diversity analysis of garlic (*Allium sativum* L.) germplasms available in Nepal based on morphological characters [Article]. *Genetic Resources and Crop Evolution, 53*(1), 205–212. DOI: 10.1007/s10722-004-6690-z

Papaioannou, C., Fassou, G., Petropoulos, S. A., Lamari, F. N., Bebeli, P. J., & Papasotiropoulos, V. (2023). Evaluation of the genetic diversity of Greek garlic (*Allium sativum* L.) accessions using DNA markers and association with phenotypic and chemical variation. *Agriculture, 13*(7), 1408. DOI: 10.3390/agriculture13071408

Paramesh, V., Ravisankar, N., Behera, U., Arunachalam, V., Kumar, P., Solomon Rajkumar, R., Dhar Misra, S., Mohan Kumar, R., Prusty, A. K., Jacob, D., Panwar, A. S., Mayenkar, T., Reddy, V. K., & Rajkumar, S. (2022). Integrated farming system approaches to achieve food and nutritional security for enhancing profitability, employment, and climate resilience in India. *Food and Energy Security, 11*(2), e321. DOI: 10.1002/fes3.321

Parameswari, K., Mulla, N., Hussain, M. A., Kumar, S. B. S., & Subha, L. (2024). Integration of AI and robotics in precision agriculture: Enhancing crop yield and resource efficiency. *Nanotechnology Perceptions, 20*(5). Advance online publication. DOI: 10.62441/nano-ntp.vi.2902

Parente-Laverde, A. M. (2020). Value chain and economic development: The case of the Colombian coffee industry. *Organizations and Markets in Emerging Economies, 11*(1), 173–188. DOI: 10.15388/omee.2020.11.29

Parreño, M. A., Werle, S., Buydens, L., Leroy, C., Roberts, S., Koirala, S., Filipiak, M., Kuhlmann, M., Brunet, J. L., Henry, M., Alaux, C., Requier, F., Piot, N., Meeus, I., Klein, A.-M., Keller, A., & Leonhardt, S. D. (2024). Landscape heterogeneity correlates with bee and pollen diversity while size and specialization degree explain species-specific responses of wild bees to the environment. *The Science of the Total Environment, 954*, 176595. DOI: 10.1016/j.scitotenv.2024.176595 PMID: 39357756

Pathak, R. (2021). Scoping the need of mainstreaming Indigenous knowledge for sustainable use of bioresources in the Indian Himalayan region. www.academia.edu

Patil, D., Chittapur, B., & Murthy, M. M. (2018). Traditional agroforestry systems and biodiversity conservation. *Bangladesh Journal of Botany, 47*(4), 785–794. https://www.researchgate.net/publication/330479881_Traditional_agroforestry_systems_and_biodiversity_conservation

Paul, A. K., Lim, C. L., Apu, M. A. I., Dolma, K. G., Gupta, M., de Lourdes Pereira, M., Wilairatana, P., Rahmatullah, M., Wiart, C., & Nissapatorn, V. (2023). Are fermented foods effective against inflammatory diseases? *International Journal of Environmental Research and Public Health, 20*(3), 2481. DOI: 10.3390/ijerph20032481 PMID: 36767847

Pe'er, G., Finn, J. A., Díaz, M., Birkenstock, M., Lakner, S., Röder, N., Kazakova, Y., Šumrada, T., Bezák, P., Concepción, E. D., Dänhardt, J., Morales, M. B., Rac, I., Špulerová, J., Schindler, S., Stavrinides, M., Targetti, S., Viaggi, D., Vogiatzakis, I. N., & Guyomard, H. (2022). How can the European Common Agricultural Policy help halt biodiversity loss? Recommendations by over 300 experts. *Conservation Letters, 15*(6), e12901. Advance online publication. DOI: 10.1111/conl.12901

Pearson, D. E., DePuy, A. L., & Kuhlman, M. P. (2025). Inspiring citizens and municipalities to initiate pollinator conservation: The urban pollinator matrix modeling tool. *Urban Ecosystems, 28*(2), 1–13. DOI: 10.1007/s11252-025-01703-9

Pedersen, N. K., Kepfer-Rojas, S., Riis-Nielsen, T., Johannsen, V. K., & Schmidt, I. K. (2025). Natural colonization in abandoned agricultural fields benefits native, insect-pollinated and bird-dispersed woody species. *Trees, Forests and People, 19*, 100755. DOI: 10.1016/j.tfp.2024.100755

Pérez-Moreno, J., Guerin-Laguette, A., Rinaldi, A. C., Yu, F., Verbeken, A., Hernández-Santiago, F., & Martínez-Reyes, M. (2021). Edible mycorrhizal fungi of the world: What is their role in forest sustainability, food security, biocultural conservation and climate change? *Plants, People, Planet*, *3*(5), 471–490. DOI: 10.1002/ppp3.10199

Peri, P. L., Dube, F., & Varella, A. C. (2016). Silvopastoral systems in the subtropical and temperate zones of South America: an overview. *Silvopastoral Systems in Southern South America*, 1–8.

Pettorelli, N., Graham, N. A. J., Seddon, N., Maria da Cunha Bustamante, M., Lowton, M. J., Sutherland, W. J., Koldewey, H. J., Prentice, H. C., & Barlow, J. (2021). Time to integrate global climate change and biodiversity science-policy agendas. *Journal of Applied Ecology*, *58*(11), 2384–2393. DOI: 10.1111/1365-2664.13985

Pilkington, A., & Meredith, J. (2009). The evolution of the intellectual structure of operations management—1980–2006: A citation/co-citation analysis. *Journal of Operations Management*, *27*(3), 185–202. DOI: 10.1016/j.jom.2008.08.001

Pouris, J., Kolyva, F., Bratakou, S., Vogiatzi, C. A., Chaniotis, D., & Beloukas, A. (2024). The role of fungi in food production and processing. *Applied Sciences (Basel, Switzerland)*, *14*(12), 5046. DOI: 10.3390/app14125046

Prasad, M. N. V. (2024). Bioremediation, bioeconomy, circular economy, and circular bioeconomy—Strategies for sustainability. *Bioremediation and Bioeconomy: A Circular Economy Approach, Second Edition*, 3–32. DOI: 10.1016/B978-0-443-16120-9.00025-X

Pretty, J., Adams, B., Berkes, F., De Athayde, S. F., Dudley, N., Hunn, E., & Pilgrim, S. (2009). The intersections of biological diversity and cultural diversity: Towards integration. *Conservation & Society*, *7*(2), 100–112. DOI: 10.4103/0972-4923.58642

Priyadarshini, P., & Abhilash, P. C. (2020). Policy recommendations for enabling transition towards sustainable agriculture in India. *Land Use Policy*, *96*, 104718. DOI: 10.1016/j.landusepol.2020.104718

Putra, P. B., Cahyono, S. A., Agus, C., Susanti, P. D., & Indrajaya, Y. (2022). The Role of Integrated Organic Cycle Farming in Tropical Agroforestry Systems for Sustainable Food Production. In *World Sustainability Series* (pp. 171–182). DOI: 10.1007/978-3-030-98617-9_10

Rabitz, F. (2015). Biopiracy after the Nagoya Protocol: Problem structure, regime design and implementation challenges. *Brazilian Political Science Review*, *9*(2), 30–53. DOI: 10.1590/1981-38212014000200010

Rachbauer, L., Granda, C. B., Shrestha, S., Fuchs, W., Gabauer, W., Singer, S. W., Simmons, B. A., & Urgun-Demirtas, M. (2024). Energy and nutrient recovery from municipal and industrial waste and wastewater—a perspective. *Journal of Industrial Microbiology and Biotechnology, 51*, 040.

Raghuvanshi, J., Agrawal, R., & Ghosh, P. K. (2019). Measuring the innovation capability of micro enterprises in India: Construct development and validation. *Benchmarking*, *26*(5), 1405–1430. DOI: 10.1108/BIJ-08-2018-0229

Rahman, S. A., Samsudin, Y. B., Bhatta, K. P., Aryal, A., Hayati, D., Cahya, M., Trihadmojo, B., Husain, I., Andini, S., Narulita, S., Rahman, M. A., Lu, X., Emang, D., Abimbola, A. I., Kustanti, A., & Larjavaara, M. (2023). The Role of Agroforestry Systems for Enhancing Biodiversity and Provision of Ecosystem Services in Agricultural Landscapes in Southeast Asia. In *In Agroforestry for Sustainable Intensification of Agriculture in Asia and Africa* (pp. 303–319). Issue April., DOI: 10.1007/978-981-19-4602-8_10

Raj, A., Jhariya, M. K., Banerjee, A., Meena, R. S., Kumar, S., Devi, A., & Poonam. (2024). Agroforestry and ecosystem services. *Agroforestry for Carbon and Ecosystem Management*, 205–221. DOI: 10.1016/B978-0-323-95393-1.00027-0

Raj, A., Jhariya, M. K., Banerjee, A., Meena, R. S., Nema, S., Khan, N., Yadav, S. K., & Pradhan, G. (2022). Agroforestry a model for ecological sustainability. *Natural Resources Conservation and Advances for Sustainability*, 289–307. DOI: 10.1016/B978-0-12-822976-7.00002-8

Rajamannar, R. (2021). *Quantum marketing: mastering the new marketing mindset for tomorrow's consumers*. HarperCollins Leadership.

Ramanatha Rao, V., & Hodgkin, T. (2002). Genetic diversity and conservation and utilization of plant genetic resources. *Plant Cell, Tissue and Organ Culture*, *68*(1), 1–19. DOI: 10.1023/A:1013359015812

Rana, D., Rana, S., & Choudhary, S. (2024). Sustainable development goals and the Indian Himalayan region. In *Advances in geographical and environmental sciences* (pp. 173–194). DOI: 10.1007/978-981-97-1685-2_10

Rana, S. K., Rawal, R. S., Dangwal, B., Bhatt, I. D., & Price, T. D. (2021). 200 years of Research on Himalayan Biodiversity: Trends, gaps, and policy implications. *Frontiers in Ecology and Evolution*, *8*, 8. DOI: 10.3389/fevo.2020.603422

Rather, Z. A., Ahmad, R., Dar, T.-U.-H., & Khuroo, A. A. (2022). Ensemble modelling enables identification of suitable sites for habitat restoration of threatened biodiversity under climate change: A case study of Himalayan Trillium. *Ecological Engineering*, *176*, 106534. DOI: 10.1016/j.ecoleng.2021.106534

Raut, R. G., Pawar, P. S., & Chavan, H. H. (2013). A community trial of Pinak® medicine in the management of snakebite cases in a rural setting of Western Maharashtra, India. *International Journal of Medicine and Medical Sciences*, *5*(4), 164. Retrieved December 07, 2023, from https://academicjournals.org/journal/IJMMS/article-full-text-pdf/293184488

Rawat, N., & Kumar, R. (2020). Agroforestry: A sustainable land use system. *ResearchGate*. https://www.researchgate.net/publication/382001712_AGROFORESTRY_A_Sustainable_Land_Use_System

Raynolds, L. T. (2000). Re-embedding global agriculture: The international organic and fair trade movements. *Agriculture and Human Values*, *17*(3), 297–309. DOI: 10.1023/A:1007608805843

Reiff, J., Jungkunst, H. F., Antes, N., & Entling, M. H. (2025). Crop productivity of Central European Permaculture is within the range of organic and conventional agriculture. *Peer Community Journal*, *5*, 5. DOI: 10.24072/pcjournal.521

Reis, F. S., Barros, L., Martins, A., & Ferreira, I. C. (2012). Chemical composition and nutritional value of the most widely appreciated cultivated mushrooms: An inter-species comparative study. *Food and Chemical Toxicology*, *50*(2), 191–197. DOI: 10.1016/j.fct.2011.10.056 PMID: 22056333

Rendón-Sandoval, F. J., Casas, A., Moreno-Calles, A. I., Torres-García, I., & García-Frapolli, E. (2023). Traditional agroforestry systems and conservation of native plant diversity of seasonally dry tropical forests. *Sustainability (Switzerland*, *12*(11). DOI: 10.3390/su12114600

Revollo-Fernández, D. A., Lithgow, D., Von Thaden, J. J., Salazar-Vargas, M. D. P., & Rodríguez de los Santos, A. (2024). Unlocking Local and Regional Development through Nature-Based Tourism: Exploring the Potential of Agroforestry and Regenerative Livestock Farming in Mexico. *Economies, 12*(6). https://www.scopus.com/inward/record.uri?eid=2-s2.0-85196781687&doi=10.3390%2Feconomies12060137&partnerID=40&md5=7fbd5dcbab6623865b01eacfa7cc7b4a

Rice, R. A. (2008). Agricultural intensification within agroforestry: The case of coffee and wood products. *Agriculture, Ecosystems & Environment*, *128*(4), 212–218. DOI: 10.1016/j.agee.2008.06.007

Risner, D., McDonald, K. A., Jones, C., & Spang, E. S. (2023). A techno-economic model of mycoprotein production: Achieving price parity with beef protein. *Frontiers in Sustainable Food Systems*, *7*, 1204307. DOI: 10.3389/fsufs.2023.1204307

Roshetko, J. M., & Bertomeu, M. (2015). Multi-species and multifunctional smallholder tree farming systems in Southeast Asia: Timber, NTFPs, plus environmental benefits. *Annals of Silvicultural Research*, *39*(ue 2), 62–69. DOI: 10.12899/ASR-1125

Rossi, L. (2025). Agroecology and Biodiversity Conservation: The Role of Organic Farming in Promoting Ecosystem Services. *Europian Journal of Organic Farming & Agri Practices e-ISSN 3051-0236*, *1*(01), 1-9.

Rousta, N., Aslan, M., Yesilcimen Akbas, M., Ozcan, F., Sar, T., & Taherzadeh, M. J. (2024). Effects of fungal based bioactive compounds on human health. *Critical Reviews in Food Science and Nutrition*, *64*(20), 7004–7027. DOI: 10.1080/10408398.2023.2178379 PMID: 36794421

Sabahy, A. A., Hendawy, S. F., Wasfy, K. I., Moursy, M. A. M., & Mohamed, R. (2024). Enhancing Water Use Efficiency and Carbon Profitability Through the Long-Term Impact of Sustainable Farming Systems. *Sustainability (Basel)*, *16*(20), 9116. DOI: 10.3390/su16209116

Sadowski, J. (2023). Total life insurance: Logics of anticipatory control and actuarial governance in insurance technology. *Social Studies of Science*, •••, 03063127231186437. PMID: 37427796

Sakib, A. I. M. (2024). Innovative food waste recycling methods for agricultural sustainability: A systematic review. *Academic Journal on Business Administration. Innovation & Sustainability*, *4*(3), 104–118.

Salgotra, R. K., & Chauhan, B. S. (2023). Genetic diversity, conservation, and utilization of plant genetic resources. *Genes*, *14*(1), 174. DOI: 10.3390/genes14010174 PMID: 36672915

Salve, A., Tiwari, C., & Baghele, L. (2022). Impact of Agroforestry Systems: A Review. *Asian Journal of Microbiology, Biotechnology & Environmental Sciences : AJMBES*, (June), 214–223. DOI: 10.53550/AJMBES.2022.v24i02.002

Sanchez, P. A. (1995). Science in agroforestry. *Agroforestry Systems*, *30*(1–2), 5–55. DOI: 10.1007/BF00708912

Sandhu, D. K., & Soni, S. K. (1989). Microflora associated with Indian Punjabi warri fermentation. *Journal of Food Science and Technology*, *26*(1), 21–25.

Santana, A. C., Santana, Á. L., Santana, Á. L., Oliveira, G. M. T., & Dos Santos, M. A. S. (2023). Bioeconomic evaluation of an agroforestry system and the potential to recover degraded areas and capitalize producers in the state of Pará, Brazilian Amazon. *Desenvolvimento e Meio Ambiente*, *61*, 439–455. DOI: 10.5380/dma.v61i0.80516

Santhapur, R., Jayakumar, D., & McClements, D. J. (2024). Formation and characterization of Mycelium–Potato protein hybrid materials for application in meat analogs or substitutes. *Foods*, *13*(24), 4109. DOI: 10.3390/foods13244109 PMID: 39767051

Santosa, F. J., & Padmaningrum, D. (2024). The economic impact of agroforestry practice in production forest areas, Central Java province, Indonesia. *Scientific Horizons*, *27*(4), 141–153. DOI: 10.48077/scihor4.2024.141

Santos, M., Cajaiba, R. L., Bastos, R., Gonzalez, D., Petrescu Bakış, A. L., Ferreira, D., Leote, P., Silva, W., Cabral, J. A., Gonçalves, B., & Mosquera-Losada, M. R. (2022). Why Do Agroforestry Systems Enhance Biodiversity? Evidence From Habitat Amount Hypothesis Predictions. *Frontiers in Ecology and Evolution*, *9*, 630151. Advance online publication. DOI: 10.3389/fevo.2021.630151

Sarkar, P. K., Dhakar, M. K., Das, B., Das, B., Sarkar, P., & Bhatt, B. P. (2024). Medicinal Plant-Based Agroforestry Systems. In *Bioprospecting of Ethnomedicinal Plant Resources* (pp. 35–54). Apple Academic Press., DOI: 10.1201/9781003451488-3

Sarkar, T., Mazumder, B., Nandi, G., & Sen, S. (2025). Reishi Mushroom (Ganoderma lucidum): Pharmacology of a potent healer in traditional Chinese medicine. *Natural Product Research*, 1–18. DOI: 10.1080/14786419.2025.2511162 PMID: 40455062

Sathvara, P. B., Anuradha, J., & Tripathi, S. (2024). Biodiversity and its importance: An introduction. In *Biodiversity Loss Assessment for Ecosystem Protection* (pp. 1–17). IGI Global. DOI: 10.4018/979-8-3693-3330-3.ch001

Satish, P., Madiwalar, A. F., Lallawmkimi, M. C., Reddy, K. J., Parveen, S. P. A., Laxman, T., Kiruba, M., & Anand, G. (2024). Agroforestry: Multifunctional Benefits and Implementation Strategies. *Journal of Geography. Environment and Earth Science International*, *28*(10), 1–12. DOI: 10.9734/jgeesi/2024/v28i10821

Sato, Y., Fukuda, H., Zhou, Y., & Mikami, S. (2010). Contribution of ethanol-tolerant xylanase G2 from Aspergillus oryzae on Japanese sake brewing. *Journal of Bioscience and Bioengineering*, *110*(6), 679–683. DOI: 10.1016/j.jbiosc.2010.07.015 PMID: 20727822

Scaife, K., Phipps, K. R., Scalise, D., & Polgár, O. (2024). Subchronic Toxicity Evaluation of Dietary Administration of a Fungal Biomass From Rhizomucor pusillus. *Journal of Applied Toxicology*. Advance online publication. DOI: 10.1002/jat.4713 PMID: 39428601

Schaeffer, E. (2017). *Industry X. 0: Realizing digital value in industrial sectors*. Kogan Page Publishers.

Secretariat of the Convention on Biological Diversity. (2011). *Nagoya Protocol on access to genetic resources and the fair and equitable sharing of benefits arising from their utilization to the Convention on Biological Diversity*. Secretariat of the Convention on Biological Diversity. https://www.cbd.int/abs/doc/protocol/nagoya-protocol-en.pdf

Secretariat of the Convention on Biological Diversity. (2020). *Global Biodiversity Outlook 5*. Secretariat of the Convention on Biological Diversity. https://www.cbd.int/gbo/gbo5/publication/gbo-5-en.pdf

Seddon, N., Smith, A., Smith, P., Key, I., Chausson, A., Girardin, C., House, J., Srivastava, S., & Turner, B. (2021). Getting the message right on nature-based solutions to climate change. *Global Change Biology*, *27*(8), 1518–1546. DOI: 10.1111/gcb.15513 PMID: 33522071

Sedkaoui, S. (Ed.). (2018). *Big data analytics for entrepreneurial success*. IGI Global.

Sekar, Santhanam & Mariappan, S. (2007). Usage of traditional fermented products by Indian rural folks and IPR. *Indian Journal of Traditional Knowledge*, *6*, 111–120.

Shahid, M., Shah, P., Mach, K., Rodgers-Hunt, B., Finnigan, T., Frost, G., Neal, B., & Hadjikakou, M. (2024). The environmental impact of mycoprotein-based meat alternatives compared to plant-based meat alternatives: A systematic review. *Future Foods : a Dedicated Journal for Sustainability in Food Science*, *10*, 100410. DOI: 10.1016/j.fufo.2024.100410

Shanmuga Sundari, M., & Rishitha, K. (2024). The Industrial Use Cases of Embodied AI Systems. In Raj, P., Rocha, A., Singh, S. P., Dutta, P. K., & Sundaravadivazhagan, B. (Eds.), *Building Embodied AI Systems: The Agents, the Architecture Principles, Challenges, and Application Domains. Information Systems Engineering and Management* (Vol. 14). Springer., DOI: 10.1007/978-3-031-68256-8_16

Shapiro, P., & Cumbelich, W. (2025). The history of Neurospora crassa in fermented foods. *Discover Food*, *5*(1), 232. Advance online publication. DOI: 10.1007/s44187-025-00547-8

Sharma, A., Patel, S. K., & Singh, G. S. (2024). Revitalizing traditional knowledge of the Himalayan Indigenous Tribes: An ethnoecological and ethnobotanical approach. In *Advances in Asian human-environmental research* (pp. 111–133).

Sharma, S., Prasad, H. K., Roy, A., & Dey, J. (2025). Fungi as a Source of Single Cell Proteins: Current Progress and Future Prospects. *Mycological Inventions for Sustainable Agriculture and Food Production*, 163-190. https://doi.org/DOI: 10.4018/979-8-3693-8201-1.ch008

Sharma, A., Bhardwaj, G., & Nayik, G. A. (Eds.). (2023). *Edible and Medicinal Mushrooms of the Himalayas: Climate Change, Critically Endangered Species, and the Call for Sustainable Development*. CRC Press. DOI: 10.1201/9781003259763

Shin, Y., Midgley, G. F., Archer, E. R. M., Arneth, A., Barnes, D. K. A., Chan, L., Hashimoto, S., Hoegh-Guldberg, O., Insarov, G., Leadley, P., Levin, L. A., Ngo, H. T., Pandit, R., Pires, A. P. F., Pörtner, H., Rogers, A. D., Scholes, R. J., Settele, J., & Smith, P. (2022). Actions to halt biodiversity loss generally benefit the climate. *Global Change Biology*, *28*(9), 2846–2874. DOI: 10.1111/gcb.16109 PMID: 35098619

Siddi, A. K., & Dalawai, R. S. (2024). A cultural drawbridge: A study of the Siddi tribe. *International Journal of Social Sciences and Humanities Review*, *11*(2), 12–20. https://www.ijsp.in/admin/mvc/upload/Vol%2011_02_03.pdf

Sima, P., Vannucci, L., & Vetvicka, V. (2018). β-glucans and cholesterol. *International Journal of Molecular Medicine*, *41*(4), 1799–1808. DOI: 10.3892/ijmm.2018.3411 PMID: 29393350

Singh, K., & Mishra, N. (2024). Biodiversity and bioeconomy: are these two faces of a single coin? *Biodiversity and Bioeconomy: Status Quo, Challenges, and Opportunities*, 3–23. DOI: 10.1016/B978-0-323-95482-2.00001-8

Singh, V. P., Sinha, R. B., Nayak, D., Neufeldt, H., van Noordwijk, M., & Rizvi, J. (2016). *The national agroforestry policy of India: Experiential learning in development and delivery phases* (ICRAF Working Paper No. 240). World Agroforestry Centre (ICRAF). https://doi.org/DOI: 10.5716/WP16143.PDF

Singh, M., Yadav, D. B., Yadav, S., Mor, V., & Poonia, B. S. (2024). Potential of conservation agricultural practices on soil quality, carbon sequestration, salinity management and productivity of rainfed areas: Conservation agricultural practices on soil quality. *Journal of Soil Salinity and Water Quality*, *16*(3), 329–344. DOI: 10.56093/jsswq.v16i3.158116

Singh, N., Gusain, A., Nigam, M., & Mishra, A. P. (2025). The pharmacological and therapeutic versatility of *Allium* species: A comprehensive exploration of bioactive constituents and biological activities. *Discover Applied Sciences*, *7*(4), 349. DOI: 10.1007/s42452-025-06800-0

Singh, P., Gargi, B., Semwal, P., & Tripathi, V. (2023). Global trends, research progress and knowledge mapping of plant–pollinator interactions through bibliometric analysis (1984 to 2023). *Biodiversity (Nepean)*, *24*(4), 200–211. DOI: 10.1080/14888386.2023.2258836

Singh, P., Gargi, B., Semwal, P., & Verma, S. (2024). Global research and research progress on climate change and their impact on plant phenology: 30 years of investigations through bibliometric analysis. *Theoretical and Applied Climatology*, *155*(6), 4909–4923. DOI: 10.1007/s00704-024-04919-5

Singh, R. D., Mody, S. K., Patel, H. B., Devi, S., Modi, C. M., & Kamani, D. R. (2014). Pharmaceutical biopiracy and protection of traditional knowledge. *International Journal of Research and Development in Pharmacy and Life Sciences*, *3*(2), 833–838. https://ijrdpl.com/index.php/ijrdpl/article/download/372/369

Singh, S., & Yadav, A. (2024). Unveiling Anthropogenic Environmental Burdens Impacting Ecosystem Services in the Himalayas. In *The Resilience of Traditional Knowledge Systems for a Sustainable Future: A Focus on Agriculture and Food Practices in the Himalayas* (pp. 221–246). Springer Nature Switzerland. DOI: 10.1007/978-3-031-56858-9_10

Singh, V., Raj, A., Jhariya, M. K., & Thakur, S. (2022). Economic evaluation of agroforestry and non-agroforestry systems in Eastern Uttar Pradesh, India. *Vegetos*, *35*(3), 810–815. https://www.scopus.com/inward/record.uri?eid=2-s2.0-85124725622&doi=10.1007%2Fs42535-022-00348-9&partnerID=40&md5=97494bb75c00eca7c8ffa64575726e4b. DOI: 10.1007/s42535-022-00348-9

Singla, A. (2020). *Protection of traditional knowledge in India with reference to Neem, Turmeric, Basmati Rice.* https://www.researchgate.net/publication/340977589_'Protection_of_Traditional_Knowledge_in_India_with_reference_to_Neem_Turmeric_Basmati_Rice'_Name_-Aditya_SinglaProtection_of_Traditional_Knowledge_in_India_with_reference_to_Neem_Turmeric_Basmati_Rice

Soccol, C. R., Da Costa, E. S. F., Letti, L. A. J., Karp, S. G., Woiciechowski, A. L., & de Souza Vandenberghe, L. P. (2017). Recent developments and innovations in solid state fermentation. *Biotechnology Research and Innovation*, *1*(1), 52–71. DOI: 10.1016/j.biori.2017.01.002

Soliman, M., & Al Balushi, M. K. (2023). Unveiling destination evangelism through generative AI tools. *ROBONOMICS: The Journal of the Automated Economy, 4*(54), 1.

Somashekar, K. S., Abhishek, G. J., Kumar, V., Tiwari, A., & Lakra, T. S. (2024). Agroecology principles, practices and their impact on sustainable food systems. *European Journal of Nutrition & Food Safety, 16*(9), 249–260. DOI: 10.9734/ejnfs/2024/v16i91544

Soni, S. K., & Sandhu, D. K. (1991). Role of yeast domination in Indian idli batter fermentation. *World Journal of Microbiology & Biotechnology, 7*(4), 505–507. DOI: 10.1007/BF00303378 PMID: 24425138

Son, J. H., Park, K. C., Lee, S. I., Kim, J. H., & Kim, N. S. (2012). Species relationships among *Allium* species by ISSR analysis. *Horticulture, Environment and Biotechnology, 53*(3), 256–262. DOI: 10.1007/s13580-012-0130-3

Srivastav, A. L., Bagherian, A., & Ghosh, D. (2025). The circular bioeconomy: Pathways to sustainability and resource optimization. *Clean Technologies and Environmental Policy, 27*(2), 815–832. DOI: 10.1007/s10098-024-03060-1

Stamets, P. (2006). Can mushrooms help save the world? *Explore (New York, N.Y.), 2*(2), 152–161. DOI: 10.1016/j.explore.2005.12.011 PMID: 16781630

Stille, L., Smeets, E., Wicke, B., Singh, R., & Singh, G. (2011). The economic performance of four (agro-) forestry systems on alkaline soils in the state of Haryana in India. *Energy for Sustainable Development, 15*(4), 388–397. DOI: 10.1016/j.esd.2011.08.002

Suárez Valdés, M. Á., Soriano Disla, J. M., Gambuzzi, E., & Castejón Martínez, G. (2024). Innovative Circular Biowaste Valorisation—State of the Art and Guidance for Cities and Regions. *Sustainability (Basel), 16*(20), 8963. DOI: 10.3390/su16208963

Sumarno, T. B., Sihotang, P., & Prawiraatmadja, W. (2022). Exploring Indonesia's energy policy failures through the JUST framework. *Energy Policy, 164*, 112914. DOI: 10.1016/j.enpol.2022.112914

Sundari, M. S., & Gangalapudi, K. P. (2025). Balancing Digital Efficiency With Human Connection in the Workplace. In Humanizing the Hyperconnected Workplace (pp. 31-68). IGI Global Scientific Publishing.

Sundari, S., Penthala, H. R., Mogullapalli, A., Sukhavasi, V., & Nayyar, A. (2024). Women and Millets: Historical Perspective. In The Role of Women in Cultivating Sustainable Societies Through Millets (pp. 196-229). IGI Global.

Surya, R., Amalia, N., Gunawan, W. B., Taslim, N. A., Ghafoor, M., Mayulu, N., Hardinsyah, H., Abdi Syahputra, R., Kartawidjajaputra, F., Rizzo, G., Tjandrawinata, R. R., Subali, D., Kurniawan, R., & Nurkolis, F. (2024). Tempe as superior functional antioxidant food: From biomechanism to future development of soybean-based functional food. *Farmatsiia (Sofia)*, *71*, 1–7. DOI: 10.3897/pharmacia.71.e116748

Szejda, K., & Parry, J. (2020). *Strategies to accelerate consumer adoption of plant-based meat: Recommendations from a comprehensive literature review*. Research Report. Washington, DC: The Good Food Institute. Available at go.gfi.org/plant-based-meat-consumer-adoption-recommendations

Takwani, C. K. (2021). *Lectures on Administrative Law* (7th ed.). Eastern Book Company.

Taleghani, A., Ayati, Z., Eghbali, S., Emami, S. A., & Tayarani-Najaran, Z. (2024). Health benefits of *Allium* spp. in metabolic syndrome: A review. *South African Journal of Botany*, *167*, 217–255. DOI: 10.1016/j.sajb.2024.01.040

Tamang, Buddhiman & Sha, Shankar & Thakur, Nagendra & Tamang, Jyoti. (2012). Haria, a traditional, rice fermented alcoholic beverage of West Bengal. *International Journal of Agriculture and Food Science Technology*. 157-160.

Tamang, J. P., Thapa, N., Bhalla, T. C., & Savitri, N. (2016). Ethnic fermented foods and beverages of India. In *Springer eBooks* (pp. 17–72). https://doi.org/DOI: 10.1007/978-81-322-2800-4_2

Tamang, B., Shukla, G., & Chakravarty, S. (2023). The urge of conserving tradition from climate change: A case study of Darjeeling Himalayan large cardamom-based traditional agroforestry farming system. *Nature-Based Solutions*, *3*, 100064. DOI: 10.1016/j.nbsj.2023.100064

Tamang, J. P. (Ed.). (2020). *Ethnic fermented foods and beverages of India: science history and culture*. Springer Nature., DOI: 10.1007/978-981-15-1486-9

Tamang, J. P., Anupma, A., & Shangpliang, H. N. J. (2022). Ethno-microbiology of Tempe, an Indonesian fungal-fermented soybean food and Koji, a Japanese fungal starter culture. *Current Opinion in Food Science*, *48*, 100912. DOI: 10.1016/j.cofs.2022.100912

Tan, M., Cui, N., Jiang, S., Xing, L., Wen, S., Liu, Q., Li, W., Yan, S., Wang, Y., Jin, H., & Wang, Z. (2025). Effect of practicing water-saving irrigation on greenhouse gas emissions and crop productivity: A global meta-analysis. *Agricultural Water Management*, *308*, 109300. DOI: 10.1016/j.agwat.2025.109300

Tata, H. L. (2019). Is smallholder farmer maintaining biodiversity in rattan agroforest? *IOP Conference Series. Earth and Environmental Science*, *298*(1), 012034. https://www.scopus.com/inward/record.uri?eid=2-s2.0-85070640436&doi=10.1088%2F1755-1315%2F298%2F1%2F012034&partnerID=40&md5=7641b3ac6f556c86d823d1ffa9f59b41. DOI: 10.1088/1755-1315/298/1/012034

Täuscher, K., & Laudien, S. M. (2018). Understanding platform business models: A mixed methods study of marketplaces. *European Management Journal*, *36*(3), 319–329. DOI: 10.1016/j.emj.2017.06.005

Tembrevilla, G., Phillion, A., & Zeadin, M. (2024). Experiential learning in engineering education: A systematic literature review. *Journal of Engineering Education*, *113*(1), 195–218. DOI: 10.1002/jee.20575

Teng, J., Hou, R., Dungait, J. A., Zhou, G., Kuzyakov, Y., Zhang, J., Tian, J., Cui, Z., Zhang, F., & Delgado-Baquerizo, M. (2024). Conservation agriculture improves soil health and sustains crop yields after long-term warming. *Nature Communications*, *15*(1), 8785. DOI: 10.1038/s41467-024-53169-6 PMID: 39389978

Teng, T. S., Chin, Y. L., Chai, K. F., & Chen, W. N. (2021). Fermentation for future food systems. *EMBO Reports*, *22*(5), e52680. Advance online publication. DOI: 10.15252/embr.202152680 PMID: 33908143

Thakur, M., Kasi, I. K., Islary, P., & Bhatti, S. K. (2023). Nutritional and health-promoting effects of lichens used in food applications. *Current Nutrition Reports*, *12*(4), 555–566. DOI: 10.1007/s13668-023-00489-6 PMID: 37581862

Thapa, S., & Tamang, J. P. (2006). Microbiological and physico-chemical changes during fermentation of kodo ko jaanr, a traditional alcoholic beverage of the Darjeeling hills and Sikkim. *Indian Journal of Microbiology*, *46*(4), 333–341.

The Economics of Ecosystems & Biodiversity. (n.d.). *TEEB Liberia*. https://teebweb.org/our-work/country-studies/enrtp2013/liberia/

The Hindu Bureau. (2025). *Deputy Commissioner stresses full and proper implementation of SCSP, TSP schemes*. The Hindu. https://www.thehindu.com/news/national/karnataka/deputy-commissioner-stresses-full-and-proper-implementation-of-scsp-tsp-schemes/article69827779.ece

The News Minute. (2025). *BJP accuses Karnataka govt of plans to use SCs/STs funds for guarantee schemes*. https://www.thenewsminute.com/karnataka/bjp-accuses-karnataka-govt-of-plans-to-use-scssts-funds-for-guarantee-schemes

Tippmann, V. (2024). Transitioning towards a Circular (Bio) economy–An Analysis of the Barriers and Potentials for an enhanced Cascade Use of Wood in Switzerland (Master's thesis, ETH Zurich).

Titova, E. S., Shishkin, S. S., & Shtykhno, D. A. (2023). Bioeconomy as One of the Ways To Sustainable Development of Russian Regions. *Federalism, 28*(1), 56–79. DOI: 10.21686/2073-1051-2023-1-56-79

Tiwari, C., & Sunny, A. (2024). To the Ones Who Listen to the Mountains: Evaluating the Role of Traditional Ecological Knowledge in Conservation Efforts in the Himalayas. In *Learning 'from'and 'with'the Locals: Traditional Knowledge Systems for Environmental Sustainability in the Himalayas* (pp. 3–22). Springer Nature Switzerland. DOI: 10.1007/978-3-031-51696-2_1

Tiwari, H., Sekar, K. C., Pandey, A., Tiwari, A., Mehta, P., Kanwal, K. S., & Arya, D. (2024). Diversity, distribution and need of urgent conservation of endemic plants in Himalaya. *Biodiversity and Conservation, 33*(8–9), 2285–2303. DOI: 10.1007/s10531-024-02815-y

Toplicean, I.-M., & Datcu, A. D. (2024). An Overview on Bioeconomy in Agricultural Sector, Biomass Production, Recycling Methods, and Circular Economy Considerations. *Agriculture, 14*(7), 1143. DOI: 10.3390/agriculture14071143

Torquebiau, E. (2024). *The agroforestry-biodiversity-climate change nexus.* 21–27.

Traditional Knowledge Digital Library. (n.d.). *Bio-piracy of traditional knowledge.* https://www.tkdl.res.in/tkdl/langdefault/common/Biopiracy.asp?GL=Eng

Tremblay, S., Lucotte, M., Revéret, J.-P., Davidson, R., Mertens, F., Passos, C. J. S., & Romaña, C. A. (2015). Agroforestry systems as a profitable alternative to slash and burn practices in small-scale agriculture of the Brazilian Amazon. *Agroforestry Systems, 89*(2), 193–204. DOI: 10.1007/s10457-014-9753-y

Tsai, S. Y., Tsai, H. L., & Mau, J. L. (2008). Non-volatile taste components of Agaricus blazei, Agrocybe cylindracea and Boletus edulis. *Food Chemistry, 107*(3), 977–983. DOI: 10.1016/j.foodchem.2007.07.080 PMID: 26065761

Tsering, J., & Rinya, K. (2020). *The hundred orchids of Sessa Orchid Sanctuary.* ResearchGate.

Tsioumani, E. (2018). Beyond access and benefit-sharing: Lessons from the law and governance of agricultural biodiversity. *The Journal of World Intellectual Property, 21*(3–4), 106–122. DOI: 10.1111/jwip.12094

Twitter, A. P. I. (n.d.). *Twitter API data.* Retrieved from https://developer.twitter.com/en/docs/twitter-api

Tzampazaki, M., Zografos, C., Vrochidou, E., & Papakostas, G. A. (2024). Machine Vision—Moving from Industry 4.0 to Industry 5.0. *Applied Sciences (Basel, Switzerland), 14*(4), 1471. DOI: 10.3390/app14041471

Udawatta, P., R., R., L., & Jose, S. (. (2019). Agroforestry and biodiversity. *Sustainability, 11*(10), 2879. DOI: 10.3390/su11102879

Ulya, N. A., Harianja, A. H., Sayekti, A. L., Yulianti, A., Djaenudin, D., Martin, E., Hariyadi, H., Witjaksono, J., Malau, L. R. E., Mudhofir, M. R. T., Mudhofir, M. R. T., & Astana, S. (2023). Coffee agroforestry as an alternative to the implementation of green economy practices in Indonesia: A systematic review. *AIMS Agriculture and Food, 8*(3), 762–788. DOI: 10.3934/agrfood.2023041

United Nations Department of Economic and Social Affairs. (2023). *The sustainable development goals report 2023.* UN. https://unstats.un.org/sdgs/report/2023/

United Nations Environment Programme. (2021). *Making peace with nature: A scientific blueprint to tackle the climate, biodiversity, and pollution emergencies.* Retrieved from https://www.unep.org/resources/making-peace-nature

United Nations. (2015). *Transforming our world: The 2030 Agenda for Sustainable Development.*

Upadhyay, S. K., Singh, G., Rani, N., Rajput, V. D., Seth, C. S., Dwivedi, P., Minkina, T., Wong, M. H., Show, P. L., & Khoo, K. S. (2024). Transforming bio-waste into value-added products mediated microbes for enhancing soil health and crop production: Perspective views on circular economy. *Environmental Technology & Innovation, 34*, 103573. DOI: 10.1016/j.eti.2024.103573

Upcraft, T., Tu, W., Johnson, R., Finnigan, T., Van Hung, N., Hallett, J., & Guo, M. (2021). Protein from renewable resources: Mycoprotein production from agricultural residues. *Green Chemistry, 23*(14), 5150–5165. DOI: 10.1039/D1GC01021B

Van Assche, K., Valentinov, V., & Verschraegen, G. (2022). Adaptive governance: Learning from what organizations do and managing the role they play. *Kybernetes, 51*(5), 1738–1758. DOI: 10.1108/K-11-2020-0759 van Rees, C. B., Waylen, K. A., Schmidt-Kloiber, A., Thackeray, S. J., Kalinkat, G., Martens, K., Domisch, S., Lillebø, A. I., Hermoso, V., Grossart, H., Schinegger, R., Decleer, K., Adriaens, T., Denys, L., Jarić, I., Janse, J. H., Monaghan, M. T., De Wever, A., Geijzendorffer, I., & Jähnig, S. C. (2021). Safeguarding freshwater life beyond 2020: Recommendations for the new global biodiversity framework from the European experience. *Conservation Letters*, *14*(1), e12771. Advance online publication. DOI: 10.1111/conl.12771

Varanini, G., Proestou, M., Goritz, N., & Feindt, P. H. (2024). Explaining low salience of environmental resilience challenges in bioeconomy strategies: A cross-regional comparative analysis. *Earth System Governance*, *21*, 100218. DOI: 10.1016/j.esg.2024.100218

Varaprasad, K., & Rao, T. V. (2024). Perspective Chapter: Agroecology-Based Natural. Updates on Organic Farming, 23.

Varela, L., Putnik, G., & Romero, F. (2024). Collaborative manufacturing and management contextualization in the Industry 4.0 based on a systematic literature review. *International Journal of Management Science and Engineering Management*, *19*(1), 78–95. DOI: 10.1080/17509653.2023.2174200

Vasiliev, D. (2022, September). The role of biodiversity in ecosystem resilience. [). IOP Publishing.]. *IOP Conference Series. Earth and Environmental Science*, *1072*(1), 012012. DOI: 10.1088/1755-1315/1072/1/012012

Veiter, L., Rajamanickam, V., & Herwig, C. (2018). The filamentous fungal pellet—Relationship between morphology and productivity. *Applied Microbiology and Biotechnology*, *102*(7), 2997–3006. DOI: 10.1007/s00253-018-8818-7 PMID: 29473099

Venkatesh, V., Brown, S. A., & Sullivan, Y. W. (2016). Guidelines for conducting mixed-methods research: An extension and illustration. *Journal of the Association for Information Systems*, *17*(7), 435–495. DOI: 10.17705/1jais.00433

Verma, K., Sharma, P., Bhardwaj, D. R., Sharma, V., & Thakur, P. (2024). Towards a greener future: Scaling up agroforestry for global sustainability. In *Green and Low-Carbon Economy*. Green and Low-Carbon Economy., DOI: 10.47852/bonviewGLCE42022481

Ververis, E., Ackerl, R., Azzollini, D., Colombo, P. A., De Sesmaisons, A., Dumas, C., Fernandez-Dumont, A., Da Costa, L. F., Germini, A., Goumperis, T., Kouloura, E., Matijevic, L., Precup, G., Roldan-Torres, R., Rossi, A., Svejstil, R., Turla, E., & Gelbmann, W. (2020). Novel foods in the European Union: Scientific requirements and challenges of the risk assessment process by the European Food Safety Authority. *Food Research International, 137*, 109515. DOI: 10.1016/j.foodres.2020.109515 PMID: 33233150

Vetter, J. (2019). Biological values of cultivated mushrooms—A review. *Acta Alimentaria, 48*(2), 229–240. DOI: 10.1556/066.2019.48.2.11

Vinaykarthik, B. C. (2022, October). Design of Artificial Intelligence (AI) based User Experience Websites for E-commerce Application and Future of Digital Marketing. In *2022 3rd International Conference on Smart Electronics and Communication (ICOSEC)* (pp. 1023-1029). IEEE.

Viola, E., Keswani, C., & Minkina, T. (2023). Impact of the sustainable agricultural practices for governing soil health from the perspective of a rising agri-based circular bioeconomy. *Applied Soil Ecology, 194*, 105199.

Violo, M. (2024). How fungi can further support the United Nation's 17 Sustainable Development Goals (SDGs). *MycoStories*. https://www.mycostories.com/post/how-fungi-can-further-support-the-united-nation-s-17-sustainable-development-goals-sdgs

Visser, F., Drouilly, M., Moodley, Y., Michaux, J. R., & Somers, M. J. (2023). Mismatch between conservation needs and actual representation of lions from West and Central Africa in in situ and ex situ conservation. *Conservation Letters, 16*(2), e12949. Advance online publication. DOI: 10.1111/conl.12949

Volk, G. M., Henk, A. D., & Richards, C. M. (2004). Genetic diversity among U.S. garlic clones as detected using AFLP methods. *Journal of the American Society for Horticultural Science, 129*(4), 559–569. DOI: 10.21273/JASHS.129.4.559

Vostriakova, V. I. (2023). Bioeconomy as a complex system of ensuring the country's sustainable development. *Bìznes Ìnform, 1*(540), 167–177. DOI: 10.32983/2222-4459-2023-1-167-177

Vuković, S., Popović-Djordjević, J. B., Kostić, A. Ž., Pantelić, N. D., Srećković, N., Akram, M., Laila, U., & Katanić Stanković, J. S. (2023). *Allium* species in the Balkan region—Major metabolites, antioxidant and antimicrobial properties. *Horticulturae, 9*(3), 408. DOI: 10.3390/horticulturae9030408

Waltz, E. (2016). Gene-edited CRISPR mushroom escapes US regulation. *Nature*, *532*(7599), 293. DOI: 10.1038/nature.2016.19754 PMID: 27111611

Wang, M., & Zhao, R. (2022). A review on nutritional advantages of edible mushrooms and its industrialization development situation in protein meat analogues. *Journal of Future Foods*, *3*(1), 1–7. DOI: 10.1016/j.jfutfo.2022.09.001

Wang, T., Zuo, Y., Manda, T., Hwarari, D., & Yang, L. (2025). Harnessing Artificial Intelligence, Machine Learning and Deep Learning for Sustainable Forestry Management and Conservation: Transformative Potential and Future Perspectives. *Plants*, *14*(7), 998. DOI: 10.3390/plants14070998 PMID: 40219066

Wang, Y., Zheng, R., Dong, W., Gao, P., & Duan, T. (2024). Conservation Agricultural Practices Increase Soil Fungal Diversity and Network Complexity in a 13-Year-Duration Conservation Agriculture System in the Loess Plateau of China. *Applied Sciences (Basel, Switzerland)*, *14*(18), 8411. DOI: 10.3390/app14188411

Wang, Z., Wang, T., Zhang, X., Wang, J., Yang, Y., Sun, Y., Guo, X., Wu, Q., Nepovimova, E., Watson, A. E., & Kuca, K. (2024). Biodiversity conservation in the context of climate change: Facing challenges and management strategies. *The Science of the Total Environment*, *937*, 173377. DOI: 10.1016/j.scitotenv.2024.173377 PMID: 38796025

Wartman, P., Acker, R., & Martin, R. C. (2018). Temperate agroforestry: How forest garden systems combined with people-based ethics can transform culture. *Sustainability (Switzerland*, *10*(ue 7)). DOI: 10.3390/su10072246

Wheeler, J. (2023). *The Digital-First Customer Experience: Seven Design Strategies from the World's Leading Brands*. Kogan Page Publishers.

Wicke, B., Smeets, E. M. W., Akanda, R., Stille, L., Singh, R. K., Awan, A. R., Mahmood, K., & Faaij, A. P. C. (2013). Biomass production in agroforestry and forestry systems on salt-affected soils in South Asia: Exploration of the GHG balance and economic performance of three case studies. *Journal of Environmental Management*, *127*, 324–334. DOI: 10.1016/j.jenvman.2013.05.060 PMID: 23810966

World Intellectual Property Organization. (2010). *Using traditional knowledge to revive the body and a community*. https://www.wipo.int/ipadvantage/en/details.jsp?id=2599

World Intellectual Property Organization. (n.d.). *Traditional knowledge*. https://www.wipo.int/tk/en/tk/

Woźniak-Gientka, E., & Tyczewska, A. (2023). Genome editing in plants as a key technology in sustainable bioeconomy. *EFB Bioeconomy Journal, 3*(February), 100057. DOI: 10.1016/j.bioeco.2023.100057

Wright, D. R., Gordon, A., Bennett, R. E., Selinske, M. J., Lentini, P. E., Garrard, G. E., Rodewald, A. D., & Bekessy, S. A. (2024). Biodiverse coffee plantations provide co-benefits without compromising yield. *Journal of Sustainable Agriculture and Environment, 3*(3), sae2.70005. Advance online publication. DOI: 10.1002/sae2.70005

Wu, J., Liu, D., Wariss, H. M., Zhang, H., Su, M., Li, W., & Han, Z. (2025). Genetic diversity and construction of core collection provides new insight for the conservation of edible *Allium galanthum* in Xinjiang. *Scientia Horticulturae, 341*, 113961. DOI: 10.1016/j.scienta.2025.113961

Wunder, S. (2005). *Payments for environmental services: Some nuts and bolts* (CIFOR Occasional Paper No. 42). Center for International Forestry Research. https://doi.org/DOI: 10.17528/cifor/001760

Xie, D.-F., Yu, H.-X., Price, M., Xie, C., Deng, Y.-Q., Chen, J.-P., Yu, Y., Zhou, S.-D., & He, X.-J. (2019). Phylogeny of Chinese *Allium* species in section Daghestanica and adaptive evolution of *Allium* (Amaryllidaceae, Allioideae) species revealed by the chloroplast complete genome. *Frontiers in Plant Science, 10*, 460. DOI: 10.3389/fpls.2019.00460 PMID: 31114591

Yadav, R. R., Negi, P. S., & Singh, J. (2021). Climate change and plant biodiversity in Himalaya, India. *DELETED, 87*(2), 234–259.

Yadav, S. P. S., Adhikari, R., Bhatta, D., Poudel, A., Subedi, S., Shrestha, S., & Shrestha, J. (2023). Initiatives for biodiversity conservation and utilization in crop protection: A strategy for sustainable crop production. *Biodiversity and Conservation, 32*(14), 4573–4595. DOI: 10.1007/s10531-023-02718-4

Yang, P., Yuan, Y., Yan, C., Jia, Y., You, Q., Da, L., Lou, A., Lv, B., Zhang, Z., & Liu, Y. (2024). AlliumDB: A central portal for comparative and functional genomics in *Allium. Horticulture Research, 11*(2), uhad285. Advance online publication. DOI: 10.1093/hr/uhad285 PMID: 38371639

Yan, J.-K., Zhu, J., Liu, Y., Chen, X., Wang, W., Zhang, H., & Li, L. (2023). Recent advances in research on *Allium* plants: Functional ingredients, physiological activities, and applications in agricultural and food sciences. *Critical Reviews in Food Science and Nutrition, 63*(26), 8107–8135. DOI: 10.1080/10408398.2022.2056132 PMID: 35343832

Yashmita-Ulman, , Singh, M., Kumar, A., & Sharma, M.Yashmita-Ulman. (2021). Conservation of plant diversity in agroforestry systems in a biodiversity Hotspot Region of Northeast India. *Agricultural Research, 10*(4), 569–581. DOI: 10.1007/s40003-020-00525-9

Yee, C. S., Asyiqin, N., Abd Rahim, M. H., Mohd Zaini, N. A., Balqis, R., Ameer, M., Ilham, Z., & Al Qadr Imad, W. A. (2025). Smart Fermentation Technologies: Microbial Process Control in Traditional Fermented Foods. *Fermentation (Basel, Switzerland), 11*(6), 323. DOI: 10.3390/fermentation11060323

Young, R. E., Gann, G. D., Walder, B., Liu, J., Cui, W., Newton, V., Nelson, C. R., Tashe, N., Jasper, D., Silveira, F. A. O., Carrick, P. J., Hägglund, T., Carlsén, S., & Dixon, K. (2022). International principles and standards for the ecological restoration and recovery of mine sites. *Restoration Ecology, 30*(S2), e13771. Advance online publication. DOI: 10.1111/rec.13771

Yuan, Y., Bai, Z., Zhang, J., & Xu, C. (2022). Increasing urban ecological resilience based on ecological security pattern: A case study in a resource-based city. *Ecological Engineering, 175*, 106486. DOI: 10.1016/j.ecoleng.2021.106486

Yusuf, J. A., & Ojedokun, R. O. (2024). The Role of Bio-Based Innovations in Circular Economy: A Biochemical and Economic Perspective. *Journal of Medical Science. Biological Chemistry, 1*(1), 21–27.

Zaki, H. O. (2022). The Impact Of Artificial Intelligence On Content Marketing. *Journal of Strategic Digital Transformation In Society, 2*(3).

Zhang, K., Wu, W., & Yan, Q. (2020). Research advances on sake rice, koji, and sake yeast: A review. *Food Science & Nutrition, 8*(7), 2995–3003. DOI: 10.1002/fsn3.1625 PMID: 32724564

Zhang, R., Farouk, M. M., Realini, C. E., & Thum, C. (2025). Hybridization in Meat-Based Dual Protein Foods: Mechanisms, challenges, and consumer insights. *Comprehensive Reviews in Food Science and Food Safety, 24*(4), e70216. DOI: 10.1111/1541-4337.70216 PMID: 40519121

Zhu, X., Liu, W., Chen, J., Bruijnzeel, L. A., Mao, Z., Yang, X., Cardinael, R., Meng, F. R., Sidle, R. C., Seitz, S., Nair, V. D., Nanko, K., Zou, X., Chen, C., & Jiang, X. J. (2020). Reductions in water, soil and nutrient losses and pesticide pollution in agroforestry practices: A review of evidence and processes. *Plant and Soil, 453*(1–2), 45–86. DOI: 10.1007/s11104-019-04377-3

Ziervogel, G., Enqvist, J., Metelerkamp, L., & Van Breda, J. (2021). Supporting transformative climate adaptation: Community-level capacity building and knowledge co-creation in South Africa. *Climate Policy*, *22*(5), 607–622. DOI: 10.1080/14693062.2020.1863180

Zou, P. X. W., & Xu, X. (2023). Mixed methods research. In *Research methodology and strategy* (pp. 89–112). Wiley., DOI: 10.1002/9781394190256.ch4

About the Contributors

Saurabh Agarwal is presently designated as a Research Professor in the Department of Computer Engineering at Yeungnam University, South Korea. He has a distinguished academic background with a Ph.D. in Computer Engineering from the University of Delhi, where his thesis focused on the design and analysis of image forensic techniques. Dr. Agarwal has accumulated extensive research and teaching experience, previously serving as a Postdoctoral Researcher at Andong National University and Kyungil University in South Korea. During his postdoctoral tenure, he led pioneering studies in image forgery detection utilizing advanced machine learning and deep learning techniques. Dr. Agarwal has held significant academic positions, including Associate Professor roles at Amity University and SRMSCET, Bareilly, India, where he instructed various computer science subjects, mentored students, and contributed to accreditation tasks. His academic contributions have been recognized with several prestigious awards, such as the Korean Research Fellowship and a Ph.D. fellowship from NSIT, Delhi. Notably, Infosys Technologies Limited highlighted his excellence in the Campus Connect Program. Dr. Agarwal's research interests encompass computer vision, image forensics, quantum computing, machine learning, and deep learning. He has published extensively, with numerous articles in high-impact international journals and conference proceedings. His recent publications include works on digital image steganalysis, forensic analysis, and image operator forensics, among others. In addition to his research, Dr Agarwal has actively participated in various training programs and workshops, including cloud computing, machine learning, and big data analytics. He has also contributed to the academic community as a reviewer for several prestigious journals and conferences. Dr Agarwal's academic qualifications include an M. Tech in Software Engineering from APJ AKTU, Lucknow, and a B.E. in Computer Science & Engineering from Barkatullah University, Bhopal. His extensive teaching experience spans over a decade, during which he has taught a wide range of subjects, mentored numerous students, and engaged in various administrative and examination duties. Dr. Agarwal has been instrumental in advancing the field of computer science through his innovative research and dedication to teaching. His work has significantly contributed to the understanding and development of image forensic techniques and machine-learning applications in digital forensics.

Lakshmi D is presently designated as a Senior Associate Professor in the School of Computing Science Engineering and Artificial Intelligence and Assistant Director, at the Centre for Innovation in Teaching & Learning (CITL) at VIT Bhopal. From 2016 to 2021 served as an Educational Research Officer, at Vishnu Educational Development and Innovation Centre (VEDIC) Shri Vishnu Educational Society, Hyderabad She has 26 years of teaching experience. She has addressed innumerable guest lectures, acted as a session chair, and was invited as a keynote speaker at several international conferences. She has conducted FDPs that cover approximately ~1,00,000 plus faculty members including JNTU, TEQIP, SERB, SWAYAM, DST, AICTE, MHRD, ATAL, ISTE, Madhya Pradesh Government-sponsored, MALVIYA MISSION TEACHER TRAINING, Central University of Gujarat and self-financed workshops across India as well as Misamis University, Philipines on Artificial Intelligence in Education: The Future and Futuristic Perspectives both theory and practical sessions. The other titles such as Modern Tools and Software in Teaching Learning Practices, Outcome Based Education, Scientific Educational Practices, Flipped Learning, Blended Learning, Activity Based Learning, Marching towards Quality and Excellence in Higher Educational Institutions, NEP 2022, Emotional Intelligence, Python Essential, Data Analytics, IoT, Machine Learning, Deep Learning, Cyber Security, AR/VR in Medicine, Architecture and Military Applications Instructional Strategies, Assessment techniques, Research Writing, etc. She has 20 international conference presentations, and 50 international journal papers and book chapters inclusive of SCOPUS & SCI. A total of 20 SCOPUS-indexed book chapters. A total of 24 patents are in various stages and 18 patents have been granted at both national and international levels. She was appointed as an external examiner for several Ph.D candidates at reputed universities. Under her supervision, one scholar has completed a Ph.D. at Annamalai University. One scholar is pursuing a Ph.D. right now at VIT Bhopal University. She has won two Best Paper awards at international conferences, one at the IEEE conference and another one at EAMMIS 2021 (First European, Asian, Middle Eastern North African Conference on Management & Information Systems). She received two awards in the year 2022 on the eve of Teacher's Day 1. Kamal Sharma Award of Academic Excellence - 2022 from Lexicon Group of Institutions, Pune for the Engaged Pedagogies. from the virtual ceremony. 2. Bharat Academic Excellence "UttamaAdyapika" Award from Braino Vision. She published a book on "Theory of Computation" in 2003, "Leading Education in the Age of Disruption" in 2021, and "Smart Cities: Disruptive

Technologies Driving the Force", in 2023. In the last three editions, she acted as a SPOC for KAVACH and the Smart India Hackathon. She has created 15 hours of the high-quality "Python Essentials" online course at VITyarthi Online Learning Platform. She has 9 SCOPUS Indexed book edits with three CRS Press, Taylor, and Francis, four IGI Global, and two IEEE Rivers.

P. Balakrishnan, an accomplished academic and researcher, is an Assistant Professor in the Department of Mechanical Engineering at SRM Institute of Science and Technology, Chennai. He specializes in flow dynamics, noise reduction, solar energy systems, and alternative fuels. Dr. Balakrishnan earned his Ph.D. from IIT Madras, focusing on jet noise reduction using coaxial swirlers. His work includes a postdoctoral tenure at SUSTech, China, where he achieved significant noise reduction in aircraft wings. With numerous publications in high-impact journals, his research has garnered international recognition, including the Young Professional Grant Award from INCE in 2016. He holds a gold medal for securing the university's first rank in his M.E., and his patented innovations contribute to noise mitigation and renewable energy technologies. Dr. Balakrishnan has also been instrumental in developing research facilities, conducting motivational lectures, and contributing to academic course development, reflecting his dedication to education and innovation.

Saurabh Chandra is an accomplished legal scholar and currently serves as an Associate Professor & Assistant Dean at the School of Law, Bennett University, India. With a robust academic foundation, he earned his Bachelor of Arts and Bachelor of Laws (B.A. LL.B.) from the prestigious Aligarh Muslim University (A.M.U.). He further specialized in Business Laws, completing his Master of Laws (LL.M.) at the renowned National Law School of India University (NLSIU), Bangalore. Dr. Chandra's academic journey culminated in a Doctorate in Law (Ph.D.) from the esteemed National Law University (NLU) Delhi. In addition to his impressive legal education from some of India's most prestigious institutions, Dr. Chandra also holds a master's degree in Management with specialization in Human Resources in Management and another in Journalism and Mass Communication. With more than sixteen years of teaching experience, he has numerous publications to his credit. His expertise is complemented by a robust portfolio of scholarly publications. Dr. Chandra brings a wealth of experience in Academic Administration, having successfully held various key administrative positions.

Oindrilla Ghosh is a Research Scholar hailing from the School of Law at Christ (Deemed to be University) Bangalore, India, who focuses her research on the health of tribal communities. Her work includes Neuroscience on Organizational Behavior publications at IGI Global and Environmental Disasters, indexed in SCOPUS and

Web of Sciences conference proceedings. She completed her early education at Loreto Day School, Bowbazar, and achieved distinction in her undergraduate (B.A/ LL.B) and postgraduate (LL.M) studies at Calcutta University. Having served as an Assistant Professor of Law at Kingston Law College, Barasat, Kolkata, Ghosh has actively participated in numerous National Conferences, International Conferences, UN Global Summits, webinars, and seminars organized by prestigious academic institutions. Her expertise lies in health perspectives, environmental studies, human behavior, and related areas. Additionally, she has conducted various workshops and lectures in educational institutions, and is engaged with NGO's to help the needy

Guneeta Kaur Gill is presently an Assistant Professor of Sociology at St. Claret College Autonomous Bangalore, India. Her research focuses on Indigenous Knowledge Systems, Ethnomedicine, Religion and Tribal Studies. Her doctorate research analysed the connections between folk healing and definitions of mental ilnesses from the perspective of tribal communities and their knowledge systems.

Vishal Jain currently serves as a Professor and Dean of Accreditation, Statutory Compliances & Institutional Development at the School of Engineering & Technology, Vivekananda Institute of Professional Studies - Technical Campus, New Delhi. His previous roles include Professor at Sharda University, Greater Noida, and Associate Professor at Bharati Vidyapeeth's Institute of Computer Applications and Management (BVICAM), New Delhi. In addition to his teaching responsibilities, Dr. Jain has organized numerous conferences, faculty development programs, seminars, hackathons, and activities for both faculty and students. His research interests encompass machine learning, information retrieval, semantic web, ontology engineering, data mining, ad hoc networks, sensor networks, and network security.

Merri Jayanti is a researcher at the Center for Environmental Research and Clean Technology, National Research and Innovation Agency (BRIN), Indonesia. She specializes in environmental hydrology and water resource sustainability, focusing on hydrological modeling, reservoir operation optimization, and sustainable drinking water systems. Her work contributes to national and international research collaborations, addressing critical water management challenges. In addition to water resource studies, she is actively exploring the role of biodiversity and agroforestry in fostering a sustainable bioeconomy. Beyond research, she engages in environmental policy and innovation projects, aiming to develop practical solutions for long-term water security and ecosystem resilience.

Binod Kumar is an undergraduate student known for his academic excellence, and his scholarly works are indexed in the Scopus database. Hailing from Ranchi,

Jharkhand, in India, Binod is pursuing a Bachelor of Law at the prestigious Radha Govind University. His passion for technology, innovation, and climate change is evident through his high academic performance and numerous projects. Binod has a keen interest in artificial intelligence, machine learning, neuropsychology, climate change, and disaster management, where he has undertaken several research projects. He is particularly fascinated by the potential of AI to solve real-world problems and is dedicated to learning and applying new techniques in these areas. He volunteers at local NGOs, teaching underprivileged children basic computer skills. With a strong foundation in law and a commitment to making a positive impact, Binod Kumar is poised to become a future leader in the legal industry

Archan Mitra is an Assistant Professor at School of Media Studies (SOMS) at Presidency University, Bangalore. He is the author of two book "Cases for Classroom Media and Entertainment Business" and "Multiverse and Media", he also has other several edited books to his credit. He has done his doctorate from Visva-Bharati Santiniketan, West Bengal in the field of "environmental informatics and communication for sustainability". In addition to that he is a certified Science Communicator and Journalism from Indian Science Communication Society (ISCOS), certified Corporate Trainer with Amity Institute of Training and Development, Certified Social Media Network Analyst. He has a strong interest in environmental communication. He was awarded certificate of merit by PRSI, Kolkata Chapter and Medal of Honor by Journalistic Club of Kolkata. He was working as a research assistant with the World Bank's "Environmental Capacity Building in Southeast Asia" project at IIM Kashipur. He was instrumental in launching the World Bank's Green MBA MOOC, he has also assisted in the research project on Uttarakhand disaster mitigation by ICSSR, the leading research on Uttarakhand disaster.

Aniket Nihal is an Indian Registered Patent Agent. He holds Master of Science (Biotechnology) from Birla Institute of Technology, Mesra; Bachelor of Science (Hons.) in Life Sciences from Central University of Jharkhand, Ranchi; & Bachelor of Laws from Ranchi University, Ranchi, India.

Sanchita Paul is a dedicated and accomplished content strategist with extensive experience in managing diverse content formats across different digital platforms. With a strong background in content strategy, editing, and UX writing, she has honed her skills in branding, social media management, and technical writing. Holding an MA in Communication and Journalism from Visva-Bharati University, Ms. Paul is proficient in tools like Google Analytics, Canva, and MS Office Suite, making her a versatile professional committed to producing high-quality, impactful content to improve brand visibility.

Anjali Roy is currently a second year Ph.D research scholar at the Department of Life Science & Bioinformatics, Assam University, Silchar. Her field of research is associated with microbial biotechnology, dealing with yeasts. She received a bachelor's degree with honors in botany from Gurucharan University, Silchar and a master's degree in botany as well from the Department of Life Science and Bioinformatics, Assam University.

Sharmistha Sharma is a Ph.D. research scholar at the Department of Life Science & Bioinformatics, Assam University, Silchar. Her interest lies in the research associated with microbial biotechnology, which deals primarily with fungi. She received a bachelor's degree with honors in botany from Gurucharan University, Silchar, and a master's degree in botany from the Department of Life Science & Bioinformatics, Assam University.

Bhupinder Singh working as Professor at Sharda University, India. Also, Honorary Professor in University of South Wales UK and Santo Tomas University Tunja, Colombia. His areas of publications as Smart Healthcare, Medicines, fuzzy logics, artificial intelligence, robotics, machine learning, deep learning, federated learning, IoT, PV Glasses, metaverse and many more. He has 3 books, 139 paper publications, 163 paper presentations in international/national conferences and seminars, participated in more than 40 workshops/FDP's/QIP's, 25 courses from international universities of repute, organized more than 59 events with international and national academicians and industry people's, editor-in-chief and co-editor in journals, developed new courses. He has given talks at international universities, resource person in international conferences such as in Nanyang Technological University Singapore, Tashkent State University of Law Uzbekistan; KIMEP University Kazakhstan, All'ah meh Tabatabi University Iran, the Iranian Association of International Criminal law, Iran and Hague Center for International Law and Investment, The Netherlands, Northumbria University Newcastle UK,

Index

A

B

C

D

E

F

G

H

I

K

L

M

N

P

R

S

T

V

W

www.ingramcontent.com/pod-product-compliance
Lightning Source LLC
LaVergne TN
LVHW062324300926
849161LV00007B/1560